Science and Philosophy in the Indian Buddhist Classics

COMPENDIUM COMPILATION COMMITTEE

Chair

Thamthog Rinpoche, Abbot of Namgyal Monastery

General Series Editor

Thupten Jinpa, PhD

Advisory Members

Geshe Yangteng Rinpoche, Sermey Monastic College
Geshe Thupten Palsang, Drepung Loseling College
Gelong Thupten Yarphel, Namgyal Monastery

Editors

Geshe Jangchup Sangye, Ganden Shartse College
Geshe Ngawang Sangye, Drepung Loseling College
Geshe Chisa Drungchen Rinpoche, Ganden Jangtse College
Geshe Lobsang Khechok, Drepung Gomang College

Science and Philosophy

in the

Indian Buddhist Classics

Volume 2

The Mind

Conceived and introduced by

His Holiness the Dalai Lama

Developed by the
Compendium Compilation Committee

Edited by Thupten Jinpa

Translated by Dechen Rochard and John D. Dunne

Contextual essays by John D. Dunne

Wisdom Publications, Inc.
199 Elm Street
Somerville MA 02144 USA
wisdomexperience.org

A translation of *Nang pa'i tshan rig dang lta grub kun btus, vol. 2*. Dharamsala, India: Ganden Phodrang Trust (Office of His Holiness the Dalai Lama), 2014.

Library of Congress Cataloging-in-Publication Data is available.
LCCN 2017018045

ISBN 978-1-61429-474-0 ebook ISBN 978-1-61429-500-6

24 23 22 21 20
5 4 3 2 1

Designed by Gopa & Ted2, Inc.
Cover art by Getty Images.
Typeset by Tony Lulek.
Set in DGP 11/14.

Printed on acid-free paper that meets the guidelines for permanence and durability of the Production Guidelines for Book Longevity of the Council on Library Resources.

Printed in the United States of America.

Contents

Preface vii

Introduction by His Holiness the Dalai Lama 1

Part 1. Mind 23

1. The Nature of Mind 39
2. Sense Consciousness 49
3. Conceptual and Nonconceptual 65
4. Valid and Mistaken 75

Part 2. Mental Factors 83

5. Distinguishing Mind and Mental Factors 97
6. Omnipresent Mental Factors 107
7. Mental Factors with a Determinate Object 111
8. Virtuous Mental Factors 119
9. Love and Compassion 125
10. Mental Afflictions 139
11. Variable Mental Factors 157
12. Mental Factors in Other Works 161
13. Substantial and Imputed Mental Factors 175
14. Alternate Presentations of Mental Factors 181

Part 3. Gross and Subtle Minds 199

15. Gross and Subtle Minds in the Shared Traditions 207
16. Gross and Subtle Minds in Highest Yoga Tantra 213

PART 4. MIND AND ITS OBJECTS 233

17. How the Mind Engages Its Objects 247

18. The Sevenfold Typology of Cognition 263

PART 5. INFERENTIAL REASONING 279

19. Reasoning and Rationality 293

20. Categories of Correct Evidence 319

21. Fallacious Inferential Evidence 337

22. Dignāga's *Drum of a Wheel of Reasons* 343

PART 6. TRAINING THE MIND THROUGH MEDITATION 351

23. How the Mind Is Trained 367

24. Calm Abiding 393

25. Analysis and Insight 423

26. Mindfulness Meditation 431

27. The Eight Worldly Concerns 445

28. Increasing Good Qualities 455

29. Concluding Topic: The Person or Self 461

Notes 467

Glossary 493

Bibliography 507

Index 517

About the Authors 553

Preface

General Editor's Note

It is a real joy to see in print this special volume presenting ideas and insights of classical Buddhist thought on the important topic of the mind and its potential for transformation. This is the second volume in *Science and Philosophy in Indian Buddhist Classics*, a unique series conceived by and developed under the supervision of His Holiness the Dalai Lama. While the first volume looked at the physical world, this volume investigates the science of the mind. Volumes 3 and 4 in the series, currently being prepared, present the philosophical insights of great Indian Buddhist thinkers.

An important premise behind the series, as His Holiness the Dalai Lama outlines in his own lengthy introduction, is the differentiation of the subject matter of Buddhist texts into the three domains of science, philosophy, and religion. He believes that while the last domain is relevant primarily to practicing Buddhists, the first two—science and philosophy as found in the classical Buddhist sources—can and should be recognized as part of global scientific and philosophical heritage. I urge the reader thus to engage first with the lengthy introduction in which His Holiness outlines in detail his vision, thought, and aspirations behind the creation of the four-volume series, and this volume on Buddhist mind sciences in particular. The opening and closing of his introduction match that of volume 1, with material unique to this volume in between.

One hallmark of ancient Indian systems of thought is their careful inquiry into the nature, function, and trainability of the human mind, with the Buddhist tradition especially excelling in this domain. Beginning with the Buddha himself, going back more than two and half millennia, the Buddhist tradition has emphasized critical inquiry into one's own mind as an indispensable path to self-mastery, spiritual transformation, and enlightenment. A famous statement attributed to the Buddha tells us that the mind is the forerunner of everything and that the person who has mastered the mind has found true peace. More systematic "scientific" and philosophical approaches emerged in Buddhist tradition a few centuries following the Buddha's death, just before the dawn of the Common Era. Scholars refer to this phase as the Abhidharma period, when, not unlike the story behind the creation of the Oxford English Dictionary, the large corpus of teachings attributed to the Buddha were compiled according to diverse systems of classification. A landmark in this Abhidharma period was the creation of the exhaustive anthology known as the *Great Explanation* (*Mahāvibhāṣa*), dated to the first or second century CE. We see in this compendium one of the first known systematic presentations of what might be called a map of the human mind. At the heart of this mind map is a distinction between the *mind* and *mental factors*, the first referring to the basic fact of awareness and the latter referring to aspects or dimensions of our mental life, defined in terms of their distinct functions. Our volume represents the broadest resource in English on this important dimension of classical Buddhist mind sciences.

Roughly speaking, our volume presents in a distilled manner four distinct aspects of classical Buddhist thought related to the discipline of mind science. First and foremost is this Abhidharma analysis of mental factors—their classifications, definitions, functions, and proximate conditions. Part 2 of our volume, the longest section, is a comprehensive presentation of these mental factors, based primarily on Asaṅga's *Compendium of Knowledge*, but including also critical comparisons with other important Abhidharma sources, such as the Theravāda Abhidhamma. This lengthy section of the volume is of special importance to contemporary psychology and neuroscience, as well as those interested in understanding the larger context of Buddhist-derived completive practices like mindfulness and compassion training. The second dimension of Buddhist mind science presented in our volume relates to what scholars refer to as Buddhist epistemology (*pramāṇa*), especially as developed by Dignāga (fifth–sixth century) and Dharmakīrti (sixth century). Parts 1, 4, and 5 of our volume presents theories and insights from this classical resource in terms of three key themes of (1) the nature and definition of the mind, including especially a distinction between perceptual and conceptual minds, (2) the mind and its object, including a typology of cognition based on understanding the manner in which the mind engages its objects, and (3) systems of reasoning that can lead to correct inference about both the world and our own mind. The third dimension of Buddhist mind sciences, presented in part 3 of our volume, pertains to the unique perspective of Buddhist tantra, which views the mind and body as ultimately nondual, and to the meditation techniques based on such a view. The fourth and final aspect of Buddhist mind sciences presented in our volume, in part 6, are the theories and methods of mental training, including especially the cultivation of calm abiding (*śamatha*) and insight (*vipaśyanā*) as well as cultivation of loving kindness and compassion.

Readers of our volume are fortunate to have in John Dunne's excellent essays preceding each of the six parts of the book insightful and highly accessible guides to the key themes, questions, and insights specific to each of the sections. Drawing on his expertise both in classical Indo-Tibetan Buddhist thought and contemporary Western thought, especially cognitive science and the philosophy of mind, these essays bring to life the key issues specific to the individual sections. They provide the larger context to the topics and valuable signposts for contemporary readers so that they can ably navigate the vast terrain covered in this volume. Crucially they provide what might be characterized as "conceptual translation" across cultures, which in addition to bringing awareness to language (the use of the word *mind* in plural form, for example) entails relating classical Buddhist concepts to their parallels in contemporary science and philosophy.

Personally, it has been a profound honor to be part of this important project. First and foremost, I would like to offer my deepest gratitude to His Holiness the Dalai Lama for his vision and leadership of this most valuable initiative of bringing the insights of great Buddhist minds to our contemporary world. In both his life and thought, His Holiness has been *the* exemplar for the depth and breadth embodied in the great Buddhist tradition.

I thank the Tibetan editors who worked diligently over many years to create this compilation, especially for their patience with the substantive editorial changes I ended up bringing to the various stages of their manuscripts. I would like to applaud and thank Dechen Rochard for taking on the challenging task of translating the entirety of this volume. Her mastery of Tibetan, expertise of the subject matter, and Western academic scholarship made her an ideal translator for this important volume. I owe a deep bow of gratitude to John Dunne for working with Dechen to refine the translation and, of course, for writing such masterful essays to introduce each of the sections. I would also like to thank our editor at Wisdom Publications, David Kittelstrom, and his colleague Mary Petrusewicz for their careful preparation of the English translation for publication.

Finally, I offer my deep gratitude to the Ing Foundation for its generous patronage of the Institute of Tibetan Classics and to the Scully Peretsman Foundation for its support of my own work, which made it possible for me to devote the time necessary to edit both the original Tibetan volume as well as this translation.

Through the publication of this volume on the Buddhist mind sciences, may the insights and wisdom of the great Buddhist masters become accessible and a source of benefit to contemporary readers across the boundaries of geography, language, and culture.

Thupten Jinpa

Translators' Note

This volume offers a concise presentation of Buddhist teachings about the mind and cognition as explained in the classical Indian commentaries composed between 200 and 1200 CE. It mainly cites treatises on Abhidharma and on logic and epistemology (*pramāṇa*), as well as contemplative manuals such as Śāntideva's *Engaging in the Bodhisattva's Deeds*, composed by masters whose breadth and depth of study is epitomized by the tradition of Nālandā University and similar monastic institutions of classical Indian civilization. This volume is the second in a series designed to express the core teachings of these Nālandā masters. It contains treasures that may initially be hard to decipher. However, with research and reflection, one can unlock them and use them as medicine for one's mind and life. It is His Holiness the Dalai Lama's heartfelt wish that these treasures may be shared with the world.

Translating this work has presented a number of challenges. First, the material is highly technical, as the quoted texts were not written for the average reader, yet this volume was compiled with the expectation of reaching a broader audience, including educated Tibetans and non-Tibetans in general as well as physicists, cognitive scientists, psychologists, philosophers, and Buddhist scholars in particular. The *geshes* who composed this text are highly trained Tibetan monks who have spent decades mastering the scriptural sources. They have expertly authored the passages linking the quotations and arranged the order of the topics in accordance with their training and without the influence of Western scholarship. This has yielded two outcomes. First, most importantly, this volume is an authentic presentation of a mainstream Tibetan interpretation of the classical Indian sources that it cites, and it covers a vast range of topics in a single work. Second, it has been written in the compressed style of scholastic Tibetan, and as a result, it contains various stylistic features that may be challenging for some audiences, such as a tendency to deploy lists of phenomena sourced from scripture in ways that may seem repetitive and to assume a preexisting knowledge of basic Buddhist doctrine. A translation of such a volume cannot avoid including these repetitions and other stylistic features, and while the translation can help readers by, for example, avoiding obscure renderings of key terms, this translation cannot simply reshape this material to mirror Western scholarship. However, Western scholars can welcome the fact that this volume offers an utterly reliable foundation from which to build bridges between the topics discussed here and their corollaries in modern cognitive science and other disciplines.

Another challenge for the translators is that this volume contains many previously untranslated passages from terse and complex Sanskrit texts, notably from the works of Dharmakīrti. To render them accurately one must know their context, which means familiarizing oneself with each of the cited texts as a whole and reading the relevant commentaries. When Sanskrit editions of texts were available, we consulted those and collated them against the Tibetan canonical versions of the same texts cited by the authors. Apart from commentaries available in both languages, we also consulted relevant subcommentaries extant in Sanskrit but not translated into Tibetan. All of this research has generally been very helpful for clarifying the meaning of the cited passages, since the Sanskrit versions of the cited texts almost always exhibit less ambiguity in meaning owing to the greater grammatical specificity of the Sanskrit language.

A further challenge concerns the interaction of intellectual traditions. In this volume we have broadly three systems of knowledge to consider: classical Indian, medieval Tibetan, and modern Western. Sanskrit is a remarkably rich language, and a single term can have as many as a dozen different meanings. Hence, when translating from Sanskrit to Tibetan several hundred years ago, Tibetan masters and their Indian collaborators had to make choices that necessarily precluded some readings of the Sanskrit original. A further difficulty occurs nowadays when translating into English. This volume treats numerous topics that have also received considerable attention within the Western intellectual and scientific disciplines. These disciplines each have their own specialized lexicons, and to enhance the accessibility of this translation for scholars within those disciplines, a broad knowledge of their disciplinary terminology was necessary to select suitable terms to convey the meaning of the Buddhist sources. This is a very difficult job. Some terms may seem suitable at first glance, but one may find after exploring them more deeply that they mean something quite different from what is being conveyed in the Buddhist literature. Hence, a great deal of circumspection is required in choosing among technical terms that already exist in English, as there is a great risk of causing confusion. But where the meaning adequately coincides, preference has been given to technical terms that already have some currency in English.

One of the greatest challenges in this volume has been to find ways of translating the various terms for mind, which often appear in the plural in Sanskrit and Tibetan sources. As noted in the introductory essay to part 1, the term *mind* is used in Buddhist texts to convey episodic mental moments that occur in a continuous stream within each living being, where one moment becomes the cause for the next. A single mental moment is often referred to as "a mind," so one person will have many minds arising and ceasing from moment to moment. Although we tried to avoid it, on many occasions we have had no other option than to use the terms *minds* and *consciousnesses* in the plural.

Translation is not univocal. As David Bellos has said, there is no single translation for any utterance of more than trivial length. Some of the texts cited in this volume have existing English translations, and we consulted them at certain stages of the work, though we did not actually use them here. We have thought very deeply about the translation choices that we made in this volume. Some of them we were fortunate enough to agree on immediately. Others we disagreed about initially, so we researched, discussed further, and eventually arrived at a consensus. Likewise, although the first volume of this series was already in print before we finalized the present volume, we have not always adopted the terminology used in the first volume. This is partly because each volume covers a different area of specialization, and when translating a topic that is treated in depth, one may need a more refined set of terms than those used when translating the same topic merely in passing. In any case, as with any translation work, we have tried to balance technical accuracy with readability for a broader audience. This is always a difficult task, and some variation is inevitable. It may be too soon to expect that these often rarefied texts can be rendered into an English version that is universally acceptable. The gradual transmission of Buddhist scientific and philosophical traditions to the West is still in its early stages. We hope that this volume will serve as a substantial contribution toward its unfolding.

As for the process of translating this volume, the first draft was completed by Dechen Rochard within the two-year time frame allotted by the Gaden Phodrang Trust. During this time Dechen was resident in Dharamsala for several months each year so as to consult with the geshe editors. Having reached the submission deadline, the text was passed on to Thupten Jinpa for his comments and suggested changes, after which it was submitted for publication and copyedited. At this point John Dunne was brought on board to write the introductory essays, which are designed to serve as an additional bridge for readers not versed in Buddhist doctrine. John's participation led to revisiting the translation, this time as a collaborative effort, with a view to making the language more suitable for a Western scientific audience. Owing to his extremely full schedule, John engaged the help of three assistants to review and research the work. As well as trying to incorporate terminology appropriate to various academic disciplines such as cognitive science, John gave special attention to citations from Sanskrit texts, with or without Sanskrit editions. During this process Dechen traveled to Madison, Wisconsin, to work in collaboration with John. They continued to consult over the internet once she returned to the United Kingdom, and together they produced the final draft.

Dechen and John are both deeply honored to have been called to work on this project. We are profoundly grateful to His Holiness the Dalai Lama for his vast and compassionate vision in conceiving of this project and for his immense energy in propelling it to completion.

In accordance with His Holiness's wishes, Dechen has been based in India for several months each year, where she has had the opportunity to meet all the members of the Compendium Compilation Committee. She is delighted to have consulted with Khen Thamthog Rinpoche, Thupten

Jinpa, Yangteng Rinpoche, Geshe Thupten Palsang, and Gelong Thupten Yarphel; and she is especially grateful to have worked closely with the four main editors of this text: Khen Rinpoche Geshe Jangchub Sangye of Ganden Shartse, Geshe Lobsang Khechok of Drepung Gomang, Geshe Chisa Drungchen Rinpoche of Ganden Jangtse, and Geshe Ngawang Sangye of Drepung Loseling. She would also like to express sincere appreciation to her old friend and fellow alumnus from Cambridge University, Thupten Jinpa, for his suggestions regarding terminology and for his patient revision of the first draft, as well as to Tenzin Tsepak, another old friend and fellow alumnus from the Institute of Buddhist Dialectics, for kindly sharing his thoughts regarding certain points of scholarship. She would like to thank her colleague Ian Coghlan for the many conversations pertinent to the translation of the first two volumes during the early years of working on this project, and Jamphel Lhundup for so kindly taking care of practical matters regarding the project and residence in Dharamsala.

John and Dechen are delighted to have collaborated on this project. We would like to thank Jeremy Manheim, Geshe Kelsang Wangmo, and Ven. Tenzin Legtsok for their skillful assistance in helping to prepare the second draft. We also wish to thank senior editor at Wisdom Publications, David Kittelstrom, for his valued suggestions regarding style and clarity of content in preparing this text for publication, as well as publisher Daniel Aitken, copyeditor Mary Petrusewicz, production editor Ben Gleason, and proofreader Megan Anderson for their helpful input at certain stages of the work.

May the translation of this work bring the teachings of the Nālandā masters to the attention of the world, and may they be understood and used to develop both inner and outer peace.

Dechen Rochard and John Dunne

Introduction

My Encounter with Science[1]

In my childhood I had a keen interest in playing with mechanical toys. After reaching India in 1959, I developed a strong wish to engage with scientists to help expand my own knowledge of science as well as to explore the question of the relationship between science and religion. The main reason for my confidence in engaging with scientists rested in the Buddha's following statement:

> Monks and scholars, just as you test gold
> by burning, cutting, and polishing it,
> so too well examine my speech.
> Do not accept it merely out of respect.

The Buddha advises his disciples to carefully analyze when they engage with the meaning of his words, just as a goldsmith tests the purity of gold through burning, cutting, and rubbing. Only after we have gained conviction through such inquiry, the Buddha explains, is it appropriate to accept the validity of his words. It is not appropriate to believe something simply because one's teacher has taught it. Even with regard to what he himself taught, the Buddha says, we must test its validity for ourselves through experimentation and the use of reason. The testimony of scriptures alone is not sufficient. This profound advice demonstrates the centrality of sound reasoning when it comes to exploring the question of reality.

In Buddhism in general, and for the Nālandā masters of classical India in particular, when it comes to examining the nature of reality, the evidence of direct perception is accorded greater authority than both reason-based inference and scripture. For if one takes a scripture to be an authority in describing the nature of reality, then that scripture too must first be verified as authoritative by relying on another scriptural testimony, which in turn must be verified by another scripture, and so on, leading to an infinite regress. Furthermore, a scripture-based approach can offer no proof or rebuttals against alternative standpoints proposed by opponents who do not accept the validity of that scripture. Even among scriptures, some can be accepted as literal while some cannot, giving us no reliable standpoints on the nature of reality. It is said that to cite scripture as an authority in the context of inquiring into the nature of reality indicates a misguided intelligence. To do so precludes us from the ranks of those who uphold reason.

In science we find a similar approach. Scientists take experimentation and the logic of mathematics as arbiters of truth when it comes to evaluating the conclusions of their research; they do not ground validity in the authority of some other person. This method of critical inquiry, one that draws inferences about the unobservable, such as atomic particles, based on observed facts that are evident to our direct perception, is shared by both Buddhism and contemporary science. Once I saw this shared commitment, it greatly increased my confidence in engaging with modern scientists.

With instruments like microscopes and telescopes and with mathematical calculations, scientists have been able to carefully analyze phenomena from atomic particles to distant planets. What can be observed by the senses is enhanced by means of these instruments, allowing scientists to gain new inferences about various facts. Whatever hypothesis science puts forth must be verified by observation-based experiments, and similarly Buddhism asserts that the evidence of direct perception must ultimately underpin critical inquiry. Thus with respect to the way conclusions are drawn from evidence and reasoning, Buddhism and science share an important similarity. In Buddhism, however, empirical observation is not confined to the five senses alone; it has a wider meaning, since it includes observations derived from meditation. This meditation-based empirical observation grounded in study and contemplation is also recognized as part of the means of investigating reality, akin to the role scientific method plays in scientific inquiry.

Since my first visit to the West, a trip to Europe in 1973, I have had the opportunity to engage in conversations with great scientists, including the noted twentieth-century philosopher of science Sir Karl Popper, the quantum physicist Carl Friedrich von Weizsäcker, who was the brother of the last West German president and also a colleague of the famed quantum physicists Werner Heisenberg and David Bohm.[2] Over many years I have had the chance to engage in dialogues with scientists on a range of topics, such as cosmology, neurobiology, evolution, and physics, especially subatomic-particle physics. This latter discipline of particle physics shares methods strikingly similar to those found in Buddhism, such as the Mind Only school's critique of the external material world that reveals that nothing can be found when matter is deconstructed into its constitutive elements, and similarly the statements found in the Middle Way school treatises that nothing can be found when one searches for the real referents behind our concepts and their associated terms. I have also on numerous occasions had dialogues with scientists from the fields of psychology and the science of mind, sharing the perspectives of the Indian tradition in general, which contains techniques of cultivating calm abiding and insight, and the Buddhist sources in particular, with its detailed presentations on mind science.

Today we live in an age when the power of science is so pervasive that no culture or society can escape its impact. In a way, there was no choice but for me to learn about science and embrace it with a sense of urgency. I also saw the potential for an emerging discourse on the science of mind. Recognizing this, and wishing to explore how science and its fruits can become a constructive force in the world and serve the basic human drive for happiness, I have engaged in dialogue with scientists for many years. My sincere hope is that these dialogues across cultures and disciplines will inspire new ways to promote both physical and mental well-being and thus serve humanity through a unique interface of contemporary science and mind science. Thus, when I engage in conversations with scientists, such as in the ongoing Mind and Life Dialogues, I have the following two aims.[3]

The first concerns expanding the scope of science. Not only is the breadth of the world's knowledge vast, advances are being made year by year that expand human knowledge. Science, however, right from its inception and especially once it began to develop quickly, has been concerned primarily with the world of matter. Unsurprisingly, then, contemporary science focuses on the physical world. Because of this, not much inquiry in science has been made into the nature of the person—the inquirer—as well as into how memory arises, the nature of happiness and suffering, and the workings of emotion. Science's advances in the domain of the physical world have been truly impressive. From the perspective of human experience, however, there are dimensions of reality that undoubtedly lie outside the current domain of scientific knowledge. It is of vital importance that the science of mind takes its place among the current fields of human investigation. The brain-based explanations in contemporary science about the different classes of sensory experience will be enriched by incorporating a more expanded and detailed understanding of the mind. So my first goal in my dialogues with scientists is to help make the current field of psychology or mind science more complete.

Not only do Buddhism and science have much to learn from each other, but there is also a great need for a way of knowing that encompasses both body and mind. For as human beings we experience happiness and suffering not only physically but mentally as well. If our goal is to promote human happiness, we have a real opportunity to pursue a new kind of science that explores methods to enhance happiness through the interface of contemporary science with contemplative mind science. It is my belief that, while acknowledging the great contribution that science has made in advancing human knowledge, our ultimate aim should always be to help create a comprehensive approach to understanding our world.

This takes us to the second goal behind my dialogues with scientists—how best to ensure that science serves humanity. As humans, we face two kinds of problems, those that are essentially our own creation and those owing to natural forces. Since the first kind is created by we humans ourselves, its solution must also be within our human capacity. In contemporary human society, we do not lack knowledge, but the persistence of problems that are our own creation clearly demonstrates that we lack effective solutions to these problems. The obstacle to solving these problems is the presence in the human mind of excessive self-centeredness, attachment, anger, greed, discrimination, envy, competitiveness, and so on. Such problems also stem from deficits in our consideration of others, compassion, tolerance, conscientiousness, insight, and so on. Since many of the world's great religions carry extensive teachings on these values, I have no doubt that such teachings can serve humanity through helping to overcome the human-made problems we face.

The primary purpose of science is also to benefit and serve humanity. Discoveries in science have brought concrete benefits in medicine, the environment, commerce, travel, working conditions, and human relationships. There is no doubt that science has brought great benefits when it comes to alleviating suffering at the physical level. However, since mental suffering is connected with our perception and attitude, material progress is not enough. Even in countries where science has flourished greatly, problems like theft and violent disputes persist. As long as the mind remains filled with greed, anger, conceit, envy, and so on, no matter however perfect our material facilities, a life of genuine happiness is not possible. In contrast, if we possess qualities like contentment and loving kindness, we can enjoy a life of happiness even without great material facilities. Happiness in life is primarily a function of the state of the mind.

If contemporary society were to pay more attention to the science of mind, and more importantly, if science were to engage more with societal concerns, including fundamental human values, I believe that this could lead to great advancement and novel outcomes. Although science has not concerned itself with the enhancement of ethics and the cultivation of basic human values such as kindness, since science has emerged as a means to serve humanity, it should never be completely divorced from the values that are of great importance to the flourishing of human society.

In Indian philosophical traditions in general, and in Buddhism in particular, one finds many techniques for training the mind, such as the cultivation of calm abiding (*śamatha*) and insight (*vipaśyanā*). These definitely have the potential to make important contributions to contemporary psychology as well as to the field of education. The mental-training techniques developed in these traditions are uniquely potent for alleviating mental suffering and promoting greater inner peace. So my second goal for my dialogues with contemporary scientists is to see how these techniques, as well as their underlying insights, can be best harnessed to the task of transforming our contemporary education system so that our society does not suffer from a deficit in basic ethics.

Today no aspect of human life is not impacted by science and technology. Science occupies a central place in both our personal and our professional lives. It is critically important that we reflect on the ultimate purpose of science, on what larger consequences and impact science can have in our world. In the early part of the twentieth century, many believed that the spread of science would erode faith in religion. Yet today, in the beginning of this twenty-first century, there seems to be a renewed interest in ethics in general and, in particular, the insights of those ancient traditions that contain systematic presentations of mind science and philosophy.

Three Domains in the Subject Matter of Buddhist Texts

In our society, all sorts of immoral acts are committed on a regular basis. We observe murder, theft, cheating, violence against others, exploitation of the weak, misuse of public goods, abuse of alcohol and other addictive substances, and disregard for societal responsibility. We also see people suffer from social isolation, from vengefulness, envy, extreme competitiveness, and anxiety. I see all these as consequences primarily of our neglect of ethics and basic human qualities such as kindness. It is essential for us to pay attention to the means that would help promote basic ethics. The profound interdependence of today's world calls us to create a society permeated by kindness.

What kind of foundation is necessary for this? Since religion-based ethical teachings are grounded in the philosophical views of their respective faith traditions, an ethics contingent on religion alone will exclude those who are not religious. If ethics is contingent on religion, it will be ignored by those who adhere to no religious faith. We do not need to be religious to see the value of kindness; we can discern it by observing our everyday life. Even animals survive by relying on the care of others.

Furthermore, impulses for empathy, kindness, helpfulness, and tolerance seem naturally present in small infants, well before the influence of religious faith begins. Looking to these innate qualities and their associated behaviors as a foundation, I have striven to promote an approach to ethics and basic human values that does not rely on the perspectives of a specific religious tradition. My reason is simply this: we can enjoy a life of peace and happiness without religion. In contrast, if we are divorced from human love and kindness, our very survival is at risk; even if we do survive, our life becomes devoid of joy and trapped in loneliness.

We can promote ethics on the basis of a specific religion, but prioritizing the perspective of one religion over others is problematic in today's deeply interconnected and global society, which is characterized by a multiplicity of religions and cultures. For an approach to the promotion of ethics to be universal, it must appeal to the fundamental values we share as human beings. If we neglect these basic human values, who can we blame for the negative consequences? Thus, when I speak of secular ethics, I am speaking of these fundamental values that are inherent to human nature, and that are in fact the very foundation of the ethical teachings of the world's religious traditions.[4]

Historically, there have been societies where respect was accorded to the perspectives of both believers and nonbelievers. For example, although the materialist Cārvāka school was the object of vehement critiques from other schools in ancient India, it was a custom to refer to the upholders of that viewpoint in honorific terms. Consonant with this ancient tradition, when India gained its independence in the twentieth century, the country adopted a secular constitution independent of any specific religious faith. This establishment of a secular constitution was not to show disrespect for religion; it was to promote peaceful coexistence among all religious faiths. One of the major forces behind the adoption of this secular constitution was Mahatma Gandhi, himself a deeply religious person. Conscious of this important historical precedent, I feel no apprehension in promoting a secular universal approach to ethics.

My own personal view is that, in general, people should remain within their own traditional religions. Changing faith can lead to difficulties for oneself, and it can also undermine the basis of interreligious harmony. With this belief I have never harbored any intention to make converts or convince followers of other religions to become Buddhists. What is appropriate for believers is to contribute to the common good by practicing those aspects of the teachings that can serve humanity as a whole. Such teachings are definitely present in all the world's main religions.

Within Buddhism, for example, I see two things with the greatest potential to serve everyone, regardless of their faith. One is the presentation on the nature of reality, or "science," as found in the Buddhist treatises, and the second encompasses the methods or techniques for training the mind to alleviate mental suffering and promote greater inner peace. In this regard it is important to differentiate among three distinct domains within the subject matter of the Buddhist sources: the presentations (1) on the natural world, or science, (2) on philosophy, and (3) on religious beliefs and practice. In general, when one speaks of religion or religious practice, it is linked with faith in a source of refuge. In this religious sense, Buddhism, too, is relevant only to Buddhists and has no particular connection to those who follow other religions and those who have no religious faith. Clearly presentations rooted in religious faith are not universally applicable, especially when we recall that among today's world population, as many as a billion human beings identify themselves as nonbelievers.

Buddhist philosophy contains aspects, such as the principle of dependent arising, that can be relevant and beneficial even to those outside the Buddhist faith. This philosophy of dependent arising can of course conflict with standpoints that espouse a belief in a self-arisen absolute being or an eternal soul, but for others, this philosophy can help expand their outlook and enable them to see things in life from multiple angles, which prevents the narrow fixation that blames everything on a single cause or condition. I see great benefit in extracting the scientific and philosophical explorations found in Buddhist texts and presenting them independently of the strictly religious teachings. This allows someone who is not Buddhist to learn about the Buddhist scientific explorations of reality as well as Buddhist philosophical insights. It also gives many people the opportunity to learn how Buddhist traditions have developed their worldview and their philosophical outlook on the ultimate nature of reality.

Take, for example, the Buddha's first teaching, the four noble truths, which is common to all Buddhist traditions. In this teaching, we can observe a clear differentiation among the "ground" (the nature of reality), the path, and the result. The statements on the nature of the four truths, e.g., "This is the noble truth of suffering," present the ground; the statements on the function of the truths, e.g., "Suffering is to be known," present the path; and the statements pertaining to the agent and the fruits of the path, e.g., "Suffering is to be known, yet there is nothing to be known," explain how the result of the path comes to be actualized. My point is that, whether the presentation is of philosophy or of ethical precepts, the fundamental approach in the Buddhist texts is to ground them in an understanding of the nature of reality.

In what is called the Mahāyāna, or Great Vehicle, too, the presentation on the two truths (conventional and ultimate) is the ground, the presentation of the two aspects (method and wisdom) is the path, and the presentation of the two buddha bodies (the form and truth bodies) is the result. All of these are grounded in an understanding of the nature of reality. Even in the case of the highest aim in Buddhism—the attainment of the two buddha bodies, or the buddhahood that is the embodiment of the four buddha bodies[5]—the potency to actualize these aims can be found in the innate mind of clear light that resides naturally within us. The presentations found in the Buddhist sources are developed on the basis of an understanding of the nature of reality. If we look at the way the words of the Buddha were interpreted in the treatises composed by the great Buddhist thinkers of the past, such as the masters of Nālandā University, there too the subject matter of the entire corpus of Buddhist texts, including those that were translated into Tibetan running into more than three hundred volumes, fall into the threefold classification of the ground, the path, and the result.

As stated, the content of the Buddhist texts can be grouped within the three domains of (1) the nature of reality, or science, (2) philosophical tenets or views, and (3) religious practice, namely the presentation of the path and the way in which the results of the path are actualized. I see great benefits if we engage with the works in the Kangyur (the scriptures) and Tengyur (the treatises) on the basis of critically examining whether their contents present science, philosophy, or religious practice.

Buddhist Presentations on Reality or Buddhist Science

In brief, the presentation of the nature of reality or science in the classical Buddhist texts can be summarized in the following four major topics: (1) the nature of the physical world, (2) the presentation of the mind, the cognizing subject, (3) how the mind engages its object, and (4) the science of logical reasoning by means of which the mind understands its object. The first topic, the nature of the physical world, as well as the presentation of the philosophical outlook and methods of inquiry underlying Buddhist science, were covered in volume 1 of the series. So in this introduction I will focus my discussion on the remaining three topics, which are the topics covered in the present volume.

The Mind, the Cognizing Subject

In general, the word *science* refers to a body of knowledge about the world obtained through a method that is verifiable by anyone who repeats the same experiment. The term can thus refer both to the body of knowledge acquired and to the method used to acquire it. In other words, *science* can refer to a specific systematic method of inquiry. For example, when a scientist investigates a particular question, he or she first develops a hypothesis. Through experiments certain results are revealed, and these findings are then subjected to confirmation by a second or a third party, such as one's colleagues. When the findings of different scientists converge, these findings come to be accepted as part of the canon of scientific knowledge. The way such discoveries are made is characterized as the scientific method. This basic feature of the scientific method seems to accord with two of the three criteria of existence proposed in the Buddhist Madhyamaka texts: that it is (1) known by a conventional valid cognition and that it is (2) not contravened by some other conventional valid cognition.[6]

Discussions of the nature of cognition or the science of mind in Buddhist sources define what is cognition, categorize the types of mind, and explore those categorized minds in detail. For example, when cognitions are differentiated, we find such twofold classifications as the division into sensory and mental cognition, which is made on the basis of whether a cognition is dependent on a physical sense faculty. There is also the twofold classification of valid versus nonvalid cognition, based on whether a given cognition is veridical. A distinction is drawn also between conceptual and nonconceptual processes, based on whether a cognition engages with its object through applying a conceptual category, such as "This is so and so," and on whether a universal image is involved in the cognition. A distinction is also drawn between the mind, which is primary, and its concomitant mental factors on the basis of whether it cognizes its object as a whole or whether it apprehends specific attributes of its object. A differentiation is made between mistaken and nonmistaken cognition based on whether the object of that awareness exists in the way that it appears to that cognition. Also, in terms of its object, a distinction is made between reflexive awareness and objective awareness. Finally, based on the ways they engage their objects, there is the sevenfold taxonomy of cognition—(1) distorted cognition, (2) doubt, (3) correct assumption, (4) indeterminate perception, (5) direct perception, (6) inferential understanding, and (7) subsequent cognition (part 4).

The Buddhist sources on mind science also explore in great detail such topics as the nature of the mental factors and their divisions, their specific functions, the process by which they arise, and their inter-relations (part 2). They also explore the nature of ignorance, and the question of how ignorance that is a distorted cognition gives rise to inappropriate attention, which in turn gives rise to afflictions like attachment and aversion, and how attachment and aversion give rise to other destructive emotions like pride, jealousy, and so on, which disturb one's mental equilibrium. Likewise, they address how attachment gives rise to mental excitation and distraction to external stimuli, and how mental laxity leads to the loss of alert awareness of the chosen object. The sources also deal with the question of how mental dullness arises, which makes the mind unserviceable and brings unclarity to the mind, as if darkness has come to settle. These sources also present ways to cultivate their counteragents, such as wisdom that helps differentiate the specific characteristics of phenomena, love, empathy, forbearance, confidence, mindfulness, and meta-awareness. In short, these texts present the techniques for enhancing qualities such as those cited and also for cultivating concentration that is characterized by single-pointedness of mind, which enhances one's capacity to sustain single-pointed attention for prolonged periods (part 6).

In brief, these Buddhist sources identify over a hundred distinct mental factors and explain how certain types of mental factors act as antidotes to other factors, and how this law of contradiction within the mental world facilitates the possibility of eliminating certain types of afflictions through enhancing the power of their counteragents. Thus this science of mind found in the Buddhist sources is something meaningful with a potential to benefit the more than seven billion human beings on Earth.

The Buddhist sources also present various levels of subtlety of consciousness. For example, consciousness during the waking state is considered grosser, consciousness in dream state comparatively subtler, and compared to that, consciousness during deep sleep is subtler still. There is also the differentiation of consciousness into the threefold category of gross, subtle, and very subtle. Within this classification, the first encompasses the five sense perceptions, the second includes the six root afflictions as well as the eighty indicative conceptions, while the third, very subtle level of consciousness, includes the minds of the four empty states.[7] Even among the minds of the four empty states, differentiations of subtlety can still be drawn. For example, compared to the innate mind of clear light, the three minds of luminosity are grosser, while the innate mind of clear light is understood to be the most subtle level of consciousness. This innate mind is characterized as energy-wind from the point of view of its movement toward an object, and as consciousness from the point of view of awareness (part 3).

An important set of topics within the Buddhist science of mind include how, given that the adventitious stains do not reside in the mind's essential nature, the essential nature of mind is that of clear light; how the continuum of this luminous and knowing reality is stable since it has no beginning; and how the qualities of the mind have the potential for limitless enhancement, for once perfected they do not require exertion of new efforts.[8] These above points are so critical that anyone who lacks deeper knowledge of these will only have partial understanding of the great Buddhist treatises.

Similarly, there are ideas and insights in Buddhist epistemological treatises, such as those of Dignāga and Dharmakīrti, that can enrich contemporary cognitive science. These include, among others, the presentations on the nature of sensory cognitions, the logical reasoning establishing how sensory cognitions are devoid of conceptualization, how thoughts engage actual reality via the medium of universals, the characteristics of universals, how general terms and cognition of universals apply to the plurality of particulars, and the various arguments put forth to demonstrate the unreality of general characteristics.

How the Mind Engages Its Object

Based on how a given object appears to the mind, a distinction is drawn between negatively characterized phenomena and positive phenomena. Similarly, based on whether the given cognition engages its object due to the object casting its form to it, there is the distinction between cognitions that engage by way of exclusion and those that engage by way of affirmation. There is also the distinction between identity and difference based on how thoughts conceive their objects, because when the object universal of a thing appears to a thought, it does so either as singular or plural. Similarly, what is essentially a single entity can be understood to possess conceptually distinct attributes, with one serving as the basis for inferring the other; for instance, being a *product* can help infer something to be *impermanent*, although that which is a product and that which is impermanent are one and the same entity. Also when a single cognition takes something as an object, one can distinguish two aspects of such cognition, the way it appears and the way it is apprehended, so a distinction can be drawn between the mind's appearing object and its object as cognized. The latter is known also as the engaged object, for when the person engages with that object just as apprehended by that cognition, he or she is not deceived. Similarly, a thing can be characterized as the object or focus of a given cognition because it serves as the basis for the mind to eliminate false conceptualizations (part 4).

These Buddhist sources also contain debates on whether sensory cognitions perceive their objects by assuming an image of the object being perceived. Those who advocate this notion of the object image state that when sensory cognitions perceive their objects, they do so by having a likeness of the object appear to the mind. If this is not the case, they argue, one will not be able to account for the diversity of perceptions that one can experience in relation to even a single object. Those who reject the notion of images argue that if sensory cognitions perceive via the medium of such images, this would mean that they do not directly perceive their objects, and furthermore, there is no evidence for the existence of external objects that are not perceptible by the senses (part 4).

The Science of Logical Reason by Means of Which the Mind Understands Its Object

The science of reasoning represents the means by which the mind engages its object, and it is like a key that helps open the doors to the secrets of reality that remain hidden from our senses. The science of logical reasoning is therefore extremely important. The basis for the application of logical reason lies in the four principles of reason, which we discussed in volume 1 (pages 10–12). The important topics covered in the Buddhist sources on logical reasoning include the characteristics necessary for a valid proof, the logical relationship between the evidence and the thesis being proven, and so on. Although, in general, one can infer the presence of something related on the basis of its relationship to something else, not all specific characteristics of that related thing can be inferred. For example, one can infer a cause from the presence of its effect, and similarly, one can draw inferences based on a shared essential nature; however, not all features of the cause or the thing can be inferred through such types of reasoning (part 5).

Also, based on the logical relationship that exists between the evidence and the thesis within a given syllogism, correct evidence is classified into three categories (effect evidence, nature evidence, and evidence consisting in nonperception).[9] Within the category of effect evidence, depending on how the syllogism is presented in line with the intention of the person to whom it is directed, five types of effect evidence are distinguished in Buddhist sources. As for nature evidence, a differentiation is made into two types, based on whether the term that states the evidence indicates an activity of a person. Within the class of evidence consisting in nonperception also, two types are identified based on whether the nonperception takes the form of the nonperception of a fact that is logically related to what is being negated or whether the nonperception is being proven on the basis of the perception of an incompatible fact. In general, in the case of something that exists but lies beyond our perception, mere nonperception cannot prove its nonexistence. Nonetheless, even here, one can demonstrate the inappropriateness of a claim on the basis of the absence of its knowledge. In contexts where a fact under discussion is something that would be perceivable were it present, here nonperception can lead to correct knowledge of its nonexistence. Thus there are two general classes of evidence consisting in nonperception (part 5).

With this second category of evidence consisting in nonperception —namely, where what is being negated is in general perceivable—there is the nonperception of something related whose presence is necessarily dependent upon the presence of its correlate. An example of this type of reasoning is where the absence of fire, on which the existence of smoke depends, is used as a basis to infer the absence of smoke. Within this type of reasoning, through the nonperception of something related, the related thing might be its cause, its universal, or its essential nature. Thus there are three kinds. In the case of evidence consisting in nonperception based on the perception of something contradictory, there is the one where the evidence is effected by the perception of something contradictory in the sense of being contrary. There is also a second type where the evidence is effected by the perception of something contradictory in the sense of being mutually exclusive without any possible alternative, as in the law of excluded middle. The Buddhist reasoning of dependent origination used to negate self-existence, for example, belongs to this second type of reasoning. Within the first kind, the perception of something contradictory in the sense of being contrary, again there are three kinds: that of essential nature, that of effects, and that of pervading universal. Within each of these three, one can further distinguish among the opposing fact itself, its effect, and its pervading universal; thus twelve distinct forms of reasoning can be differentiated.[10] In fact, there can be many different ways in which this type of reasoning can be further distinguished (part 5).

Further, there is the reasoning in the form of a consequence that reveals a contradiction within the opponent's position. Although this type of consequence-revealing reasoning on the whole negates viewpoints primarily through demonstrating a logical contradiction within the opponent's position, there are some that also implicitly present correct logical syllogisms to establish a given thesis. Those that do not imply such a correct syllogism negate their opponent's viewpoint by leveling objections.[11]

In brief, the science of logical reasoning offers a multiplicity of avenues to critically analyze reality, and through these diverse avenues, insight into the ultimate nature of reality can be discerned. When one engages with reality repeatedly through these avenues of inquiry, the cognitive power of one's mind will come to be enhanced, leading to the development of a balanced view when it comes to understanding the ultimate nature of reality.

Buddhist Philosophy

Philosophy represents the summation of the conclusions about the nature of reality developed through critical inquiry. In *Science and Philosophy in the Indian Buddhist Classics*, philosophy will be treated in volumes 3 and 4, but I will touch on it briefly here. In Buddhism, works explicitly presenting philosophical views evolved early. We see this with the appearance of the *Questions of King Menander* before the Common Era, the Abhidharma treatises starting around the first century CE, the six philosophical treatises of Nāgārjuna shortly thereafter,[12] and so on. There also appeared, in Buddhism's classical era, treatises in which the principal views of both Buddhist and non-Buddhist Indian schools were presented together in a single work and critically examined. For example, Bhāvaviveka composed his *Blaze of Reasoning* in the fifth century, and in the eighth century Śāntarakṣita authored his *Compendium of Reality*.

Buddhism's basic philosophy is encapsulated in what are known as the view of the four "seals," or axioms: all conditioned things are impermanent, all contaminated things are characterized by suffering, all phenomena are empty and devoid of selfhood, and nirvāṇa is peace. *Impermanence* refers to the fact that things, right from their birth, do not remain static even for a single moment. This is because things do not depend on some third condition for their disintegration; the very causes that produce them also make them susceptible to disintegration. We can see this truth of impermanence for ourselves if we contemplate deeply the gross changes we observe in things. The statement that "all contaminated things are characterized by suffering" indicates how our existence is bound to a causal nexus of undisciplined states of mind that keeps it under their power. As for the statement "nirvāṇa is peace," Dharmakīrti identifies this with the possibility of eliminating pollutants from the mind. He establishes the existence of such a state of freedom through reasoning so that we do not need to rely on faith alone to explain it. The teaching on no-self relates principally to the ultimate nature of reality, namely that things do not exist the way they appear to.

All Buddhist schools reject the existence of a self that is eternal, unitary, and autonomous. Yet many Buddhist schools assume that what we call "self" or "person" must nonetheless exist in some form. We find the assertion that the self exists on the basis of the aggregates, with some proposing that all five aggregates constitute that person and others positing the mind alone to be the person. Some, recognizing that the six types of consciousness are unstable like bubbles in water, assert eight classes of consciousness and posit foundational consciousness (*ālayavijñāna*) to be the real person. Others, seeing faults in identifying the person with the aggregates, assert a self (or person) that is neither identical to nor different from the aggregates.

As we can see, there is a divergence of interpretations and subtleties among the various Buddhist schools with respect to the meaning of no-self. The Tibetan tradition relies chiefly on the interpretation of the Perfection of Wisdom sūtras by Nāgārjuna and his disciples. In this view, the meaning of no-self is understood by way of dependent origination. Two types of selflessness are differentiated from the perspective of their bases (persons and phenomena), but there is no difference in subtlety in what is negated; in both contexts, it is independent existence. The very fact that things are dependently originated establishes that they are devoid of self-existence. When we think, for example, in terms of the designator and the designated, the knower and the known, the agent and the act, and so on, we can see the utter mutuality and contingency of these things. If the table in front of us, for example, were to exist objectively without depending on conceptual designation, the table itself could provide the criteria of what constitutes a table from its own side. This is not the case. We have no choice but to accept that what we call "table" is posited by the mind.

What we see is a mutual dependence. The objective world exerts constraints on the mind, and the mind in turn exerts constraints on the objective world. Take the simple example of a handwritten letter *a*. So many factors converge that are part of its dependent reality. There is, for example, the shape of the letter, the pen that wrote it, the ink used to write it, the paper on which it is written, the person who wrote it, the intention of the writer, the convention that established this letter, those who accept this as a letter, and the cultural environment in which this letter has a meaningful usage. Without these, its existence as a letter is simply impossible. The nature of all things is exactly like this. Therefore things are explained as having a nature of dependence requiring so many other factors for their existence. This is why Madhyamaka thinkers such as Candrakīrti speak of how things are unfindable when subjected to ultimate analysis and of how their existence can only be posited as designated by the mind. This view is strikingly similar to explanations found in contemporary physics about how nothing can be found to possess reality when analyzed at the subatomic level.

Another important philosophical view in Buddhist texts is that of the two truths. We find the language of "two truths" in the non-Buddhist Indian philosophical schools as well. In Buddhism, all four schools of thought equally accept the notion of two truths, but what constitutes these two varies from school to school.[13] Between the two Mahāyāna schools, for example, there is not much difference in the way the Cittamātra (Mind Only) and Madhyamaka (Middle Way) schools define the two truths. Nonetheless there is a substantial difference in the specific examples they give for those two truths.

In brief, *ultimate truth* pertains to the ultimate nature of things while *conventional truth* relates to perspectives rooted in the apparent world. Both the Madhyamaka and Cittamātra schools explain ultimate truth in terms of emptiness. Cittamātra speaks of the emptiness of external reality or the emptiness of subject-object duality, while Madhyamaka speaks of the emptiness of real existence of everything, even the minutest particles of matter. Conventional truth encompasses the entirety of the everyday reality we perceive—the natural world, the beings who inhabit it, arising and disintegration, progress and decline, cause and effect, happiness and suffering, good and bad, and so on. In short, the flowerpot we see in front of us is conventional truth, while its absence of objective existence—that this pot cannot be found when sought through ultimate analysis—is its ultimate truth.

That pot is empty at the very moment it is perceived, and it can be perceived while simultaneously being empty. Madhyamaka thinkers explain this by saying that the two truths have the same nature but are conceptually distinct. When the Buddhists speak of the way things exist, they maintain that we need to transcend both extremes—the extreme of reification and the extreme of denial—and view things simply as they are.

Buddhist Religion

Generally speaking, although aspects of the Buddhist tradition that fall under religion are connected with faith, the basic framework of Buddhist religious practice is grounded in the principle of causality, which is part of the laws of nature. For example, the impulse to shun pain is part of our natural disposition, and our existence as conditioned beings is the basis for the arising of suffering. Therefore Buddha taught the reality of *suffering* as the first truth of our existence. Since suffering necessarily arises from a cause, he identified the second truth as the *origin* of suffering. These two truths pertain to the cause and effect of suffering. What is the cause of suffering? Its ultimate source is explained as ignorance, and since this ignorance can be brought to an end, the Buddha taught the third truth, the *cessation* of suffering and its origin. Since such a cessation must also have a cause, the Buddha taught the truth of the *path*, the means of attaining such a cessation. There is thus a cause-and-effect pair of truths pertaining to the attainment of freedom. Clearly the foundation of Buddhist practice described in the four noble truths is the natural law of cause and effect.

When Dharmakīrti introduces the truth of cessation, he demonstrates the possibility of bringing an end to ignorance, the cause of suffering. Nowhere does he speak of the need to demonstrate the truth of cessation by relying on scriptural authority. Furthermore, Dharmakīrti offers a profound explanation of suffering and its origin in terms of the sequence of the twelve links of dependent origination, and cessation and the path in terms of the reverse order of the twelve links.[14] Since happiness and suffering are characteristics of sentient experience, no account of them can be divorced from sentient experience. Therefore Dharmakīrti also offers an extensive account of cause and effect as it relates to the inner world of experience. Furthermore, when one speaks of Dharma (religion) in Buddhism, its true meaning must be understood in terms of the attainment of nirvāṇa. The term *Dharma* refers to the means and the path that lead to nirvāṇa as well as the scriptures taught by the teacher, the Buddha, that present this path.

Having now distinguished three domains of subject matter in the Buddhist sources—(1) scientific presentations about the natural world, (2) philosophy, and (3) religious practice—we might ask from what sources the presentations in this series on the first two dimensions, science and philosophy, are developed. Among the Buddhist classics available in Tibetan, we have the two canonical collections introduced above. The precious collection of the Kangyur contains translations of the Buddha's words as embodied in the "three baskets" (Tripiṭaka), containing both sūtra and tantra teachings. The precious collection of the Tengyur contains the treatises of great masters such as the seventeen Nālandā masters that include Nāgārjuna and Asaṅga, the two trailblazers prophesized by the Buddha.[15]

The Tibetan translations that comprise the Kangyur and the Tengyur are the largest body of Indian Buddhist texts extant today anywhere. Today modern scholars who engage in objective studies of Indian Buddhist sources state that these Tibetan collections not only contain the largest number of texts but also represent the best translations and most comprehensive Buddhist canon. Many of these works composed in classical Indian languages, especially Sanskrit, were entirely lost in their original language through changes of history and environmental conditions. Only a few of the great works remain in original Sanskrit. In the Pali canon, we find scriptures associated with the Theravāda tradition but not of other Buddhist schools, such as the Mahāyāna sūtras and tantras. Although a great number of Buddhist texts were translated into Chinese, modern researchers say that because of the character of the Chinese language, those translations tend to be looser and do not match the rigorous correspondence, both in terms and meaning, found in the Tibetan translations. Today, therefore, the Tibetan language is the storehouse for the Buddha's scriptural teachings in their entirety. By offering access to the complete system of the Indian Buddhist tradition encompassing all three vehicles—the shared teachings, the Mahāyāna, and the Vajrayāna—there is simply no alternative to the literary heritage of the Tibetan language.

On This Compendium of Buddhist Science and Philosophy

Unlike the world's other major religions, the Buddhist tradition's canons contain an extremely large number of texts. Even in the case of the part translated into Tibetan, there are more than five thousand individual texts in over 320 large volumes. The size of the collection means that it would be difficult for a person to read the entire collection even once. So a tradition emerged, from the period of the trailblazers, to extract the essential points from this vast body of scripture and present them in accord with the interests and capacities of the aspirants, in accessible formats such as compendiums and manuals. For example, Nāgārjuna composed the *Compendium of Sūtras*; Śāntideva too composed a compendium of sūtras[16] as well as the *Compendium of Training*; the glorious Atiśa composed the *Extensive Compendium of Sūtras* (*Mahāsūtrasamuccaya*); and the trailblazer Asaṅga composed his *Compendium of Knowledge*. Similarly, Dignāga wrote his *Compendium of Valid Cognition* (*Pramāṇasamuccaya*), bringing together the essential points of numerous works he had authored previously, such as his *Analyses*[17] and his short verse texts. All of these various compendiums proved to be of tremendous benefit to subsequent students of Buddhism.

Taking these precedents as our inspiration, I recognize that in today's time, too, presentations based on the words of the excellent teacher, the Buddha, pertaining to the basic nature of reality as well as associated philosophical concepts can be a source of benefit to humanity, irrespective of whether one is Buddhist or non-Buddhist, religious or not religious. My aspiration has been to see the creation of these compendiums in a format consistent with the approach of contemporary academic scholarship. This way these presentations can benefit many people. So several years ago I discussed this vision with others and tasked a group of scholars to initiate the project. Today, this group has completed the work of creating compendiums on the presentations on the nature of reality and on philosophy, the first two domains within the threefold division of the subject matter of Buddhist texts. With great efforts the compilers have gathered a vast number of citations from authoritative sources relevant to these two domains. Thus my wish to see such compendiums on science and philosophy from the Buddhist classics—wherein the presentations on these two domains are explained separately in their own rights—has today become a reality. I offer my appreciation to the compendium editors as well as to those senior scholars who have advised the editors. I also thank the translators who have rendered these volumes into other languages.

Given that the scriptures and their commentarial treatises in the Buddhist classics are so vast and profound in their meaning, it is conceivable that there are shortcomings in these volumes in the form of omission, over-reading, or even error. At the very least, what the editors have achieved is a series that demonstrates with clarity that there exists within the subject matter of the Buddhist texts three distinct domains of science, philosophy, and religious practice. If, in the future, there should be a need for additional material or deletion of some elements, the structure is now in place so that such modifications can be made easily.

In conclusion, I would like to share my hope that these volumes on the presentations on the nature of reality and their associated philosophical concepts from the Buddhist sources will make an important contribution to our collective human knowledge by offering the gift of a new set of insights. I pray that these volumes become a source of great benefit to many people.

The Buddhist monk Tenzin Gyatso, the Dalai Lama
Introduction translated into English by Thupten Jinpa

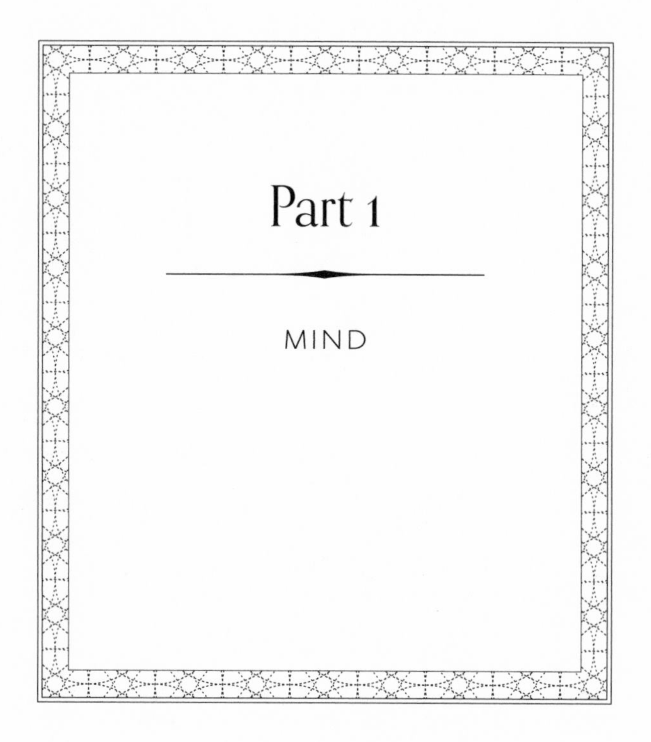

Part 1

MIND

The Path of Knowledge

From their earliest days, Buddhist traditions have emphasized the importance of the mind. Traditional accounts maintain that Prince Siddhārtha, who went on to become the historical figure we call the Buddha, the "Awakened One," experienced a kind of spiritual crisis some time in his twenties. The crux of that crisis was the problem of *duḥkha*, a term that is usually translated as "suffering," but which also points to a subtler, more elusive sense of persistent dissatisfaction. It is said that the young prince left his life of leisure and privilege and set out to solve the problem of suffering, and he encountered a panoply of options offered by various spiritual guides. Among the traditions he encountered, some urged spiritual seekers to manipulate their bodies through physical practices, including severe austerities. Such approaches, in various ways, see the problem of suffering as primarily a physical—and not a mental—issue, and they thus propose that the problem of suffering can be solved through manipulating the external world, the body, or both. Traditional accounts maintain that young Prince Siddhārtha plunged into such practices, and eventually, owing to the emaciation brought on by his physical austerities, he could press his hand on his belly and easily feel his backbone underneath. This hyperbolic image of the gaunt prince starkly conveys the strength of his determination to find a physical resolution to the problem of suffering and dissatisfaction.

In the end, however, Siddhārtha turned away from this more physical approach and adopted a perspective that appears to have been increasingly influential in his time. The physical world and the body itself must be maintained, but the problem of suffering cannot be eliminated through purely physical manipulations. Instead, suffering arises from a fundamental distortion in how we experience the world, such that we live in a state of perpetual ignorance (Skt., *avidyā*) or confusion (*moha*). Thus, to relieve suffering, one must remove that fundamental confusion by counteracting it with wisdom (*prajñā*), which "sees things as they truly are" (*yathābhūtadarśana*).

By interpreting suffering as a problem of ignorance, Siddhārtha had embarked on a spiritual path that, in common with some other Indian traditions, came to be known as a "path of knowledge" (*jñānamārga*). For any path to knowledge, including all Indian Buddhist traditions, the fundamental goal of philosophy and contemplative practice is to uproot the confusion that underlies all suffering. Buddhist accounts often focus on exactly what constitutes ignorance—the foundational cognitive distortion that lies at the root of suffering—since identifying ignorance properly enables one to cultivate its antidote. Different levels of philosophical analysis interpret ignorance differently, and at the most basic level, ignorance concerns the cognitive distortions that induce the sense that one has a fixed and completely autonomous personal identity. Subsequent volumes in this series will examine not only that basic level but also the more finely grained accounts found at higher levels of analysis, such as the radical anti-essentialism found in Madhyamaka philosophy. For all these levels of analysis, the key point is that the defect that produces suffering and dissatisfaction is within the mind itself, and Buddhist theorists from the earliest period onward were thus obliged to engage in an extensive and robust inquiry into the nature of the mind. In that endeavor, they explored the processes of cognition, the contours of affective states, the constituents of reliable knowledge, the meditative methods for transformation, and so on. This volume, compiled by a skilled team of Tibetan scholars, presents the Indian Buddhist account of the mind and its workings, and in this first part, our authors focus on the nature of the mind itself, along with key issues related to cognition. Before examining some key issues in part 1, let's clarify the sources and methods for this account.

Sources and Methods

As noted in Thupten Jinpa's introductory essays in the first volume of this series, the main sources for this compilation come from Buddhist works written originally in Sanskrit, although some are now only available in Tibetan translation. The present volume on mind cites dozens of sources, including discourses (*sūtras*) attributed to the Buddha, but three genres are especially central: the Abhidharma corpus, the *pramāṇa* or epistemological literature, and manuals for contemplative practice. Also key to the first volume of this series, the Abhidharma corpus emerges from the earliest period of Buddhist history, and these texts are especially concerned with giving a systematic account of the fundamental constituents of the mind. One of the most enduring Abhidharma contributions is the account of "mind and mental factors" that informs part 2 of this volume, but Abhidharma sources are often cited in other contexts as well, with the *Treasury of Knowledge* (*Abhidharmakośa*) by Vasubandhu (ca. fourth century) a frequent source.

Unlike the Abhidharma, the epistemological literature emerges later in Buddhist history, and it is associated especially with the Indian theorists Dignāga (ca. fifth–sixth century) and Dharmakīrti (ca. sixth century). For analyses of perception, inferential reasoning, concept formation, and other cognitive processes, Dharmakīrti's works in particular are a key source. Our authors also cite the epistemological texts for their influential theories on the nature of mind.

The great Buddhist authors of India also wrote treatises (Skt., *śāstra*) that sometimes defy easy categorization, but many of these can fit under the general rubric of "contemplative manuals," in that even when straying into abstruse philosophy, they remain centrally concerned with providing instructions, theories, or explanations for effective contemplative practice. Among the many such sources cited by our authors, one of the most frequent is *Engaging in the Bodhisattva's Deeds* by Śāntideva (ca. seventh–eighth century), which is exemplary in the way it interweaves straightforward instructions for contemplative practice with philosophical analysis.

Traditionally, all of these sources are organized according to the hierarchical schema of the "four schools" of Buddhist philosophy: the Vaibhāṣika, Sautrāntika, Yogācāra, and Madhyamaka. The Madhyamaka school is traditionally considered the "highest" because its critiques of essentialism are said to yield the most accurate account of what truly exists, but the models of mind and cognition endorsed by the Madhyamaka are drawn almost entirely from the three lower schools. The third school, Yogācāra, is often considered a form of philosophical idealism, such that the existence of matter is critiqued. While the Yogācāra school provides some key sources for the models of mind and cognition presented in this volume, it is the two lower schools—the Vaibhāṣika and the Sautrāntika—that provide the analytical perspective for the vast majority of the content articulated here. Setting aside the finer distinctions between the two lower schools, the most relevant aspect of their perspective is that they endorse a form of dualism, such that the world consists of irreducible physical stuff—generally conceived as infinitesimally small particles of matter—and irreducible mental stuff, the various components of consciousness and cognition.

Methodologically, our authors adopt a traditional, pragmatic approach to this wide-ranging material. Since the higher levels of analysis provided by the Madhyamaka and Yogācāra schools are often counterintuitive, and since the basic models of mind, affect, and cognition can be adequately articulated without appealing to those higher levels of analysis, our authors primarily choose to present this material from the lower level of analysis assumed by the Vaibhāṣika and Sautrāntika schools. Higher levels of analysis are occasionally invoked, but this basic level, where mind and matter are kept distinct, is the default perspective for most explanations, in part because it does not stray too far from our ordinary intuitions. Still, as will become apparent, those ordinary intuitions are often called into question even at this lowest level of analysis, and this is precisely why that more basic level is pragmatically useful for explanatory purposes, since a critique that strays too far from our ordinary intuitions might prove challenging in an unnecessarily distracting way.

The Nature of Mind

The first part of this volume presents an account of mind drawn from Indian Buddhist sources, and this immediately raises some difficulties for readers accustomed to notions of mind in Western philosophy, psychology, and cognitive science. One key issue is that the term *mind* in Western contexts suggests a single entity that endures over time and has various capacities, dispositions, or features. In contrast, the Buddhist sources cited by our authors maintain that mind is *episodic*, such that a mind (Skt., *citta*) is a continuum (*santāna*) of mental moments, each moment causally emerging from the previous moment and acting as a cause for the subsequent moment. Each mind is thus a unique moment of consciousness (*jñāna*) or awareness (*saṃvitti*). The analysis of the nature of mind is thus actually an analysis of what, in many Western contexts, would be a moment of mind or a "mind event." In a way that can be additionally confusing, Buddhist authors will often speak of plural "minds" that pertain to the same person at different points of time or in different contexts, such as the mind in a moment of visual consciousness or the mind in a moment of one-pointed concentration. Both of these minds could be in the same continuum, such that in referring to someone named Jane, one could speak of "Jane's many minds," a locution that seems odd in Western contexts. In our translation, we have tried to avoid using the plural *minds* as much as possible, but it is important to note that, even in the singular, the term *mind* refers to a single mind event—that is, a discrete moment in a mental continuum.

Turning now to the nature of mind, our authors focus on the most widely cited account—namely, that mind is clear (Tib., *gsal ba*) and aware (*rig pa*). Here, the term *clear* renders two distinct Sanskrit terms that evoke the phenomenal character of mind and also one of its essential properties. In relation to the Sanskrit term *prakāśa*, the Tibetan translation is most accurately rendered as "luminous," in the sense that the mind "illuminates" or presents contents, much as a lamp illuminates whatever is nearby. Unlike a physical lamp, however, the mind does not depend on proximity to "illuminate" whatever is present in a moment of consciousness; a concept of the Eiffel Tower, for example, can be presented in consciousness without any need for one to be in Paris. A second key feature is that, even in exceptional cases where a moment of consciousness has no cognitive content, a mind or moment of consciousness still includes the phenomenal character of presenting or illuminating, even though there is no content to be illuminated.

The mind is also clear in that it is transparent (Skt., *prabhāsvara*). Here, the term *clear* points to a fundamental property of mind. Water, for example, is by nature clear, in that even when it is murky, the impurities that obscure it can be removed, and its natural clarity or transparency will return. Likewise, the mind is clear in that no particular object (such as the Eiffel Tower) or affective state (such as anger) is essential to a moment of consciousness. This point is especially crucial for Buddhist approaches to personal transformation and behavior change, since it means that the dysfunctional habits that produce suffering and dissatisfaction can be transformed, precisely because they are not essential to the mind itself. This means in particular that ignorance—the fundamental cognitive distortion that underlies suffering—is not an essential property of the mind, and it is therefore possible to remove that distortion without putting an end to consciousness itself.

While the mind is clear—or perhaps "luminously clear," to capture the two senses encompassed by that term—it is also aware (Tib., *rig pa*; Skt., *saṃvit*). In general, this means that a mind or moment of consciousness has an epistemic character; that is, the mind does not simply illuminate, it also does so in an informative way. More specifically, at the level of analysis deployed by our authors, a mind or moment of consciousness is *about* its object, and it presents that object in a way that is relevant to action that engages that object. Our authors point out that mind, by virtue of being aware, is distinct from matter, which lacks this character. Yet two points here are crucial. First, while the mind is distinct from matter, it nevertheless depends on a material "basis" for it to function. In other words, the mind is necessarily embodied, and the constraints posed by a particular embodiment—such as the capacities of one's sensory organs—must be taken into account when examining cognitive processes and other aspects of the mind. Second, while the mind is intrinsically aware by virtue of presenting objects as relevant to embodied action, it is not necessarily the case that a moment of awareness—that is, any given mind event—provides reliable information about its object. It is for this reason, in part, that Buddhist theorists are so concerned to distinguish the many varieties of awareness.

Varieties of Cognition: The Case of Sensory and Mental Consciousness

Much of part 1 is devoted to examining different ways of categorizing mind, and of particular note here is the distinction between sense consciousness and mental consciousness. In this context, the use of the term *minds* in the plural does not apply just to the continuum of mind events. We can also properly say that a sentient being in any given moment may have multiple minds. This points to an important feature of the Buddhist account presented by our authors—namely, that it adopts what could be called a modularity thesis, similar to some contemporary accounts in cognitive science. Modularity itself is a complex topic, and contemporary notions encompass a variety of competing approaches. In simple terms, *modularity* means that distinct cognitive processes are executed by distinct mental modules that are "encapsulated," in that they can function independently of other modules. Strong forms of modularity often connect these functions to specific brain regions, and they may posit modularity even at complex levels of processing. Weaker forms of modularity assert that modules operate at more basic levels of processing, and such theories may not maintain a strong correspondence between a module and any localized brain region. From the perspective of this highly simplified account of modularity, the Indian Buddhist theorists assert a weak form of modularity, especially in regard to the five forms of sense consciousness.

In brief, an instance of sense consciousness emerges initially when the physical sense faculty contacts the object and thereby becomes its *dominant condition* (Skt., *adhipatipratyaya*). With additional conditions in place, the mental processes required to produce, for example, the first moment of a visual consciousness can occur simultaneously with the processes that produce the first moment of any other sense consciousness, such as an auditory or olfactory one. In this way, multiple sense "minds" can (and usually do) arise simultaneously, and this suggests that sense consciousness exhibits at least a weak form of modularity. For at the low level of processing required to produce the first moment of a sense consciousness, the five kinds of sense consciousness operate independently of each other.

Our authors point out that the low level of processing that produces an initial moment of sense consciousness is not sufficient to induce an action that engages with a sensory object. To facilitate action toward a sensory object, it must be conceptualized or categorized, and this can only occur when the data from sense consciousness moves into mental consciousness. At that point, on the basis of additional conditions such as desires and goals, a concept that facilitates goal-oriented action can occur. Now, however, the modularity related to sense consciousness no longer applies, because only one mental consciousness can occur at any given time. When conceiving sense objects, mental consciousness depends on the lower-level processing provided by sense consciousness as well as other mental processes such as memory.

One important aspect of this articulation of six types of consciousness—mental consciousness and the five forms of sense consciousness—is that it reflects the commitment to developing models of cognition that do not require an autonomous self (Skt., *ātman*) or perceiver (*bhoktṛ*) as the agent of a cognitive act. Drawing on the work of Dignāga and Dharmakīrti, our authors point out that the sense of subjectivity in any moment of consciousness is simply a momentary "phenomenal form" or "image" (*ākāra*) that emerges simultaneously with the image or representation of the object. This "image" of subjectivity thus has no causal role or agency in that moment of visual perception; it instead reflects a basic structural feature—the subject-object relation—that characterizes any moment of consciousness bearing on an object.

Epistemic Reliability and the Suspicion of Concepts

The emphasis on building cognitive models that set aside any notion of an autonomous, unchanging "self" acting as the agent of cognition returns us to the basic problem of suffering and dissatisfaction. As noted above, Buddhist theorists in India maintain that the fundamental problem of suffering is caused by ignorance, a fundamental cognitive distortion whose most basic manifestation is precisely this sense of an autonomous, unchanging "self" as the agent of actions, the perceiver of perceptions, the controller of the mind-body system, and so on. For these theorists, this distortion can only be solved by cultivating a form of wisdom that counteracts it. A basic theory here is that two cognitive states can stand in opposition to one another such that one necessarily inhibits the other from arising. An additional claim is that, when a nondeceptive (Skt., *avisaṃvādi*) cognition—one that is epistemically reliable—comes into conflict with an unreliable one, the reliable cognitive state, if sufficiently robust, will always inhibit the unreliable one. Moreover, the dispositions that cause the unreliable or distorted cognition to arise can eventually be eliminated by using contemplative techniques to immerse oneself in the experience of the reliable cognition. This basic model, discussed at length in part 6, applies to all cases where one seeks to eliminate cognitive distortions, and it applies especially to the cultivation of wisdom as a means to uproot ignorance.

Given the central role played by epistemic reliability or nondeceptivity in practices designed to transform a practitioner, it is no surprise that this issue surfaces repeatedly in Buddhist analyses of cognitive processes. The Indian Buddhist tradition contains an enormous amount of material simply on the question of epistemic reliability, especially in the context of a *valid cognition* (*pramāṇa*), which is both reliable and also a motivator of action. The foundational question of epistemic reliability leads to many other nuanced and subtle inquiries that produce the taxonomies in this section of part 1. One intriguing distinction that emerges in these taxonomic analyses is the notion that epistemic reliability can still apply to cognitions that are "mistaken" (*bhrānta*). Well-formed inferences, for example, are always epistemically reliable, but since they are necessarily conceptual, they are also mistaken. Here, epistemic reliability is rooted in the way that a cognition facilitates effective action in relation to an object, and in part this means that a cognition need only be accurate in regard to the causal capacities of the object relevant to one's goal-oriented action. Thus, if I infer from billowing smoke that a fire is occurring in a particular location, my conceptual cognition of fire can enable me to take effective action—whether I seek warmth or want to douse the fire. Yet that conceptualization of "fire" itself is also mistaken, precisely because it is conceptual.

The notion that conceptual cognitions are necessarily mistaken—even when they are epistemically reliable—reflects an overall suspicion of conceptuality that characterizes Indian Buddhism from its earliest days, but the technical account in part 1 draws especially on Dharmakīrti and other Buddhist epistemologists. For these theorists, conceptual cognitions are always mistaken in two ways. First, the object that appears phenomenally in my awareness, known as the conceptual "image" (*pratibimba*) of the object, is taken to be identical to the functional thing that I seek to act upon as the *engaged object* (*pravṛttiviṣaya*) of my action. In other words, the phenomenally presented object "fire" in my conceptual cognition does not have the causal properties of an actual fire—the thought of a fire cannot burn wood. Yet our cognitive system creates a fusion (*ekīkaraṇa*) of this phenomenal appearance with the engaged object to which the conceptual image of "fire" refers.

Conceptual cognitions are also mistaken in another way: they take the categories presented in conceptualizations as truly real. Buddhist epistemologists say those categories are actually constructed through the process of concept formation. For example, the conceptualization of fire, when taken as referring to a real, causally efficient thing, presents that thing as bearing the same fundamental properties—some essence that constitutes a thing as "fire"—as all other things that can be categorized as "fire." This projection of our categories into the world, however, is false for these theorists, since they maintain that all instances of things that we call "fire" are entirely unique. Instead, we construct concepts and categories through a process of exclusion (*apoha*), whereby the cognitive system forms categories primarily by excluding what is incapable of or irrelevant to the desired causal outcomes.

This brief excursus into the questions of reliability, error, and conceptualization demonstrates the finely grained and insightful analyses typical of the material found throughout this volume. Much more could be said about the taxonomies presented in part 1, but as is perhaps already evident from the approach to conceptualization discussed above, many of these materials point to a key issue: the primacy of direct perception, or what might be called a kind of "empirical stance."

Empiricism and Buddhist "Science"

In his first essay in volume 1 of this series, Thupten Jinpa speaks eloquently about the notion of "Buddhist science" and the ways we might understand that term. Readers are encouraged to consult that essay for an appreciation of the methodological and theoretical commitments that Buddhist authors hold, and the way we might answer the question "Is this science?" Here, deferring to Jinpa's lengthier discussion, I will just examine two issues that point to the type of "yes and no" answer that he gives.

One way to understand what we mean by *science* concerns the scientific method and the various commitments that it entails. In terms of the process of implementing research, an idealized and simplified account of *science* would involve: (1) formulating a theory, (2) generating hypotheses based on the theory, (3) testing hypotheses through experimentation, and (4) revising or confirming the theory based on the results of experiments. This abbreviated and idealized account excludes many issues, such as the way institutional and cultural pressures might prevent theory revision even in the face of contrary experimental evidence. Yet even setting aside these issues, this idealized process does point to one challenge for the notion of Buddhist science: theory revision.

This volume focuses especially on Indian Buddhist sources written in Sanskrit, and these texts emerged over more than a millennium of inquiry. We might wonder how exactly these theories and models were formulated. For example, what role did actual phenomenological inquiry play, especially if investigated with contemplative techniques? Were there attempts to test hypotheses with experimentation? Were textual accounts concerned primarily with claims made in other texts, or did they employ empirical observations to rebut textual critiques? The short answer to these and related questions is that we do not know. Certainly, just as with modern science, the need to defend one's published views would drive many responses and adjustments to theories, but we do not know how much observation and even experimentation went into the Indian Sanskrit texts. However, one point is clear: substantial theory revision about key issues such as the nature of concept formation has not occurred for many centuries. And if we take science to require an ongoing process of theory revision, then we would likely conclude that "Buddhist science" is a highly contentious term.

Science, however, may also be characterized as embodying a particular commitment to a rigorous and ongoing inquiry into the nature of the world and the beings within it, where our notion of "rigor" requires us to base our knowledge claims in experience itself. This "empirical stance," to borrow a term from Bas van Fraassen, requires us to set aside our theory commitments, our texts and publications, and our assumptions about the possible in favor of a kind of careful and disciplined observation that is rooted in the evidence of the senses. This attitude is precisely what underlies the Buddhist suspicion of conceptuality and theorization as an end in itself. And while it too can be—and indeed, has been—idealized in the context of Buddhist inquiries into the mind and its processes, the principled commitment to this empirical stance, which requires setting aside the authority even of the Buddha's own words, might be enough to say that, yes, we can properly speak of Buddhist science. As you read through this volume, that empiricist spirit may not be evident on every page, but if you keep it in mind, perhaps you will agree that Buddhist science is an appropriate rubric for the detailed materials curated by our skilled authors.

John Dunne

Further Reading in English

For a concise overview of the major schools of Indian Buddhist philosophy, see John D. Dunne, "Buddhism, Schools of: Mahāyāna Philosophical Schools of Buddhism," in *Encyclopedia of Religion*, 2nd ed., edited by Lindsay Jones (Detroit: Macmillan Reference USA, 2005).

On the notion of mind in Buddhism see Christian Coseru, "Mind in Indian Buddhist Philosophy," in *The Stanford Encyclopedia of Philosophy*, edited by Edward N. Zalta (Metaphysics Research Lab: Stanford University, 2017). https://plato.stanford.edu/archives/spr2017/entries/mind-indian-buddhism/. See also chapter 2 of Chakravarthi Ram-Prasad, *Indian Philosophy and the Consequences of Knowledge: Themes in Ethics, Metaphysics and Soteriology* (Aldershot, England: Ashgate Publishing, 2007).

For issues around epistemic reliability and the suspicion of concepts, see chapter 1 of Paul Williams, *Buddhist Thought: A Complete Introduction to the Indian Tradition*, 2nd ed. (London: Routledge, 2012). For this issue in relation to the problem of language, see Jonathan Gold, *Paving the Great Way: Vasubandhu's Unifying Buddhist Philosophy* (New York: Columbia University Press, 2014). And for the account in the epistemological tradition, see chapter 4 of John D. Dunne, *Foundations of Dharmakīrti's Philosophy* (Boston: Wisdom Publications, 2004).

For an accessible account of the empirical orientation of Buddhist thought, see chapter 2 of the Dalai Lama, *The Universe in a Single Atom: The Convergence of Science and Spirituality* (New York: Morgan Road Books, 2005). An account of scientific empiricism is given by Bas C. van Fraassen in *The Empirical Stance* (New Haven, CT: Yale University Press, 2002).

1

The Nature of Mind

IN THE FIRST VOLUME of this series, we presented objects of knowledge in general, and physical entities in particular, including the process of arising and dissolution of the universe and living beings—this is the "objective" side. Now we will present the mind—the "subjective" side. In general *mind* or *consciousness* refers to inner experience. This includes feelings that are pleasant or painful, states of mind that are happy or miserable, emotional experiences such as fear of danger, anger toward those who inflict harm, affectionate attachment to close relatives, and compassion when observing suffering sentient beings. It includes sense consciousness—such as a visual consciousness that sees a vase filled with beautiful flowers, or an auditory consciousness that hears the sounds of music or singing. It also includes cognitions that remember previous experiences, as in "I remember this" or "I thought this," and cognitions that consider reasons and think "If this is the case, then that must also be the case," and so on. Whatever position one holds—that the mind is material or immaterial—in general what we call *consciousness* [2] is known to exist based on experience, so it does not necessarily need to be proved through reasoning. Nevertheless, there are many types of subtle mental states that must be proved through reasoning.

Someone may ask, "What is the difference between consciousness and physical things?" In general, matter and consciousness are differentiated by whether they are obstructive, established as clear and aware in nature, merely experiential in nature, and capable of knowing objects. Many earlier Buddhist masters identify the essential nature of consciousness as the opposite of obstructive physical things: lacking obstructivity, consciousness has the quality of clarity (in that external and internal things can appear to it), it has the quality of knowing its object, and it has the nature of mere experience.

Thus, when identifying the nature of the mind, the sūtras teach that the mind is difficult to catch hold of: being intangible, it is weightless; like a firebrand spinning around, it does not rest; like the constantly moving waves of the ocean, it is unstable; like a forest fire, it ignites all deeds of body and speech; like a great river, the mind forcefully draws along a vast number of internal movements of awareness; and so forth.

In general, the mind is the main factor involved in accomplishing the goals that living beings desire. The mind, unlike physical things, is difficult to identify. Yet if we train our mind, then in reliance on mindfulness, meta-awareness, heedfulness, and so on, we will attain both temporary [3] and final happiness. Therefore, in the Buddhist tradition, the system of analyzing the mind in great detail flourished from the earliest times. This is also the main reason why Buddhist texts have extensive explanations about psychology. For example, in the Buddha's own *Dhammapada* it states:

> Do not commit any evil,
> practice supreme virtue,
> thoroughly tame your own mind;
> this is the teaching of the Buddha.[18]

Both the happiness and suffering that arise in relation to beings' desired goals finally depend on the functioning of the mind—for it is owing to the force of having one's mind tamed or not tamed that temporary or lasting happiness and suffering arise. The essence of the Buddha's teaching is said to be the thorough taming of the mind. The *Collection of Aphorisms* also says:

> Taming the mind is excellent.
> A tamed mind leads to bliss.[19]

A tamed mind is the highest excellence, and it is on the basis of a well-tamed mind that happiness arises. [4] *Engaging in the Bodhisattva's Deeds* also makes the same point:

> "In this way, all fear and
> immeasurable suffering

arise from the mind,"
so taught the Speaker of Truth.[20]

Generally in the context of Buddhist texts, the terms *cognition* (which is the translation of the Sanskrit word *buddhi*), *consciousness* (the translation of *jñā[na]*), and *awareness* (the translation of *samvitti*) are all treated as coextensive or synonymous. The nature of *cognition* is stated to be awareness, and the nature of *consciousness* is said to be clear (or luminous)[21] and aware. "Clear" here expresses the essential nature of consciousness, and "aware" expresses its function. "Clear" also indicates: (1) that consciousness is beyond the nature of matter, which is characterized as tangible and obstructive, so it is clear in nature; (2) that just as reflections appear in a mirror, any internal or external object whatsoever—good or bad, pleasant or unpleasant—can appear in consciousness, so consciousness is luminous in that it illuminates objects; and (3) that the essential nature of consciousness is not contaminated by the stains of mental afflictions such as attachment, so its nature is clear or luminous. Thus several meanings of the word *clear* are stated in the texts.

That consciousness is devoid of the property of obstructivity is explicitly stated in the sūtras. For example, the *Teaching on the Undifferentiated Nature of the Sphere of Reality* says: "Is this mind blue? Is it yellow?"[22] Also [5] the *Chapter on the Thirty-Three* says: "The mind has no form, it cannot be pointed out, it has no obstructivity, and it is not representable."[23] Passages such as these explain how the mind lacks color and shape, cannot be visually pointed out, and has no obstructivity.

As noted above, consciousness is posited to be luminously clear in that any internal or external object whatsoever—good or bad, pleasant or unpleasant—can appear in it. This is explained as follows. Just as sunlight or electric light illuminates forms, any external or internal phenomena can be illuminated by—that is, they can appear to—consciousness; it is what makes objects known or manifest. Yet the illumination of sunlight or electric light and the "luminous clarity" that is the essential nature of consciousness are utterly dissimilar, for sunlight and electric light illuminate only what is in their immediate vicinity and cannot illuminate other things. The way that consciousness illuminates objects is not restricted in that way.

Many Mahāyāna and Hīnayāna scriptures teach that afflictions such as attachment present within the mind are adventitious, and that the nature of the mind is luminous clarity. For example, a sutta of the Theravāda school in the *Numerical Discourses* says: "O bhikkhus, this mind of luminous clarity is afflicted due to adventitious afflictions. Ordinary beings who have not heard this do not thoroughly understand reality just as it is. Therefore, not having heard, ordinary beings do not cultivate their minds. [6] Thus I have spoken."[24] The *Perfection of Wisdom Sūtra in Eight Thousand Lines* says: "Thus this mind is devoid of mind, for the mind's nature is luminous clarity."[25] And the *Descent into Laṅkā Sūtra* says: "The mind's nature is luminous clarity."[26]

The *Buddha Nature Sūtra* says:

> Just as where the most precious gold exists,
> one completely purifies it then makes use of it,
> likewise, I see all sentient beings continuously
> being harmed for a long time by mental afflictions,
> so I teach the Dharma as a means to purify
> the nature of what are "adventitious afflictions."[27]

Also, *Ornament for the Light of Wisdom That Engages the Object of All Buddhas Sūtra* says: "Mañjuśrī, the mind is luminous clarity by nature, not afflicted by nature; therefore, since all the afflictions are adventitious, all the afflictions are removable. [7] Anything that is just luminous clarity by nature is unafflicted."[28]

Among the treatises, Maitreya's *Sublime Continuum* says:

> Like the purity of a jewel, space, or water,
> it is always undefiled in nature.[29]

Just as the nature of gold is not rusted, the nature of space is not clouded, and the nature of water is not muddied, so the nature of the mind is not stained. The same text states:

> The luminously clear nature of the mind
> is unchanging, like space, because
> it does not become defiled by adventitious stains,
> attachment, and so on, which arise from an improper way of
> thinking.[30]

That is, just as empty space does not change by virtue of being inside different containers, so the ultimate nature of the mind, which is merely the lack of true existence of the mind, cannot change into something else on account of the afflictions such as attachment, which are adventitious stains. According to this text, therefore, the "nature of the mind" is the absence of true existence of the mind; [8] and the meaning of "adventitious stains" is that they do not exist truly as the essential nature of the mind, which is luminously clear and aware. Moreover, if the stains within the mind were truly existent, they would have to exist without depending on anything; hence, it would be impossible to remove them from the mind by cultivating their antidotes. However, since they lack true existence, it is possible to generate antidotes for the stains and remove them from the mind. The text cited above is arguing for all this.

Also, Dharmakīrti says in his *Exposition of Valid Cognition*:

> The nature of the mind is luminous clarity;
> the stains are adventitious.[31]

The stains of attachment and so on, which are the basic causes that give rise to suffering, can be removed in reliance on their antidotes; and since those stains do not impinge on the nature of the mind, the nature of the mind is luminous clarity, as discussed above.

Also, according to one interpretation, the line "The nature of the mind is luminous clarity" points to the fact that, just as reflections appear in a mirror, the quality of consciousness is that internal and external objects can appear to it. Moreover, those who accept reflexive awareness interpret the *luminous clarity of mind* as not just meaning that moments of consciousness illuminate their objects; instead, consciousness also illuminates itself, which is to say that it is by its very nature self-luminous.

Exposition of Valid Cognition says:

> Therefore, according to us, the mind itself is luminous,
> since it is by nature luminous clarity; [9]
> and something else [namely, an object] is also luminous
> in that it is illuminated by projecting its form into the mind.[32]

The mind, the subjective, is luminous because it is by nature itself luminous clarity. Something else—namely an object such as a visible form—becomes the appearing object of that consciousness by means of something like the transference of its image—that is, visible form and so on—into that luminously clear consciousness. Therefore it is said that forms and so on clearly appear to that consciousness.[33] In that case, consciousness has both the quality of illuminating, in the sense of being illuminating itself, and the quality of illuminating its object, in the sense that the image (*rnam pa*) of the object appears. In some contexts, when "luminously clear" and "aware" are discussed, the mind may be additionally called "empty" in the sense of being naturally empty of obstructivity, for those texts identify the nature of consciousness as having three qualities—emptiness, luminous clarity, and awareness.

The second quality of consciousness mentioned above is *awareness*, which expresses the function of consciousness. *Exposition of Valid Cognition* says:

> Consciousness has the attribute of apprehending its object;
> it apprehends it in the way that it exists;
> and by virtue of being existent, the nature of the object
> is to produce consciousness.[34]

Consciousness apprehends its object or operates by way of cognizing its object, [10] which must indicate that the foremost unique attribute of consciousness is to apprehend its object or to cognize its object.

Śāntarakṣita speaks about the essential nature of consciousness in terms of having the quality of cognizing its own nature, which is the opposite of the nature of matter or something having a material form. *Ornament for the Middle Way* says:

> Consciousness arises in a way that is opposite
> to anything in the nature of matter;
> that which is nonmaterial in nature
> is that which knows its own nature.[35]

In the *Ornament for the Middle Way Autocommentary* related to this stanza, it says:

> First, according to those who posit consciousness to be without a dualistic image [of object and subject], consciousness does not meet with an appearance of its object but is involved merely in cognizing itself. This means that, by its nature, it cannot experience an object other than itself. Consciousness is thus posited to have the nature of reflexive awareness because it is itself luminously clear by nature and because it is opposite to the nature of chariots and so on, which lack awareness. Since consciousness is reflexively aware of itself, it does not depend on some other cognizer [for consciousness itself to be known], unlike the case of blue and so on. Thus what we mean by *awareness* is that it is not unaware.[36] [11]

As for the meaning of those passages, Kamalaśīla's *Commentary on Difficult Points of the Ornament for the Middle Way* says: "The concise meaning is as follows. The functioning of reflexive awareness is the very opposite of the nature of material things, such as chariots or groups of parts; that is, awareness does not depend on any other illuminator [in order to occur]; and with that function in place, one thus engages in activities."[37]

According to a literal reading of Śāntarakṣita's root text and commentary and Kamalaśīla's *Commentary on Difficult Points of the Ornament for the Middle Way*, they both uphold the position asserting reflexive awareness, in that they explain the essential nature of consciousness to be the opposite of physical phenomena. However, even according to those who do not accept the notion of "reflexive awareness," Śāntarakṣita's reasoning, from a general point of view, points to a hugely important difference between matter and consciousness. For example, to know something physical such as a red lotus flower or, to use Śāntarakṣita's example, something external such as a chariot, it must depend on some other nonphysical phenomenon, namely, consciousness. Thus, in general, whereas consciousness can be aware of consciousness, a physical thing cannot be aware of a physical thing. Therefore obstructive physical phenomena do not have the quality of knowing objects. [12]

If, in general, matter and consciousness are fundamentally different with respect to their essential natures—in such terms as whether they are obstructive, exist by nature as clear and cognizing, and are in the nature of mere experience—then the question arises, which of these two must be considered fundamental in reality? Also, what is the nature of the relationship between matter and consciousness? Now the proponents of the Cārvāka ancient Indian philosophical school posit physical phenomena to be fundamental and inner mental phenomena to be mere attributes of physical phenomena—like the shadow of a physical object, like the light of an oil lamp, or like the potency of beer. However, the general position of classical Buddhist thinkers is that material phenomena and consciousness are equally fundamental in reality.[38] They maintain that just as physical phenomena cannot be reduced ultimately to some kind of substance or stuff of consciousness, likewise mental phenomena cannot be reduced ultimately to some kind of material substance such as the subtle elements.

Furthermore, although anything that is not consciousness, such as something physical, cannot be the substantial cause of consciousness, both matter and consciousness are mutually dependent on each other. Hence one must definitely accept that the two have a cooperative causal relationship. It is very clear, for example, that the sense consciousnesses would not arise without their causal bases, the physical sense organs. Also, most types of mental consciousness, having been drawn forth by a sense consciousness, [13] are indirectly dependent on the physical sense faculties. Likewise, according to Buddhist philosophical views, many specific worldly environments have a karmic relationship with the beings living in them. In particular, the texts of the highest yoga tantra[39] present the essential nature of the inner wind and of the mind to be that of indivisibility; thus even in the case of a subtle mental consciousness, it is understood to be inseparable from its medium, which is the wind. Therefore one must accept that form and consciousness are always inseparable. Nevertheless, it is not necessarily the case that inasmuch as there is consciousness, it must be contingent on the physical sense faculties as well as states of the brain, a fact that was explained earlier when discussing the relationship between the body and the mind.[40]

Someone may ask, "Although *consciousness* is defined as that which is luminously clear and aware in the sense of being beyond the physical, what exactly is the support or basis of consciousness?" In general, consciousness is based primarily on the wind element, which is located throughout every part of the body, upper and lower, from the crown of head to the soles of the feet. In particular, among the five sense consciousnesses, the eye consciousness is located in the physical organ that is the site of the eye faculty.[41]

Likewise, the ear consciousness is located in the physical organ that is the site of the ear faculty, and similarly for the rest, up to the body consciousness, which is located in the physical organ that is the site of the body sense faculty. The first four sense consciousnesses are located at individual parts of the body, and the fifth, the body sense consciousness, is located at every part of the body. The mental sense consciousness has both gross and subtle types: the gross is based on the gross life-sustaining wind (*prāṇa*), and the subtle [14] is based on the subtle life-sustaining wind, which the texts of the highest yoga tantra describe as being located at the heart area.

In the case of a human being, the heart area, the location of the life-sustaining wind that is the basis of the very subtle mental consciousness, is situated between the breasts, close to and directly in front of the spine. In between the upper and lower parts of the knot formed by the middle, left, and right channels is located the indestructible drop, which is radiant. The gross life-sustaining wind, in its basic state, leaves and enters the nostrils and stays around the heart channel, but it does not remain in the central channel, or *dhūtī*, of the knot at the heart channel. The special abode of the five-branched wind (the life-sustaining branch being the most important) is the heart area (which is the principal place of their arising and subsiding) and the lotus of the heart channel.[42] The ways in which the five sense consciousnesses function with regard to the five objects—form, sound, and so on—should be understood in accordance with how they are explained in Āryadeva's *Lamp on the Compendium of Practices*, which was discussed in the first volume of this series. [15]

The Ways of Categorizing Cognition

To understand the presentation of mind, early Buddhist texts offer many ways of categorizing cognition. These are included within the sevenfold typology of cognition, the threefold division of cognition, and various twofold divisions of cognition. First, the sevenfold typology of cognition consists of: (1) direct perception, (2) inference, (3) subsequent cognition, (4) correct assumption, (5) indeterminate perception, (6) doubt, and (7) distorted cognition.[43] The way that these cognitions engage their objects will be explained later.

Second, the threefold division of mind consists of: (1) conceptual consciousness, which has a universal as its observed object; (2) unmistaken nonconceptual consciousness, which has a unique particular as its observed object; and (3) mistaken nonconceptual consciousness, which has a clear appearance of a nonexistent as its observed object. Examples of these are, in the first case, a conceptual consciousness perceiving a pot; in the second case, a direct perception perceiving a form; in the third case, a distorted nonconceptual consciousness to which a snow mountain appears yellow. The threefold division is made from the point of view of the observed object and represents principally the Sautrāntika viewpoint.

Third, the various twofold divisions are: (1) sense consciousness versus mental consciousness, (2) conceptual versus nonconceptual consciousness, (3) valid versus nonvalid consciousness, (4) mistaken versus unmistaken cognition, (5) those that engage via exclusion versus those that engage via affirmation, [16] and (6) mind and mental factors. These six divisions are listed in the texts presenting such categories. The twofold division of sense consciousness versus mental consciousness is made from the perspective of whether the relevant cognition needs to be based on a physical sense faculty as its dominant condition. The twofold division of conceptual consciousness versus nonconceptual consciousness is made from the perspective of whether a linguistic referent appears in that cognition. The twofold division of valid consciousness versus nonvalid consciousness is made from the perspective of whether a cognition is newly realizing or is deceptive or nondeceptive. The twofold division of mistaken cognition versus unmistaken cognition is made from the perspective of whether the object of a cognition exists as it is perceived. The twofold division of engagement via exclusion versus engagement via affirmation is made from the perspective of whether a cognition engages its object by parsing its object into components. Finally, the twofold division of mind and mental factors is made from the perspective of whether a given cognition is primary or secondary, and whether it apprehends the entity itself or the specific attributes of the object.

2

Sense Consciousness

Sense Consciousness versus Mental Consciousness

There are six categories of consciousness: eye, ear, nose, tongue, and body consciousness, plus mental consciousness. The first five are the sense consciousnesses, each of which arises in dependence on a physical sense faculty that is its own uncommon dominant condition. The last is the mental consciousness, which arises in dependence on the mental sense faculty that is its dominant condition. Form is perceived by the eye consciousness, sound by the ear consciousness, smell by the nose consciousness, taste by the tongue consciousness, and touch by the body consciousness. [17] The five sense consciousnesses arise through the functioning of their respective dominant conditions, the five sense faculties, and their respective objective conditions, the five sense objects of form and so on. The uncommon dominant condition of the mental consciousness is the mental sense faculty, but it is not a physical sense faculty. Therefore a distinction is made between the sense consciousnesses and the mental consciousness from the point of view of their uncommon dominant conditions. Regarding this, Dignāga's *Compendium of Valid Cognition* says:

> [A sense consciousness] is named based on a sense faculty
> because this is its uncommon cause;
> its experienced object is a whole
> because that [sense consciousness] is produced by many
> objects together.
>
> Something's having many qualities
> cannot be realized through a sense faculty.

> The object of a sense perception is to be known for oneself;
> it is an entity that cannot be indicated with words.
> This includes mental perception of an object and of attachment and so on.[44]

Sense consciousness and *mental consciousness* are not named from the point of view of their objects but from the point of view of their uncommon dominant conditions. For example, although the sound of a drum arises from using a hand or a stick and so on, it is not referred to as the sound of those things; the drum is its uncommon cause, and it is hence called "the sound of a drum." [18]

To give some other examples, when you view a garden containing many types of flowers, even though there are different kinds of flowers in that garden, they appear to your eye consciousness as a single mass. If the mind cognitively selects a particular variety of flower and apprehends it, that cognition has now gone beyond the bounds of sense consciousness and has entered the domain of mental consciousness. Also, if a visible form appears to eye consciousness at the same time as a melodious sound appears to ear consciousness, then the object toward which an inclination—the wish to hear or the wish to see—is more strongly habituated will be the only object that the sense consciousness comes to apprehend and determine on that occasion; other objects will appear to other sense consciousnesses at that time, but they will not be ascertained. In this way, those inclinations—wishing to hear or wishing to see—are forms of mental consciousness.

All kinds of things can appear to the mental consciousness—past, future, concepts, designations, and so on—that cannot appear to a sense consciousness. When a present object such as blue manifestly appears to the eye consciousness, if the mind does not attend to it, such as through the wish to see blue, then that eye consciousness will not realize that it is blue even though the blue clearly appears to that eye consciousness. On the other hand, if the mind focuses on it, then that eye consciousness can induce the ascertainment "This is blue."

Also, when your eye consciousness is looking at a form, your mind may be distracted toward some other object. However, while your eye consciousness is in contact with the form that is its object, it is indeed aware of that form. But from that awareness, any cognition of that object as good or bad can only be a mental consciousness. [19] As in an Abhidharma treatise too, Dignāga's *Compendium of Valid Cognition Autocommentary* says: "Someone with eye consciousness cognizes blue but does not think 'This is blue.'"[45]

The sense consciousnesses totally depend on there being an external object of focus nearby, and they can arise only through the functioning of the objective condition, their own observed object. But the mental consciousness need not depend on the presence of an external observed object. So although an eye consciousness seeing a flower will not arise if there is no flower to be the observed object, a mental consciousness perceiving a flower can still arise through remembering or conceiving it, even when no flower is in the vicinity. Dharmakīrti's *Ascertainment of Valid Cognition*, for example, states:

> This is not possible for a sense consciousness because it arises through the causal capacity of its object. Since it arises through the causal capacity of its object, it conforms to the character of that [external object].

Also:

> Conceptual cognition is a mental consciousness. It arises, without depending on the proximity of a causally efficient object, through the latent potencies for that conceptualization. It apprehends an object that is not restricted to a sense faculty, and it does so through some relation to a sensory experience, either together or separately.[46]

The first excerpt says that since a sense consciousness arises through the causal capacity of its observed object, it cannot apprehend its object in a manner that associates words and their referents. [20] The second says that a conceptual cognition does not depend on the proximity of an observed object's causal capacity, and that it apprehends its object without (necessarily) ascertaining the universal of an object that has been apprehended by a sensory perception. Through these points, one can clearly understand the difference between a sense consciousness and a conceptual mental consciousness.

In brief, when experiential consciousness engages an object such as a visible form or a sound, it does so in one of two ways: primarily in relation to the body or primarily in relation to the mind. In the former case, the consciousness is called *sense consciousness*, and in the latter case it is called *mental consciousness*. For example, on the basis of a sense consciousness seeing a form, hearing a sound, and so on, a mental consciousness evaluates the object, as in "It is this" or "It is like this" and so on. When the eyes are closed, the eye consciousness remains inactive, but the mental consciousness is still present and can engage with its object. Similarly, although sense consciousness does not operate during the dream state, even in the dream state the concept-based mental consciousness can engage in various activities such as undergoing experiences that are feelings of pleasure, pain, and so on. Thus our own experience demonstrates that there are two kinds of consciousness—sense consciousness and mental consciousness.

Indeed many types of cognitive events must be understood primarily in terms of mental consciousness, as in the following examples. For instance, thoughts such as "There are flowers over there" and "These are flowers" apprehend merely the entities themselves. Thoughts like "These flowers smell sweet" principally apprehend the object's distinctive attributes. Likewise, thoughts may involve a mind seeing beautiful flowers in a dream, a recollection of flowers seen earlier, or a cognition that apprehends flowers even when the eyes are closed. There are also emotions, such as fear, love, or torment. [21] There are also afflictive mental states, such as pride, hatred, or attachment; and their converse too, such as the meditative states of calm abiding (*śamatha*), special insight (*vipaśyanā*), and so on. All of these examples must be construed primarily as instances of mental consciousness.

Having discussed the difference between sense consciousness and mental consciousness, the essential nature of each can now be presented as follows. A *sense consciousness* is defined as a consciousness that depends on a physical sense faculty that is its uncommon dominant condition. A *mental consciousness* is defined as a consciousness that depends on a mental sense faculty that is its uncommon dominant condition.

When sense consciousness is categorized, there are five types: eye consciousness, ear consciousness, nose consciousness, tongue consciousness, and body consciousness. A consciousness that depends on the eye sense faculty as its dominant condition is called *eye consciousness*—for example, one that is seeing the color white. A consciousness that depends on the ear sense faculty as its dominant condition is called *ear consciousness*—for example, one that is hearing the sound of running water. A consciousness that depends on the nose sense faculty as its dominant condition is called *nose consciousness*—for example, one that is smelling the scent of sandalwood. A consciousness that depends on the tongue sense faculty as its dominant condition is called *tongue consciousness*—for example, one that is experiencing the taste of sugar. A consciousness that depends on the body sense faculty as its dominant condition is called *body consciousness*—for example, one that is experiencing a soft object of touch. [22] *Unraveling the Intention Sūtra* says:

> Viśālamati, it is on the basis of and abiding in this appropriating consciousness that the group of six consciousnesses—eye consciousness, ear, nose, tongue, body, and mental consciousness—arise. The eye consciousness arises in connection with consciousness in dependence on forms and the eye sense faculty. Immediately following the eye consciousness, there arises a conceptual mental consciousness with the same duration and same object of experience.[47]

Likewise, a division into a sixfold group of consciousnesses is clearly taught in the *Descent into Laṅkā Sūtra*, the *Pile of Precious Things Collection*, the *Great Play Sūtra*, and so on.

Mental consciousness includes, in addition to the conceptual cognitions, direct mental perceptions that are concept-free and have arisen on the basis of a sense consciousness acting as its immediately preceding condition. It also includes such direct mental perceptions as those that directly realize the truth about a thing's way of being, such as impermanence, on the basis of progressive engagement in learning, reflection, and meditation, whereby what is perceived gradually becomes clearer and clearer. As in the case of directly realizing impermanence, there are many examples of mental consciousness that is perceptual.

In general, classical Buddhist thinkers have divergent opinions as to whether there exists a type of concept-free mental consciousness that arises from a sense consciousness acting as its immediately preceding condition [23] and, like the sense perception, takes an external object as its observed object. Those who assert the existence of such a mental direct perception rely on the following sūtra passage: "Monks, the knowledge of material form is of two types—one based on the eye [sense faculty] and the other on the mental [faculty]."[48] Dharmakīrti, the great Buddhist master of epistemology, says in *Ascertainment of Valid Cognition:*

> A mental consciousness that arises
> from a sense consciousness acting as the immediately preceding condition
> and that apprehends the object immediately preceding it
> is also called *direct perception.*

> This verse says that a mental consciousness that is produced by a sense consciousness acting as the immediately preceding condition, with the immediately preceding moment of its own object acting as a supporting condition, is also an instance of direct perception.[49]

This passage clearly states that a mental consciousness can be produced when a sense consciousness apprehending something external as its object acts as the immediately preceding condition; in that case, the mental consciousness must be posited as a direct perception. When analyzed precisely, this excerpt says that the very sense consciousness that induces the mental consciousness is both its immediately preceding condition and its dominant condition. [24] Since it enables the arising of that object-cognizing mental perception, such a sense consciousness immediately preceding it must be its dominant condition; and since it generates that very mental perception as a clearly cognizing experience, such a sense consciousness must be its immediately preceding condition.

A question arises: "Does such an object-cognizing directly perceiving mental consciousness apprehend an object that has already been perceived by the sense consciousness inducing it, or does it apprehend an object not yet perceived by that preceding sense perception?" In the first case, that mental consciousness would be a subsequent cognition. In the second case, a further question arises: "If seeing a material form does not depend on the presence of eyes, there would surely be occasions when even a blind person could see forms?" A response is given in *Exposition of Valid Cognition:*

> Thus, in arising from the sense consciousness
> that is its immediately preceding condition,
> the mental consciousness apprehends a different object;
> therefore a blind person does not see.[50]

Such a consciousness apprehends the next moment of the observed object of the sense consciousness that is its dominant condition. Therefore, when there is such a directly perceiving mental consciousness, the fallacy of a blind person seeing forms does not occur. [25]

The Three Conditions of a Sense Consciousness

To explain the three conditions necessary for a sense consciousness to arise, we may ask, "Given that consciousnesses are unlike physical things, what are their special causes and conditions?" The two categories of consciousness—sense consciousness and mental consciousness—each have particular causes and conditions. A sense consciousness has three conditions, and a mental consciousness has either two or three conditions. The definition of a condition is: that which assists a cause in producing its results. For example, the condition of sprinkling water helps a seed give rise to a sprout. Generally speaking, *cause* and *condition* are mutually inclusive; they just have different names. There are four types of conditions: causal condition, objective condition, dominant condition, and immediately preceding condition. Vasubandhu's *Treasury of Knowledge Autocommentary* says: "Where are these stated? In the sūtras it says, 'There are four conditions: causal condition, immediately preceding condition, objective condition, and dominant condition.' So it is taught."[51]

The definition of the objective condition of a sense consciousness is: that which directly and principally gives rise to a sense consciousness bearing an image of the object. The definition of an observed object of a sense consciousness is: that which directly and principally gives rise to a sense consciousness bearing an image similar to itself. [26] Thus there is some question about whether the *objective condition of a sense consciousness* and the *observed object of a sense consciousness* are synonymous. The definition of the dominant condition of a sense consciousness is: that which directly and principally and gives rise to a sense consciousness able to apprehend its object. The eye sense faculty is posited as the *uncommon* dominant condition of the eye consciousness that is its effect, and the mental sense faculty is posited as the *common* dominant condition of the eye consciousness that is its effect. The definition of the immediately preceding condition of a sense consciousness is: that which directly and principally gives rise to a sense consciousness that is clear and cognizing in nature.

To illustrate the three conditions, based on the example of an eye consciousness: its objective condition is a visible form; its dominant condition is the eye sense faculty; and its immediately preceding condition is a prior moment of consciousness attending to a visible form. Likewise, the three conditions of an ear consciousness are: its objective condition is a sound; its dominant condition is the ear sense faculty; and its immediately preceding condition is a prior moment of consciousness attending to a sound. The way of positing the three conditions of the remaining sense faculties such as nose consciousness is just the same. A sense consciousness arises when all three conditions are satisfied: externally there is a nearby objective condition—form, sound, and so on; internally there is a dominant condition—a physical eye sense faculty and so on; and there is an immediately preceding condition—a prior moment of consciousness. Dharmakīrti's *Drop of Reasons* says:

> How the eye consciousness arises in dependence on the eye [sense faculty] and so on should be understood as follows. On the basis of its similar immediately preceding condition, an eye consciousness has the nature of perceiving; [27] on the basis of the eye sense faculty, along with having the nature of perceiving, an eye consciousness also arises with the restriction that it has the capacity to apprehend visible forms; and on the basis of its object, an eye consciousness bears an image similar to that object. Although the direct effect [in this case, an eye consciousness] is not several different entities, it has different attributes based on causes that are different entities; and although the causes are different, the particular object they give rise to is a single entity. Due to not requiring any further causal mediation, different causal potentialities may be ready to give rise to an effect having distinct attributes in accord with the nature of those potentialities. In that case, due to momentariness, those causal potentials that require no further mediation are called

> the "basis for establishing the nature" of the effect of that complete causal collection. Thus, from the complete collection [of those three kinds of conditions], a single entity arises having different qualities—namely, that it has the nature of perceiving, is restricted to apprehending visual forms, and bears an image of its object.[52]

In this way, each of the sense consciousnesses apprehending forms and so on arises in dependence on three conditions. *Exposition of Valid Cognition* says:

> A sense faculty, an object, and an awareness,
> or a prior moment of attention,
> from such a complete collection of causes and effects
> does the related arise, but not otherwise.[53] [28]

A sense consciousness arises as the effect of a combination of the dominant condition (a sense faculty), the objective condition (one of the five sensory objects such as visible form), and the immediately preceding condition (a prior moment of attention). For example, the arising of an eye consciousness apprehending blue requires a complete causal collection that includes the objective condition, blue, and the dominant condition, the physical eye sense faculty to which that blue itself appears; it must arise in dependence on those.

In the case of a blind man, a visible form that is the outer condition may be present right in front of him, but such visible forms are not perceivable to his eye sense faculty, which is the dominant condition, so an eye consciousness does not arise.

Each of the three conditions of a sense consciousness imprints unique traces in the mindstream. From traces of the objective condition arises the consciousness bearing the image of the object. From traces of the dominant condition arises the sense consciousness with the ability to apprehend the object. From traces of the immediately preceding condition arises the consciousness with a clear and cognizing nature. For example, a sense consciousness to which blue appears arises having the image of the object, blue, which is caused primarily by the blue that is its objective condition. A sense consciousness to which blue appears arises with the ability to apprehend its object, blue, which is caused primarily by the eye sense faculty. And a sense consciousness to which blue appears arises with a clear and cognizing nature, which is caused primarily by the immediately preceding condition.

The dominant condition determines whether a consciousness belongs to one of the sense consciousnesses or to the mental consciousness. The eye consciousness apprehends visible forms, [29] but it does not apprehend sounds and so on; the ear consciousness apprehends sounds, but it does not apprehend visible forms and so on. This is due to the uniqueness of their dominant conditions.

There are three conditions for a directly perceiving mental consciousness apprehending a form and of a directly perceiving mental consciousness apprehending a sound and so on. The dominant condition for such a directly perceiving mental consciousness is the factor that, based on the previous moment of directly perceiving sense consciousness, automatically produces a directly perceiving mental consciousness as its result. The immediately preceding condition of such a directly perceiving mental consciousness is the factor that, based on the previous moment of directly perceiving sense consciousness, produces a mental consciousness that is clear and knowing as its result. Thus the dominant condition and the immediately preceding condition of such a directly perceiving mental consciousness are actually the same, but they are conceptually distinguishable. The objective condition of such a directly perceiving mental consciousness is the second moment of an object similar in type to that apprehended by the last moment of the directly perceiving sense consciousness which induced that directly perceiving mental consciousness.

This explanation of the three conditions is presented in accordance with the standpoint of the Sautrāntika school. This way of presenting the three conditions of perception is shared, in general, as a common position of the Buddhist tenet systems. As for the Vaibhāṣika tradition's unique view on the topic, which is presented in Vasubandhu's *Treasury of Knowledge*, it will be explained in a later section.[54]

The Cittamātra Approach to External Objects

The Cittamātra tradition's way of positing the objective condition is presented in Dignāga's *Investigation of the Object*:

The nature of a knowable thing, though internal,
is such that it appears as if external; [30]
it is the object because it has the nature of consciousness;
it is therefore its condition as well.[55]

The essential nature of an object of knowledge is internal consciousness; even though there is no external object, the object, which is actually internal, appears as if it were external. Just that is the objective condition. Since the eye consciousness apprehending blue takes on the appearance of blue, that appearance as blue to the eye consciousness is its appearing object; moreover, it is also its condition. In the Cittamātra tradition, this is posited as the *metaphorical* objective condition or the objective condition that appears. The *actual* objective condition, according to their viewpoint, is as stated in *Investigation of the Object*:

Owing to potentialities placed, [the previous sensory awareness] is, in sequence, [the objective condition].[56]

Each preceding sense consciousness has the potential to cause the subsequent sense consciousness to assume the object's image. Since that potential is the condition for placing the image of the object in the subsequent sense consciousness, [31] it is said to be, in a succession of moments, the objective condition for the subsequent sense cognition. That, in this view, is the actual objective condition. Thus the causal potential within the preceding sense consciousness is posited as the objective condition for its resultant, subsequent sense consciousness. It is not posited as the objective condition by virtue of being both the object of the subsequent sense consciousness as well as its condition. Instead, it is so posited because it is the main condition that causes a subsequent sense consciousness to arise with an image of its object. Therefore, in this view, out of these two types of objective condition—the objective condition that appears and the objective condition that is the potentiality—the former is the metaphorical objective condition and the latter is the actual objective condition.

The reason why Dignāga's own *Investigation of the Object Autocommentary* says that the objective condition must have two attributes is given as follows. According to the system claiming that an external object, such as blue that appears to a sense consciousness, is itself the objective condition, it is so because it is both the object of that consciousness as well as its condition—and this is what it means to have two attributes. According to Dignāga's own system, which maintains that the causal potentiality is the objective condition, what it means for the objective condition of a sense consciousness to have two attributes is that the potentiality is what deposits the image of the object within the sense consciousness, and it is a condition of such a cognition. Therefore, if something is the objective condition of a sense consciousness, it must be what makes the sense consciousness arise as bearing the image of the object. That said, one should understand that although something could be what makes the sense consciousness arise as bearing the image of the object, it does not need to be the object of that sense consciousness. [32]

Thus among the three conditions of consciousness, the immediately preceding condition gives rise to this instance of clear and cognizing consciousness in the present moment. And the potentiality deposited by the present moment of clear and cognizing consciousness gives rise to the consciousness in the next moment and so on, continuously from one moment of clear and cognizing consciousness to the next.

The way earlier and later moments in the continuum of consciousness flow like this implies the topic of past and future lives, so we will offer a brief discussion of that topic.

The Proof of Past and Future Lives

Any conditioned thing whatsoever must arise from both a substantial cause and a cooperative condition. The consciousness within one's present continuum does not arise from a discordant cause or from no cause at all; it arises from a concordant cause. And that cause has both (1) a substantial cause that primarily produces the entity itself and (2) a cooperative condition that, although not the primary producer of the thing itself, acts as a complementary condition to help the substantial cause to produce its effects. When one considers in this way how the first moment of consciousness in this life must arise from both a substantial cause and a cooperative condition, then one will establish the existence of past and future lives by means of correct reasoning.

When one thinks carefully about this in the case of a person, it is obvious that in regard to the form aggregate and the consciousness aggregate, the form aggregate that emerges as the gross body is composed of atoms, and it has evolved from the atomically composed egg and sperm of the parents. The bodies of the mother and father too have arisen from other bodies, specifically from the egg and sperm of their own parents, and so on. When one inquires and analyzes how one body arises from an earlier one in a constant continuum in this way, it becomes clear that there is no starting point. [33] As Āryadeva says in *Four Hundred Stanzas*:

> In the case of any effect,
> a first cause is not perceivable.[57]

Just as the substantial cause of a gross physical thing must be a continuum of what is of a similar type as itself, in the same way the earliest moment of mental consciousness in this life must arise from a continuum of what is of a similar type as itself. Moreover, if something that is not consciousness, such as a form that is atomically established, were the substantial cause of consciousness in this life, then the fallacy would occur of two things with utterly different natures turning into each other. Dharmakīrti's *Exposition of Valid Cognition* says:

> It can also be proved because what is not consciousness
> is not the substantial cause of consciousness.[58]

If consciousnesses did not need to arise from previous mental phenomena of similar type but could arise in dependence on physical elements, then distinct mental states such as love, wisdom, and attachment could increase or decrease merely through the force of certain physical elements increasing or decreasing. However, this is not the case. Instead, these qualities are seen to increase and decrease out of particular familiarity with previous mental states of similar type. [34]

Someone may say, "Although one has a mind from an earlier continuum of mind, it is part of the minds of one's parents, and this is the substantial cause of one's own mind." In that case, if the parents were skilled in handicraft and so on, then their children would be too. But there is no such invariable relation, as we can confirm by direct observation. Also, children from the same parents have very different mental attitudes. That objections such as these refute the above assertion is stated by Dharmottara in his *Proof of the Afterlife*:

> Suppose someone says, "Even if it has another mind preceding it, the child's mind is born from the mother's mind." This is not a correct statement. For the existence of an excellent intelligence [in a child] is the effect of a prior habituation whose presence or absence is the determining factor. It cannot be considered to have some other cause. Among those born of the same mother and similarly endowed with good physique, youth, and fully functioning sense faculties, one may be very intelligent and the other not.[59]

In brief, the mind of this present life must have a previous continuum that is its substantial cause, and since some material entity or one's parent's minds have been ruled out through the process of elimination, the conclusion is that it is the preceding moment of one's own mental continuum that serves as the substantial cause of the first moment of the newborn's mind. It is this reasoning that proves the existence of one's mind in a former birth. Thus it similarly follows that an earlier mind [35] within one's own continuum must precede the mind of that former birth. This proves that there is no past or future temporal limit to one's own or another's mind, that there is no beginning to one's series of births.

As for the existence of future lives, just as the mind of one's past lives is connected with subsequent moments of that mental continuum, right up to the present lifetime, similarly with respect to the final mind of this life at the point of death. Given that all the conditions are complete for it to link itself to future moments of similar types, it is established that the mind at the point of death will connect to a later mind. The existence of later lives is proven in this way.

When one understands how the self of this life migrates to the next rebirth and how the consciousness of this life is connected to the consciousness of the next rebirth, then one can know that it is perfectly suitable for one type of being to be reborn as another type of being, and one can understand how such rebirth occurs. According to Buddhist thinkers, there is no permanent self that resides within the aggregates of this present life that then leaves the present life's aggregates and enters the aggregates belonging to the next life. It is also not the case that rebirth occurs without causes, or through the design of a creator, or from a total cessation of the continuum of consciousness. Rather, when the consciousness of this life has separated from each of the gross aggregates, and the appearances of this life have totally vanished, whereupon the substantial continuum of this very consciousness [36] becomes connected to the consciousness of the next rebirth, this is what is meant as the self of that person taking a rebirth.

A more extensive presentation of the arguments in Buddhist texts to prove past and future lives will be taken up in greater detail in a later volume in this series.

3

Conceptual and Nonconceptual

GENERALLY SPEAKING, the Tibetan term usually rendered as "conceptuality" (*rtog pa*)[60] has multiple senses in translations of the early Buddhist texts. For example, the *Treasury of Knowledge* says:

> Inquiry and analysis are coarse and subtle.[61]

When engaging the object, *inquiry* (*rtog pa*, *vitarka*) is a mental factor that engages it in a coarse way and *analysis* (*dpyod pa*, *vicāra*) is a mental factor that engages it in a subtle way. In *Distinguishing the Middle from the Extremes*, however, it says:

> The conceptual constructions (*kun rtog*, *parikalpa*) that pertain to the unreal
> consist of minds and mental factors of the three realms.[62]

Minds and mental factors that are mistaken in having a dualistic appearance and are included within the levels of any of the three realms are said to be *conceptual constructions*. Dignāga's *Compendium of Valid Cognition* says: [37]

> Direct perception is free from conceptualization
> that attaches a name, a type, and so on.[63]

Here Dignāga defines *conceptualization* (*rtog pa*, *kalpanā*) in terms of those cognitions that apprehend their objects by way of applying a linguistic "name" or a "type"—that is, a class—to the object. In other words, the object is qualified by some form of universal. Of these different senses, we will use *conceptuality* here in accordance with how it is defined in the Buddhist epistemological treatises of Dignāga, Dharmakīrti, and their followers. This usage of the term also occurs in the sūtras. For example, *Descent into Laṅkā Sūtra* says: "Mahāmati, conceptual thought is what finds expression in words such as 'This is an elephant, a horse, a chariot, a pedestrian, a living being, a woman,' and so on. Thus a conceptualization is that which illuminates the warrant for applying a term to an object, as when one thinks 'It is this kind and not another.'"[64]

As explained above, sense consciousness is nonconceptual, whereas mental consciousness may be either conceptual or nonconceptual. Certain types of consciousness directly cognize whatever object they engage, and based on this they are said to be nonconceptual, such as an eye consciousness seeing a vase. Also, there are many types of consciousness that cannot perceive the object directly but apprehend it by way of taking as their observed object something that partially resembles the actual object. Such types of cognition are referred to as conceptual. An example is a cognition that remembers yesterday's meal. Although yesterday's meal does not appear directly to a memory occurring now, there is an appearance in that memory that resembles yesterday's meal. [38] And it is via this appearance that yesterday's meal is taken as the object of that cognition and is remembered. Furthermore, although an eye consciousness seeing a vase is nonconceptual, when the eyes are closed after having seen the vase, the appearance of the vase persists in the mental consciousness, and such a mental consciousness is conceptual. Regarding this, Dharmakīrti's *Exposition of Valid Cognition* says:

> A cognition connected to concepts
> does not have a clear appearance of the object.[65]

And:

> Any cognition that has a clear appearance [of the object]
> is nonconceptual.[66]

When an object appears to a consciousness mixed with an object universal,[67] then it is a conceptual consciousness. If a cognition directly perceives its object unmixed with a universal, then it is a nonconceptual consciousness.

In *Ascertainment of Valid Cognition*, Dharmakīrti gives a very clear account of the difference between conceptual and nonconceptual consciousness:

> Conceptual cognition is a mental consciousness. It arises, without depending on the proximity of a causally efficient object, through the latent potencies for that conceptualization. It apprehends an object that is not restricted to a sense faculty, and it does so through some relation to a sensory experience, either together or separately.[68]

Conceptual minds do not rely on the proximity of an observed object's causal capacity [39] but arise owing to the power of past habituations or latent potencies of language-based thought processes. *Ascertainment of Valid Cognition* says:

> An attribute, an attribute-possessor, the relation,
> and customary conventions—having apprehended these separately,
> an object is cognized in that way
> by associating them, and not otherwise.

> If one understands the attribute, the attribute-possessor, their relationship, and the customary conventions posited, then by associating them, one apprehends the attribute-possessor as qualified by some attribute. Such is the case in the expression "staff holder." That cognition does not happen otherwise, because one does not have that cognition if one does not cognize the things [i.e., attribute and attribute-possessor], their relationship, and the conventions expressing them.[69]

Here the following kinds of cognition are recognized as conceptual: those that involve predication, such as a thing and its property; the relation between horns and the horned animal; and relations involving customary conventions, such as the notion of a chief and his deputies. In brief, all cognitions that combine objects through mutual association involving an attribute and an attribute-possessor are said to be conceptual mental cognitions.

The same text says:

> It is not possible for [a direct perception such as eye consciousness to apprehend something] as the possessor of a universal, or a quality, or an activity. This is so because the distinct things [e.g., the universal and its possessor] and their relationship do not appear [in direct perception], so it cannot associate them; and this is so also because they are not perceived that way [i.e., as distinct but related], just like water and milk [are not perceived as distinct when mixed together]. And even in the case of cognizing a distinct thing [in a way that categorizes it], that cognition:
>
> > . . . depends upon recollecting a linguistic convention, and
> > by nature it associates
> > [the present object] with what has been previously experienced. [40]
> > How could that cognition occur in eye consciousness,
> > which lacks any apprehension of past and future?
>
> Within this [eye consciousness] there is no ability to make associations because, given that it arises through the proximity of an object, there is no analysis involved. If it were to analyze, then there would be no difference between sense consciousness and mental consciousness. If there were no difference between them, various consequences would absurdly follow, such as it would both apprehend and not apprehend past and future; it would both categorize and not categorize its object; it would both infer and not infer its object; and it would both depend on and not depend on the presence of the object.[70]

A nonconceptual sense consciousness arises from its three conditions having been met, whether one wants it to arise or not. For example, when the three conditions of an eye consciousness seeing a cow have been met and that eye consciousness arises, even if one were to deliberately try to think it is a horse, one will still see the cow. Conversely, with respect to a thought conceiving something as a cow, if one were to deliberately think it to be otherwise, it is possible to change that thought of a cow. Such a difference between conceptual and nonconceptual cognition is stated in *Ascertainment of Valid Cognition*:

> Moreover, if the thought of an attribute and so on were to arise, then even if [the three conditions for sensory perception] have been fully met, that thought can deliberately be changed, [41] as is the case with any conceptual consciousness. The content of conceptual consciousness can be changed through deliberate thought, but this is not so for sense consciousness. Once the three conditions have been met, even if one sets aside the thought of the object as a cow and instead thinks of it as a horse, one still sees a cow.[71]

In general there is the way in which conceptual thought is defined in the manner of the earlier quotation from the *Compendium of Valid Cognition*: "Direct perception is free from conceptualization that attaches a name, a type, and so on."[72] Also, Dharmakīrti's *Exposition of Valid Cognition* says:

> Any consciousness that apprehends a linguistic referent
> is a conceptual cognition of that object.[73]

A consciousness is said to be a conceptual cognition regarding an object if the consciousness that apprehends the object takes that object's universal or linguistic referent as its observed object. Also, *Ascertainment of Valid Cognition* says: "What is conceptual cognition? Conceptual cognition is an expressive cognition. A conceptual cognition is one that has an appearance that is suitable to be associated with words."[74] Conceptual cognition is a cognition whose phenomenal content appears or is apprehended to be associable with language. *Drop of Reasoning* makes the same point.

So the definition of a conceptual cognition is: a construing awareness (*zhen rig*) that apprehends *word* and *referent* as suitable to be associated. The significance of the phrase "suitable to be associated" in this definition is explained as follows. Some conceptual cognitions, such as one in the mind of a preverbal infant, take as their objects either the object universal or the word universal, but they do not take a combination of both as their object. [42] Therefore such cognitions do not apprehend word and referent as associated. Nevertheless, those cognitions apprehend them as suitable to be associated. Alternatively, some texts say that what is meant by "word referent" (*sgra don*) is the mere universal that is apprehended by the mind, and its meaning should not be understood as the separate terms *word universal* and *object universal*.[75]

There is a twofold categorization of conceptual cognitions: those that concur with their objects and those that do not. A conceptual cognition whose engaged object exists is called a "conceptual cognition that concurs with its object." An example is the conceptual cognition apprehending a pot. A conceptual cognition whose engaged object does not exist is called a "conceptual cognition that does not concur with its object." An example is the conceptual cognition apprehending a rabbit's horns. Here, a pot is asserted to be the engaged object of a conceptual cognition apprehending a pot, and the horns of a rabbit is the engaged object of a conceptual cognition apprehending a rabbit's horns.

Also, when conceptual cognitions are categorized according to their purpose, there are two divisions: a conceptual cognition associating a word and a conceptual cognition associating a referent. A conceptual cognition associating a word is one that associates the *word* from the time of learning the linguistic convention with the *referent* at the time of applying the convention as a classificatory one. An example is "This dappled thing is a cow." The difference between a linguistic convention and a classificatory convention is as follows. When a word is first applied to an object, that is a *linguistic convention*, and when that convention is used on later occasions, it is called a *classificatory convention*.

A conceptual cognition that connects a subject of predication with a predicated quality is called a "conceptual cognition associating a referent." An example is the conceptual cognition "This person is a staff holder." It cognizes by applying the predicated quality, "staff holder," to the subject of predication, "this person." Dignāga's *Compendium of Valid Cognition Autocommentary* says: [43]

> Suppose someone asks, "What is this so-called conceptual cognition?" It is that which associates a name, a class, and so

> on [with an object]. In the case of arbitrary words, an object is expressed as qualified by a name, such as "Ḍittha." In the case of class words, the object is expressed as qualified by a class, such as "cow." In the case of quality words, the object is expressed as qualified by a quality, such as "white." In the case of action words, the object is expressed as qualified by an activity, such as "cooking." And in the case of substance words, the object is expressed as qualified by a substance, such as "staff holder" or "horned."[76]

Here "name" indicates a conceptual cognition associating a word, and "class" indicates a conceptual cognition associating a *referent*. A conceptual cognition thinking "This person is Ḍittha" is a conceptual cognition associating a *word*. A conceptual cognition thinking "This conglomerate of characteristics such as a hump and so on is a cow" is a conceptual cognition associating a referent by way of a *class*. A conceptual cognition thinking "The color of a puṇḍarīka flower is white" is a conceptual cognition associating a referent by way of an *attribute*. A conceptual cognition thinking "Rice is cooking" is a conceptual cognition associating a referent by way of an *activity*. A conceptual cognition thinking "This person is a staff holder" or "This yak is horned" is a conceptual cognition associating a referent by way of a *substance*. [44]

Also in some texts three types of conceptual cognitions are explained: conceptual cognitions that rely on a linguistic convention, conceptual cognitions that superimpose something else onto their objects, and conceptual cognitions of that which is hidden. A conceptual cognition that arises in dependence on associating a name or linguistic convention is a conceptual cognition that relies on a linguistic convention. An example is the conceptual cognition that arises in dependence on the statement "This bulbous thing is a pot" (which expresses the proper use of the word or verbal convention for *pot*). Superimposing something else onto an object, or apprehending it to be other than what it actually is, is a conceptual cognition that superimposes something else. An example is the conceptual cognition that, with inappropriate attention, superimposes attractiveness onto an unattractive object.[77] A conceptual cognition focused on something that is hidden to one is a conceptual cognition of that which is hidden. An example is a conceptual cognition apprehending a pot to be impermanent. This interpretation of types of conceptuality is presented in Dharmakīrti's *Exposition of Valid Cognition*:

> [Two types of] conceptual cognition—one based on a convention (*saṃketa*)
> and one that superimposes another object—
> sometimes cause the error [of seeming to be perceptual]
> because they immediately follow a perception.
>
> A cognition such as a recollection,
> being the conceptual cognition of a remote object,
> is dependent upon convention (*samaya*),
> and it does not apprehend a perceptual object.[78]

The nature of a nonconceptual cognition is to be a moment of awareness that is free from any construing awareness that apprehends word and referent as suitable to be associated. There are two types: nonconceptual consciousness that concurs with the object and nonconceptual consciousness that does not concur with the object. A nonconceptual consciousness whose engaged object exists is called a nonconceptual consciousness that [45] concurs with the object; examples include an eye consciousness apprehending a pot and an eye consciousness apprehending a pillar. A nonconceptual consciousness whose engaged object does not exist is called a nonconceptual consciousness that does not concur with the object; examples include a sense consciousness to which a snow mountain appears blue and a sense consciousness to which one moon appears as two.

In brief, a conceptual mental cognition and a nonconceptual sense cognition are different in various ways, including whether the cognition's object appears clearly; the cognition apprehends word and referent as suitable to be associated; the cognition depends on a linguistic convention; when this or that object appears in cognition, it appears to be fused with an object universal; a subject and its properties appear separately; the cognition apprehends a subject and its properties as mutually associated; the cognition qualifies its object with a class, a quality, or an action; the cognition is caused by remembering a linguistic convention; the cognition is qualified by time; the cognition apprehends earlier and later times in combination; or the cognition occurs due to the proximity of its observed object. [46]

4

Valid and Mistaken

Valid versus Nonvalid Forms of Cognition

The purpose of the presentations on valid cognition found in the scriptures is to help beings accomplish their desired goals. The various sufferings that exist in the world, which are universally unwelcome, occur on the basis of ignorance of the nature of things in reality. If one can find the valid means of knowing, then in reliance on them one can engage in correct norms of what is to be affirmed and what is to be rejected. In this way, one will be able to accomplish the desired goals. Dharmakīrti's *Drop of Reasoning* says: "Since correct cognition is the prerequisite to the fulfillment of everyone's aims, I will explain it."[79]

The Sanskrit word for "valid cognition" is *pramāṇa*, which is composed of two parts, *pra* and *māṇa*. Although the prefix *pra* can have many meanings, in this case it means "first." The second part, *māṇa*, means "to measure, comprehend, realize, or know." Therefore valid cognition is that which first, or newly, comprehends something that was not realized before. For example, *Exposition of Valid Cognition* says:

> It also reveals an object not known before.[80]

Alternatively, the syllable *pra* can mean "excellent" or "supreme," which in Sanskrit is *paramārtha*. Here too, the syllable *māṇa* means "to measure." So valid cognition is that which is nondeceptive and thus supremely knows its object. *Exposition of Valid Cognition* explains:

> Valid cognition is nondeceptive cognition.[81] [47]

Thus the definition of valid cognition is: a newly acquired and nondeceptive cognition. The part of this definition that says "newly" excludes subsequent cognition from being valid cognition. The part that says "nondeceptive" excludes correct assumption, doubt, and distorted cognition from being valid cognition. The part that says "cognition" excludes the physical eye sense faculty and so on from being valid cognition. *Exposition of Valid Cognition* says:

> Valid cognition is nondeceptive cognition.
> [Regarding its object's] ability to perform a function,
> it is nondeceptive.[82]

And:

> It also reveals an object not known before.[83]

That is to say, consciousness that accords with the object's real way of existing is nondeceptive. Also, since it cognizes an object for the first time, or determines it on its own accord, it is characterized as "newly realizing."

Alternatively, valid cognition can be defined as: a consciousness that is nondeceptive regarding its comprehended object, which it has determined on its own accord. For example, *Ascertainment of Valid Cognition* says:

> Right cognition has two aspects: direct perception and inferential understanding. These two are right because, when we act having determined an object through one or both of these two, we are not deceived with regard to that object's functioning.[84]

On this point—the definition of valid cognition—the views of the Buddhist schools from the Sautrāntika up to the Svātantrika-Madhyamaka are in harmony. [48]

In Candrakīrti's *Clear Words*, however, Dignāga's definition of valid cognition and views on direct perception and so on are refuted on the grounds that they are set forth on the basis of presupposing inherently existent phenomena. In Candrakīrti's own system, valid cognition is posited only on the basis of being nondeceptive in agreement with what is commonly accepted in everyday conventions of the world. Therefore being "nondeceptive" alone exhausts the meaning of what a valid cognition is, and there is no need to add the qualification "newly realizing" to the definition. So we need to understand from statements such as these that there are unique epistemological views in Candrakīrti's system.

In general, although the word *valid* can be applied to valid cognition, valid person, and valid scripture, the last two are not valid in their own right; so to be *valid* is synonymous with being a *valid cognition*. As for the two types of valid cognition—direct perception and inferential understanding—this will be explained in the context of the seven types of mind in chapter 18.

Buddhist epistemological texts differentiate the two kinds of effects of valid cognition: *mediated* effects of valid cognition and *unmediated* effects of valid cognition. This is to emphasize the point that every genuine goal desired by living beings is accomplished as either a direct or an indirect effect of valid cognition.

The definition of a nonvalid cognition is: that which is not a newly acquired and nondeceptive cognition with respect to its object. Instances of nonvalid cognition include a subsequent cognition[85] realizing sound to be impermanent and a thought holding sound to be permanent. Although a subsequent cognition realizing sound to be impermanent is nondeceptive regarding sound being impermanent, thus satisfying the meaning of *nondeceptive*, it does not newly realize sound [49] to be impermanent and thus does not satisfy the meaning of *newly realizing*. A thought holding sound to be permanent, being a mind that distortedly superimposes an incorrect attribute on sound, does not satisfy either the meaning of *nondeceptive* or of *newly realizing*. Among the seven types of cognition to be discussed later, distorted cognition, doubt, correct assumption, subsequent cognition, and indeterminate perception are all nonvalid cognitions.

As for Prāsaṅgika-Madhyamaka thinkers like Candrakīrti, since they maintain that valid cognition does not need to be newly realizing, and they define valid cognition as consciousness that is merely nondeceptive in accordance with what is commonly accepted in the world, subsequent cognition too qualifies as valid cognition because it is nondeceptive with regard to its principal object. Also, both conceptual and nonconceptual subsequent cognition are accepted to be valid direct perception because the former understands its object not in dependence on reasoning but through experience. For example, without relying on reasoning, you can know, in a mundane sense, a person whom you met before, so you can claim, "I know that person through direct perception." Therefore both conceptual and nonconceptual forms of direct perception are accepted in this system. Candrakīrti's refutation of Dignāga's epistemology in general—as well as his refutation of Dignāga's views on perception in particular—and Candrakīrti's own unique epistemological views will be addressed in detail in volume 4. [50]

Mistaken versus Unmistaken Cognition

There are two ways in which a cognition may be mistaken: it may be mistaken only with regard to its appearing object or it may be mistaken with regard to its engaged object as well as its appearing object. For example, a conceptual cognition apprehending a pot is mistaken only with regard to its appearing object, and a conceptual cognition apprehending a pot to be permanent is mistaken with regard to its engaged object as well as its appearing object. In the case of a conceptual cognition apprehending a pot, its appearing object appears to be a pot, but it is not a pot,[86] so this mind is mistaken with regard to its appearing object. In a conceptual cognition apprehending a pot to be permanent, not only does a pot *appear* as permanent, but the pot is *apprehended* to be permanent too, just as it appears. Therefore this mind is mistaken with regard to both its appearing object and its engaged object; thus it is known as a distorted cognition.

The definition of a mistaken cognition is: a mind that is mistaken with regard to its appearing object. Moreover, a mistaken cognition, while it must be mistaken with regard to its appearing object, may be either mistaken or unmistaken with regard to its object as cognized.[87] There are two types of cognition that are mistaken: mistaken conceptual cognition and mistaken nonconceptual cognition. Examples of these are, in the first case, a conceptual cognition apprehending a pot and, in the second case, a sense consciousness to which a white snow mountain appears blue. (The latter is categorized as a distorted cognition.) Conceptual cognitions are mistaken regarding their appearing object in that an object universal appears to be the object. Distorted cognitions are mistaken regarding their engaged object in that they hold their engaged object to exist in a way it does not. [51]

A mistaken cognition may be mistaken regarding its object due to the influence of either a temporary cause of error or a deeper cause of error.

There are four types of temporary causes of error: (1) one that exists in the basis, (2) one that exists in the object, (3) one that exists in the location, (4) one that exists in the immediately preceding condition. First, an example of a nonconceptual cognition with a temporary cause of error in the basis is the visual cognition of two moons that occurs when the eye is affected by an eye disorder. The main condition causing this mind to be mistaken comes about through the functioning of the *basis*—that is, the visual sense organ—so the cause of error is said to exist in the basis. Second, an example of a nonconceptual cognition with a temporary cause of error in the object is a sensory cognition in which a disc appears when a fan spins around very quickly. Third, an example of a nonconceptual cognition with a temporary cause of error in the location is the sensory cognition in which trees appear to be moving, as can occur when someone travels in a boat. Fourth, an example of a nonconceptual cognition with a temporary cause of error in the immediately preceding condition is a sensory cognition in which the ground appears to be red due to the mind being disturbed by anger. *Drop of Reasoning* says: "A direct perception is a cognition in which no error has been induced by an eye disorder, fast spinning, boat travel, mental disturbance, and so on."[88]

There are two types of conditions that cause such errors: conditions that distort the physical sense faculties and conditions that distort the mental sense faculty. The first has two types: external conditions that distort the physical sense faculties and internal conditions that distort them. The first type includes things like mirrors, voices inside a cave, summer sunlight on a pale sandy area nearby, and so on. [52] These, respectively, cause the apprehension of a reflection as a face, echoes as speech, a mirage as water, and so on. Second, internal conditions that distort the physical sense faculties include such things as an eye disorder, jaundice, infectious disease, and so on; these cause sensory cognitions to be mistaken.

There may also be errors regarding shape, color, number, measurement, and so on. For example, a circle of light appearing when a firebrand is spun around very fast is an error regarding *shape*; a white conch shell appearing yellow is an error regarding *color*; trees appearing to be moving as an effect of traveling in a boat is an error regarding *activity*; one moon appearing as two is an error regarding *number*; falling hairs appearing when there are none is an error regarding *nature*; sunshine at midnight appearing in a dream is an error regarding *time*; a large object appearing small when seen from a distance is an error regarding *measurement*. Among those, a dream consciousness is a mental consciousness and the rest are sensory cognitions.

Also, Asaṅga's *Levels of Yogic Deeds* speaks of other causes of error: "One errs in five ways. What are those five ways? (1) Erring with regard to discernment, (2) erring with regard to number, (3) erring with regard to shape, (4) erring with regard to color, and (5) erring with regard to action."[89] In this context, erroneous discernment [53] is, for example, when one sees Devadatta and thinks it is Yajñadatta. "Erring with regard to action" is a synonym for "erring with regard to function."

As for deeper causes of error, these are as stated in *Ornament for the Middle Way*:

> From the ripening of latent potencies
> within a beginningless mental continuum,
> projected images appear, but since they are mistaken,
> they are in nature like illusions.[90]

According to sources such as the above, these are stable causes of error that have arisen since time without beginning. Since one is deeply habituated to viewing oneself and others as existing in an independent and autonomous way, then when anger, for example, arises toward people such as enemies whom one dislikes, the people who are the objects of that mental state appear to be independently and objectively unattractive. Similarly, *Descent into Laṅkā Sūtra* says:

> Just so, Mahāmati, spiritually immature beings have minds affected by misconceptions that proliferate owing to various factors from time without beginning; they have mental states completely consumed by the fires of attachment, anger, and delusion.[91]

As for the second type, conditions that distort the mental sense faculty include dreams, intoxication, and medication, because these cause the mental consciousness to be mistaken. [54]

In the Buddhist epistemological texts, mistaken cognition is referred to also by the term *fallacious perception*. What might be the difference between these two terms? Mistaken cognition and fallacious perception are in fact synonymous. The definition of a fallacious perception is: a cognition that is mistaken with regard to its appearing object. Dignāga's *Compendium of Valid Cognition* says:

> Mistaken cognition, conventional consciousness,
> inference, that which arises from inference,
> recollection, and desire, along with [perceptions distorted by an] eye disorder
> are fallacious perceptions.[92]

Here fallacious perception is presented in terms of seven categories, in the following order: (1) mistaken conceptual cognition, such as a conceptual cognition apprehending a mirage to be water, (2) conventional conceptual cognition, such as a conceptual cognition apprehending a gross composite object or a continuum, (3) conceptual cognition of an inference, such as the conceptual cognition apprehending a reason, (4) conceptual cognition that arises from an inference, such as an inferential understanding based on reasoning, (5) conceptual cognition that is recollection, such as a conceptual cognition remembering a past thing, (6) conceptual cognition that aspires to something, such as a conceptual cognition wishing to actualize something in the future, (7) nonconceptual fallacious perception, such as a sense consciousness to which falling hairs appear when the eye is afflicted by a disorder, and so on.

A conceptual cognition apprehending a reason is called an *inference*; this is because the thought apprehending a reason is the cause of its resultant inferential understanding based on reasoning, so the name of the result is applied to its cause. In Dharmakīrti's *Exposition of Valid Cognition*, however, these seven types of fallacious perception are subsumed into four classes: (1) a mind that is based on a linguistic convention, (2) a mind that superimposes something else onto its object, (3) a mind that has a hidden object, [55] and (4) a cognition that arises from an impaired sense faculty as its basis. The first three of these fallacious perceptions are conceptual, whereas the fourth is nonconceptual.[93]

The definition of an unmistaken cognition is: a cognition that is not mistaken with regard to its appearing object. According to the Sautrāntika and the Sautrāntika-Madhyamaka systems, direct perception is unmistaken. So for them, a visual cognition of blue is an unmistaken cognition.

Conceptual cognitions mistake their object's universal for the object itself, thus they are necessarily mistaken cognitions, no matter whether they are mistaken or unmistaken regarding their object as cognized. If a nonconceptual consciousness is unmistaken regarding its object as cognized, then it must be unmistaken regarding its appearing object. Therefore if a cognition is unmistaken, it must be a nonconceptual consciousness. Also, since directly perceiving subsequent cognitions and indeterminate perceptions are also unmistaken cognitions, unmistaken cognitions need not necessarily be valid cognitions.

According to the Cittamātra, the Yogācāra-Madhyamaka, and the Prāsaṅgika-Madhyamaka systems, visual cognitions apprehending blue and so on are mistaken cognitions because they are distorted by a deeper cause of error. As for the Prāsaṅgika-Mādhyamikas, they consider all conceptual and nonconceptual cognitions within the mindstreams of ordinary beings to be mistaken cognitions, yet they also maintain that all instances of cognition directly apprehend their own appearances. Subtle distinctions such as these that are drawn by the Prāsaṅgikas will be explained in volumes 3 and 4. [57]

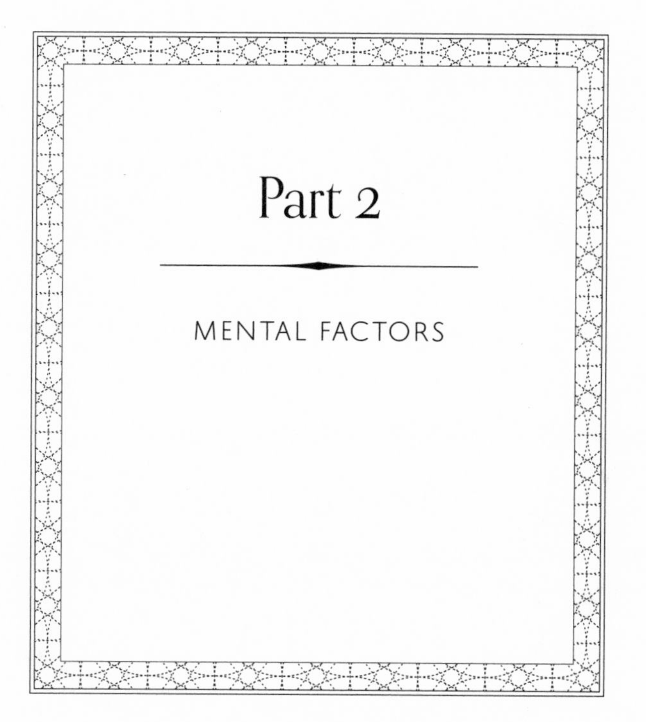

Part 2

MENTAL FACTORS

Categories of Mind

As discussed in part 1, the Indian Buddhist account of mind argues for a causal continuum of "minds," such that each unique mind or mind event in the continuum arises from the previous moment and also acts as a cause for the arisal of the next unique mind event. In this way a mind or moment of consciousness is causal in nature, but this causal process is not simply a matter of one mind moment producing the next. Instead, at least some other fundamental elements known as *dharmas* must be involved. More specifically, a core set of dharmas categorized under the general rubric of *mental factors* (Skt., *cetasika* or *caitta*) must be in place for a mind moment to occur, and often many other mental factors are also active. In this account, consciousness thus involves a complex interaction of multiple cognitive and affective elements, and from the earliest days of Buddhism until roughly the end of the first millennium, Indian Buddhist theorists articulated multiple finely grained accounts of these mental factors and their various functions. To introduce this part of our volume on mind, I will begin by examining some of the basic motivations that inform the analysis of mental factors. The key features of the most prominent account will then draw our attention, and I will conclude with some reflections on some revealing aspects of this model, including the absence of any category for emotions.

Abhidharma Motivations

While other topics treated in this volume draw on the wide range of sources introduced in part 1, the topic of mind and mental factors is based almost entirely on the Abhidharma literature, and the general motivations of Abhidharma analysis are clearly reflected in this material. In particular, two overall motivations underlie the detailed analysis of the mental factors that contribute to a moment of consciousness. First, a dominant, if implicit, role is again played by the critique of an enduring, autonomous self (Skt., *ātman*) that seems to act as an agent of cognition, an experiencer of affective states, a controller of the mind-body system, and so on. In ways that were already evident in part 1, the history of Abhidharma accounts of mind can be read in part as an attempt to construct models that explain how mental processes can work in the absence of any such essentialized self. The taxonomic analyses of mental factors, however, play another important role: they provide a mapping of the mind-body system that serves to direct contemplative practices whose goal is, quite literally, to search for precisely the kind of autonomous, controlling self whose existence is so strongly suggested by our ordinary intuitions—though from the Buddhist perspective these intuitions are both false and the ultimate cause of suffering. By carefully parsing the constituents of consciousness into discrete elements, the Abhidharma taxonomy of mental factors enables practitioners to search for this alleged self in a systematic way. Of course, from the Abhidharma standpoint, the conclusion of that search is that no such self can be found, but without a thorough taxonomy of the elements of experience, one might suspect that some key candidate for such a self has somehow been overlooked. By providing what purports to be an exhaustive map of all the possible candidates, the Abhidharma analysis enables one to make one's search conclusive.

Another motivation for these detailed taxonomies emerges from concerns with transforming inner lives and behaviors. Buddhist practices aim to radically transform individuals by uprooting ignorance, the mental distortion that creates the illusion of an autonomous self. Yet this final goal is understood to be preceded by a long period of preparation that involves intermediate goals, especially the attainment of mental capacities and behaviors conducive to contemplative practice. A widespread model maintains that the wisdom or insight required to uproot ignorance can only emerge with the suitable level of meditative concentration (Skt., *samādhi*), and since the requisite concentration cannot be developed in a chaotic mind, practitioners must first cultivate a lifestyle that embodies ethics (*śīla*). The claim here is that a mind filled with disturbing, non-virtuous mental states such as hatred will necessarily be unstable, and the Abhidharma analysis of mental factors is thus also concerned with a detailed account that distinguishes nonvirtuous (*akuśala*) mental states from virtuous (*kuśala*) ones. Through that analysis, practitioners learn to identify these states and can apply various techniques for reducing the nonvirtuous and enhancing the virtuous.

Along these same lines, the account of mind and mental factors also provides a model for understanding how specific practices affect cognitive and affective processes in beneficial ways. A straightforward example is the role that the mental factor called *aspiration* (*chanda*) plays in attentional processes, such that in the absence of aspiration—which involves an intense interest directed toward an object—sustained attention can easily degrade. Another example is *mindfulness* (*smṛti*), also a key mental factor in practices that seek to enhance attention. The detailed analysis of the capacity of this mental factor to, for example, inhibit attention capture or distraction clearly connects to some of the contemplative techniques discussed in part 6. In this way, the account of mental factors is best read with an eye to the way that these analyses enhance the psychological and behavioral changes that are targets of Buddhist practice.

Asaṅga's Model: Omnipresent and Determinate Factors

Indian Buddhist theorists developed various and somewhat divergent accounts of the mental factors, and given the usual importance accorded Vasubandhu's *Treasury of Knowledge*, one might expect that text to play a central role here as well. Nevertheless, for this particular topic, Tibetan scholarship has instead focused on the model presented by Asaṅga (ca. fourth century) in his *Compendium of Knowledge* (*Abhidharmasamuccaya*). This text is often called the "higher Abhidharma" because it is inspired by a higher level of philosophical analysis in contrast to Vasubandhu's *Treasury of Abhidharma*, which is "lower." Our authors do discuss Vasubandhu's approach along with several other models, but the focus of their treatment is Asaṅga's model. Here, they follow the "mind and mental factors" (Tib., *sems dang sems byung*) literature, a genre in Tibet that involves extracting from Abhidharma texts—especially Asaṅga's—the material specifically relevant to this topic. Asaṅga's list of mental factors is extensive, but he identifies a set of particularly important "omnipresent" (Skt., *sarvatraga*) factors that must occur with any moment of consciousness, and he also introduces a crucial category of mental factors "with a determinate object" (*pratiniyataviṣaya*). Let us examine some key features of these categories in turn.

Omnipresent Factors

Centuries before Asaṅga, Buddhist theorists were already identifying the mental factors that they deemed common to all moments of consciousness, all mind events. All of these accounts assume a distinction between what Tibetan scholarship calls the main mind (Tib., *gtso sems*) and the mental factors that occur with it. In brief, the *main mind* is simply the fact of awareness of some object within any moment of consciousness. The *omnipresent mental factors* are the features of that moment of consciousness that must be present for consciousness to occur. This list reveals important insights about what is minimally required for an object to be presented in awareness. Based on the paradigmatic case of sense perception, traditional lists often begin with *contact* (Skt., *sparśa*), which concerns the relationship between a physical sense organ and the mind. Three factors—intention (*cetanā*), attention (*manasikāra*), and discernment (*saṃjñā*)—have to do with the minimal attentional features required for a moment of awareness. *Attention* orients the mind toward the object, and *discernment* is what functions to "grasp the mark" (*nimittagrahaṇa*), in effect, selecting the object in distinction from other objects. Finally, the mental factor of *feeling* (*vedanā*) accounts for the hedonic tone of the awareness of the object as pleasant, unpleasant, or neutral. The inclusion of feeling as necessary for any awareness points to the causal role that hedonic tone plays in Buddhist accounts of behavior, inasmuch as a positive or negative tone plays a significant downstream causal role in one's continued engagement with the object through behaviors focused on obtaining what is pleasant and avoiding what is unpleasant.

Our authors present a succinct and clear account of these omnipresent mental factors, and there is no need to repeat their efforts here. Instead, we might consider some intriguing implications of this model. In particular, it is noteworthy that one of Asaṅga's contributions is to narrow the list of omnipresent factors to just five items. In his *Treasury of Knowledge*, Vasubandhu (traditionally considered to be Asaṅga's half-brother) bundles many more factors into the "omnipresent category," taking the aforementioned *aspiration* and other factors such as *concentration* to be omnipresent. By removing these and other factors from the omnipresent category, Asaṅga seems to be pointing to a quite minimal moment of consciousness. To understand Asaṅga's possible motivations, we might consider the notions of *phenomenal* and *access* consciousness.

Developed by the philosopher Ned Block, the distinction between phenomenal and access consciousness roughly parallels the Buddhist distinction between nonconceptual and conceptual consciousness. As Block puts it, "Phenomenal consciousness is experience; the phenomenally conscious aspect of a state is what it is like to be in that state." Access consciousness, by contrast, is characterized by its "availability for use in reasoning and rationally guiding speech and action" (Block 1995, 228). According to Block's distinction, in an instance of phenomenal consciousness, we are aware of some content or object but without the ability to categorize or name what we are aware of. For example, I may have a vague feeling of unease as I am closing my car door, but I do not fully realize that this feeling is emerging from my awareness that the car keys are lying on the dashboard. In contrast, if an instance of access consciousness were active in that moment, I would know explicitly, "I left my keys on the dashboard!" Phenomenal consciousness still has downstream causal effects—my sense of unease may prompt me to wonder whether I have forgotten something—yet it does not involve explicit categorization or conscious action toward the object. It may be that Asaṅga was attempting to develop a similar distinction, perhaps because he was writing in an age when the Buddhist epistemologists were beginning to articulate a sharper distinction between conceptual and nonconceptual awareness.

Mental Factors with a Determinate Object

When Asaṅga shortens the list of omnipresent factors, he does so by creating a new, separate category: mental factors "with a determinate object." Here, Block's distinction between phenomenal and access consciousness again seems relevant, in that the cognitive states that involve these mental factors appear to be those required for "access," including conscious categorization of the object and voluntary action toward it. Although Asaṅga's account is vague at points, it seems that the presence of at least some of the factors in this category accounts for the kinds of determinate cognitions that may follow upon an initial, indeterminate moment of awareness, such as the first moment of sense consciousness. Certainly, the term *determinate object* itself indicates that the presence of these mental factors is what enables conceptual engagement with the object presented in cognition.

Although not explicitly thematized in this way by traditional sources, all of the mental factors with a determinate object relate to holding an object in attention, perhaps in ways similar to contemporary scientific notions of working memory, as suggested by Georges Dreyfus (2011). Asaṅga's model does not specify how many factors must be present simultaneously in order for an object to be held in attention, but one insight in the formulation of this category is that the list begins with *aspiration* and *resolution* (*adhimukti*). Both of these factors mark a kind of goal-oriented interest in the object. If we assume that one or both of these must be present for an object to be held in attention, then this model intriguingly suggests that goal-oriented interest is a key ingredient in this form of attention.

Beyond the goal-oriented dimension of attention provided by aspiration and resolution, the mental factor of mindfulness is even more clearly attentional in nature. In part 6, we will see that the term *mindfulness* plays a broad role in Buddhist contemplative practice, but in the specific, technical context of mental factors, this term points especially to the capacity to block distraction. In other words, the mental factor called *mindfulness* does not account for one's orienting toward, selecting, or focusing on an object; instead, when an object is being held in attention, mindfulness is what prevents attention from being captured by a competing stimulus. In this way, the original Sanskrit term *smṛti,* which literally means "remembering" or "recollecting," might best be understood as a metaphor: just as "remembering" something means that we are not forgetting it, so too, when the mental factor *smṛti* is in place, we are not "forgetting" or losing track of the object due to our attention being captured by some other object. I will have more to say about this and other aspects of mindfulness in the introduction to part 6.

The last two mental factors in this category, *concentration* and *wisdom* (*prajñā*), at first glance may seem to apply uniquely to the realm of meditative practices designed to cultivate them. And our authors do indeed acknowledge the importance of these factors in the contemplative context, not least because the term for the factor *wisdom* is identical to the term for the insight that uproots fundamental ignorance. Nevertheless, these factors do not just concern contemplation. Elsewhere, in the context of a "lack of meta-awareness" (*asamprajanya*), Asaṅga again uses the term for wisdom, *prajñā*, but notes that, in that context, it occurs with afflictive mental states that produce suffering and that it is thus involved in nonvirtuous behavior. And even the high level of one-pointed attention involved in the mental factor of concentration need not be necessarily virtuous. And as it turns out, for Asaṅga and many other (but not all) Abhidharma theorists, even the mental factor of mindfulness can occur in "afflicted" (*saṃkliṣṭa*) or nonvirtuous mental states. The upshot is that, while this category of mental factors is clearly associated with the cultivation of highly focused states of extraordinary attention, the cultivation of refined attention in itself does not guarantee that the mind will be in a wholesome state. This is undoubtedly one reason Abhidharma theorists are so concerned with properly distinguishing virtuous mental factors from those that they refer to as "afflictions."

Virtuous Factors, Mental Afflictions, and Compassion

As we have seen, Asaṅga's model—the most dominant one for the Tibetan tradition—lays out the minimal set of omnipresent mental factors that must be present in any moment of consciousness, and he articulates a set of factors that must be present, at least in part, for more stable attentional states to occur. At the same time, however, a moment of consciousness can (and usually does) include many other mental factors, and these are parsed into three categories: virtuous factors, nonvirtuous factors called *mental afflictions* (*kleśa*), and variable factors that can be either virtuous or non-virtuous. Let us examine some key features of this part of Asaṅga's model.

Overall, the distinction between virtuous and nonvirtuous factors is based largely on the functions and effects of these processes and states. In the long term, virtuous factors produce *sukha*—which can be translated as happiness, pleasure, or well-being—because they establish *karma* or mental conditioning that creates such results. In contrast, nonvirtuous states have the long-term effect of *duḥkha*—suffering, pain, or dissatisfaction—because they establish the conditioning that leads to *duḥkha*. In nearer terms, these virtuous and nonvirtuous factors also support more immediate downstream effects. For example, Asaṅga notes that the virtuous factor nonhatred (*adveṣa*), which inhibits the arisal of hatred, functions in a way that prevents negative or destructive behavior. The nonvirtuous state of anger (*dveṣa*), in contrast, functions to induce unpleasant mental states and to induce negative, destructive behavior.

While mental factors are often described in terms of their function, phenomenological features also appear in their descriptions. In particular, mental afflictions—the general rubric for nonvirtuous states—are said to arise in such a way that the mind is disturbed or unsettled (*apraśamita*) in various ways, including by distraction and agitation. Here again, we see the influence of concerns about contemplative practice in these accounts of mental factors. As noted above, Buddhist theories of spiritual transformation maintain that, while only wisdom can uproot the deepest causes of suffering and dissatisfaction, one must also train in meditative concentration, since the cultivation of wisdom requires that capacity. Likewise, the contemplative training in concentration will not succeed with a chaotic mind; one must also train in ethics—that is, one must cultivate virtue and reduce nonvirtue—precisely because mental afflictions, or nonvirtues, cause the mind to be unsettled in a way that inhibits the cultivation of attention.

Along similar lines, our authors also examine several virtues central to Tibetan Buddhist contemplative practice—love, forbearance, and compassion—even though these virtues do not figure explicitly in Asaṅga's list of virtues. Here, our authors cite a beloved verse from the work of Śāntideva, "All those who are happy in the world are so from wanting others to be happy; all those who suffer in the world are so from wanting themselves to be happy." As suggested by this verse, love, forbearance, and compassion are crucial because they enable practitioners to radically reorient their self-focused, self-cherishing attitudes and motivations in such a way that their mental life and manifest behavior become focused on the welfare of all others impartially, and through that shift in perspective, practitioners achieve their own highest spiritual goals, including authentic happiness. This reorientation toward concern for others, however, can be effected by beginning with the kind of innate, spontaneous loving kindness and compassion we feel, in a biased way, toward our loved ones and others whom we experience, in some sense, as extensions of ourselves. In this section, our authors thus explore some of the contemplative techniques that use our own innate capacity for loving kindness and compassion as a springboard toward the development of unbiased, universal forms of these capacities.

Variable Lists and the Question of Emotions

As we have just seen, loving kindness and compassion are key virtues targeted by Buddhist contemplative practice, but they do not figure explicitly in Asaṅga's list. Their absence points to some important issues around the categorization of mental factors, and our authors end this part with an inventory of lists found in Buddhist sources other than Asaṅga's work. These include not only other Abhidharma sources but even scriptural materials, such as the *Sūtra on the Application of Mindfulness* (which is exactly *not* the same as the well-known Pali text, the *Satipaṭṭhānasūtta*). The mere presence of so many alternative accounts—without coming to any final reconciliation—points, first of all, to the instrumental nature of these lists: they are motivated by the transformational goals of Buddhist practice. The lists also make sense as responses to the variety of goals and historical contexts that characterize the Buddhist traditions. This does not make these accounts somehow "unscientific"; after all, numerous historians and philosophers of science have articulated the way that the concerns and cultural contexts of modern science also shape its models, theories, and outcomes. Seeing the various lists of mental factors as responsive to motivations and contexts in this way also helps to explain one of their striking features: the absence of any category for emotions.

Lisa Feldman Barrett and others have argued that contemporary approaches to the science of emotions reflect age-old cultural assumptions that inform the creation of such a category, especially by setting "emotions" in opposition to "reason." When traced back to Greek philosophers such as the Stoics, the Western cultural heritage of emotion as a category often characterizes emotions as clouding reason, even to the point that the complete elimination of emotion is seen as a worthy goal of rational, philosophical practice. Scientific work first championed by Antonio Damasio has shown, however, that this perspective is deeply flawed: without emotions, our capacity for reason becomes severely impaired. Nevertheless, the divide between reason and emotions persists as a cultural motif, in part because it suggests a path for cultivating a virtuous life where the pursuit of "higher" reason is what keeps in check the base, even animalistic, "passions" that motivate nonvirtuous acts.

As we have seen, a crucial distinction for Buddhist theorists is between virtuous and nonvirtuous mental factors. The question of whether a mental factor might be called an emotion is simply not relevant. We have seen that a nonvirtuous mental state is one that, on the one hand, is disturbed or unsettled, and on the other hand, produces suffering and dissatisfaction, at least in the long term. Additionally, mental afflictions such as anger are also rooted in a fundamental confusion about the nature of personal identity, or even about reality itself. Thus even purely cognitive factors such as inquiry, which could scarcely be categorized as an emotion, manifest as nonvirtuous when they perpetuate that fundamental confusion. And of course, ignorance—the root of that confusion—also is not what we would call an emotion, yet it is seen as the most destructive mental affliction.

Thus, in contrast to the divide between reason and emotion, the Buddhist approach is far more concerned with the distinction between wisdom and ignorance. And this interest may help to explain the absence of nearly all the "basic emotions" identified by Paul Ekman and other scientists who champion the notion of universal, transcultural emotional states. Ekman's list consists of six such emotions: anger, disgust, fear, happiness, sadness, and surprise. In the most influential account, the one by Asaṅga, only anger appears as a mental factor, and while at least one other list (in the *Sūtra on the Application of Mindfulness*) includes fear and sadness, none include all six of Ekman's basic emotions. From this, we might conclude that these Buddhist theorists were not very skilled at their observation of mental states. But given the increasing importance within the contemporary science of emotions of the role that culture and context play in our conceptualization of emotions, a more likely conclusion might be that Asaṅga and other Buddhists theorists have not simply missed something. Instead, perhaps what they offer is an intriguing, alternative approach to categorizing the facets of our mental lives.

John Dunne

Further Reading in English

For an overall account of the Abhidharma context, see chapter 1 of Jan Westerhoff, *The Golden Age of Indian Buddhist Philosophy* (Oxford: Oxford University Press, 2018).

Asaṅga's model is presented by Evan Thompson and Georges Dreyfus in "Asian Perspectives: Indian Theories of Mind," in *The Cambridge Handbook of Consciousness*, edited by Philip David Zelazo, Morris Moscovitch, and Evan Thompson (Cambridge: Cambridge University Press, 2007).

The importance of compassion within Buddhism and in broader contexts is discussed extensively by Matthieu Ricard in *Altruism: The Power of Compassion to Change Yourself and the World* (New York: Little, Brown and Company, 2015).

The record of a historic encounter between Buddhism and science on the topic of emotions is recounted by Daniel Goleman in *Destructive Emotions: How Can We Overcome Them? A Scientific Dialogue with the Dalai Lama* (New York: Bantam Books, 2004). For an account of emotions as constructed, see Lisa Feldman Barrett, *How Emotions Are Made* (New York: Houghton Mifflin Harcourt, 2017).

5

Distinguishing Mind and Mental Factors

CONSCIOUSNESS, as explained above, has the nature of being clear and aware. On the basis of primary function, Buddhist texts distinguish between a moment of consciousness—a *mind*—and the *mental factors* that are facets of that mind.

As for the mind in general, Asaṅga's *Compendium of Knowledge* says: "What is posited as the aggregate of consciousness? Whatever is mind and mentality, that too is consciousness."[94] In Sanskrit, *mind* is *citta*, *mentality* is *manas*, and *consciousness* is *vijñāna*, and these three are equivalent, or synonymous. These three terms have different etymological interpretations. *Mind* refers to the six engaging consciousnesses within the continuum of a living being, which accumulate (Skt., *cit*) the latent potencies of experiences, habits, and so on. *Mentality* is that which functions to cognize (*man*) its object. *Consciousness* is that which knows (*jñā*) differentiated (*vi*) aspects of its object. Alternatively, *mind* connotes the accumulation of latent potencies, *mentality* connotes the basis, and *consciousness* connotes what is based on it. [58]

Also, in the texts of the Cittamātra school, which assert the notion of a foundation consciousness (or "store consciousness"), there are contexts where the three—mind, mentality, and consciousness—are presented as having different referents. In such texts, *mind* refers to the foundation consciousness (*ālayavijñāna*) that is the base for the seeds of actions, *mentality* refers to the afflictive mental consciousness (*kliṣṭamanas*) that grasps the foundation consciousness as "me," and *consciousness* refers to the six engaging consciousnesses that cognize their respective objects. *Descent into Laṅkā Sūtra*, for example, says:

> The *mind* is the foundation consciousness;
> what thinks of that as "me" is *mentality*;
> that which apprehends objects
> is known as *consciousness*.[95]

In Asaṅga's *Compendium of the Mahāyāna*, too, it says: "It is not correct to think, as this opponent does, that mind, mentality, and consciousness are synonymous, with merely the words being different, because one can see that *mentality* and *consciousness* have different referents. Therefore *mind* too will have a different referent."[96]

Regarding the distinction between mind and mental factors, *Distinguishing the Middle from the Extremes* says:

> That which sees the object is consciousness;
> what sees its attributes, the mental factors.[97]

Both the root text and the autocommentary make this point clear. There are two ways of understanding the referent of the word *it* in the line "its attributes, the mental factors." [59] One is to take it as referring to the mind (that is, attributes of the mind), and the other is to read it as referring to the object of the mind. According to the first explanation, "mental factors" refers to attributes of the main mind[98] or to attributes of its functions. According to the second reading, "mental factors" refers not merely to cognizing the object itself but to cognizing the specific attributes of the object as well.

The first interpretation draws the distinction between minds and mental factors not in terms of their object but in terms of the cognizing subject's specific functions; it defines mental factors on the basis of the different aspects of the function of the main mind or consciousness. Consider, for example, a visual consciousness and its concomitant mental factors apprehending a visible form. Here the sense consciousness's mere perception of its object, the visible form, constitutes the primary consciousness. The mental factors accompanying it, on the other hand, not only perceive the form but are each characterized in terms of specific functions, such as turning the mind toward the object, not forgetting an object already cognized, and so on. Therefore a main mind is a cognizing agent that is characterized only in terms of perceiving its object without needing to be qualified in terms of something else. A mental factor is a cognizing agent that perceives the very same object and engages that object principally by way of different attributes, such as function and so on. [60]

The mind and its accompanying mental factors are concomitant in sharing five similar features, therefore there is no difference between them insofar as both cognize the object itself and its attributes. Nevertheless, the main mind and mental factors other than feeling, such as discernment, *experience* their object not through their own power but through the power of that concomitant factor of feeling; also, the main mind and any mental factors other than discernment *distinguish* their object not through their own power but through the power of the factor of discernment. The same applies to all the remaining mental factors. For example, if the main mind has the aspect of apprehending a pot, the aspects of all its accompanying mental factors—feeling, discernment, intention, contact, and attention, as well as any others there may be—are equally that of apprehending the pot. Likewise, just as the aspect of the mental factor feeling is that of experiencing an object that is beneficial, harmful, or neutral, so the aspect of the main mind and any other mental factors such as discernment, intention, faith, and so on must be that of experiencing the object that is beneficial, harmful, or neutral. *Clarifying the Meaning of the Treasury of Knowledge* by Yaśomitra, also known as Jinaputra, says:

> "Same basis, object, aspect, time, and substance" indicates that whatever basis gives rise to the mind, that very basis gives rise to feeling, discernment, intention, and so on. Likewise, whatever object [61] gives rise to the mind, that very object gives rise to feeling and so on. Whatever aspect gives rise to the mind, that very aspect gives rise to feeling and so on—in that if the mind's aspect is blue, then feeling and so on, concomitant with it, arise having only the aspect of blue. And at whatever time the mind arises, feeling and so on arise only at that time.[99]

The second interpretation of the phrase "its attributes, the mental factors" says that the mind cognizes the object itself, such as a visible form, sound, and so on, while the mental factors cognize the object's attributes. The meaning of "the object itself" and "its attributes" is as follows. A visible form, for example, is the object itself. That visible form's size, attractiveness, and so on are the object's attributes. Likewise, a sound is an object itself, and its melodiousness or harshness and so on are its attributes. The six types of consciousness that perceive the object itself are main minds. The fifty-one mental factors such as feeling that perceive the object's attributes such as attractiveness or unattractiveness, the perception of which gives rise to pleasant or unpleasant states and so on, are mental factors. For example, when the visual consciousness apprehends a visible form, its accompanying feeling apprehends its qualities, such as whether it is beneficial or harmful; discernment apprehends the object's distinguishing mark, to which various classificatory conventions will be applied; contact apprehends the form's attributes, such as pleasant or unpleasant; [62] attachment and hatred apprehend it as attractive, unattractive, and so on. This same point applies to the remaining factors and their aspects. Asaṅga says in *Levels of Yogic Deeds*: "Here any painful, pleasant, or neutral characteristics are cognized by contact; any beneficial, harmful, or neutral characteristics are cognized by feeling; any characteristic that is the distinguishing mark for a classificatory convention is cognized by discernment."[100]

Thus, according to either of the two explanations, the way in which a single mind has significantly different accompanying mental factors should not be understood in the manner of a king surrounded by an entourage of many different ministers. For it says in a sūtra, "The consciousnesses of sentient beings occur as distinct individual streams of awareness."[101] Also, *Exposition of Valid Cognition* says:

> Two conceptual consciousnesses are never observed
> together.[102]

So to view the mind and its concomitant mental factors as substantially different would contradict many such passages that recur again and again in the sūtras and śāstras. [63]

One might ask, "Well then, how does one main mind have diverse accompanying mental factors?" When we look at a single flower, for example, we have to posit a general mental experience in relation to it. This basic awareness of the object is the main mind. There is then a *feeling* that arises simultaneously with it, which experiences a pleasant, unpleasant, or neutral sensation. *Discernment* distinguishes the attributes of the object, such as "This is blue and that is yellow." *Intention* moves the mind and the mental factors toward the object, just as a magnetic stone moves iron. *Attention* directs the mind to a specific object among several objects of the mind. *Contact* causes the mind to encounter the object. *Mindfulness* causes the mind to hold the object without forgetting it. *Concentration* causes the mind to stay placed on the object and so on. In brief, mind and mental factors must be differentiated and identified in terms of either the functions of the mind or the functions of the attributes of the object. When remembering an object such as a visible form, although both the main mind and its accompanying factor of mindfulness are concomitant in observing the same object, a visible form, they engage it differently. The main mind does so by cognizing the object itself—the visible form—whereas mindfulness does so by causing the object not to be lost or forgotten. [64]

Thus both of these two explanations accord with the above-quoted text from *Distinguishing the Middle from the Extremes*, as well as Vasubandhu's *Commentary on Distinguishing the Middle from the Extremes*, which says: "The mere object is seen by consciousness; the object's attributes are seen by mental factors, such as feeling and so on,"[103] and Sthiramati's *Explanatory Commentary on Distinguishing the Middle from the Extremes*, which says:

> In the line "The mere object is seen by consciousness," the word *mere* is used to exclude its attributes; therefore it perceives the thing itself but does not apprehend its attributes. Also, "the object's attributes are seen by mental factors, such as feeling and so on" because that entity's attributes are perceived as that by those mental factors. Here any state—pleasure, pain, mental bliss, and so on—is apprehended as an attribute of that thing by feeling. Any distinction between the meanings of distinguishing marks of applied conventions, such as woman or man, is apprehended by discernment. Later it states, "One must apply this analysis likewise to the others."[104] [65]

In the *Treasury of Knowledge* it says:

> Mind, mentality, and consciousness
> are synonymous; mind and mental factors
> are concomitant in sharing the same five features,
> including basis, object, and aspect.[105]

The definition of a main mind is: a consciousness that is concomitant with its accompanying mental factors in sharing the same five features. When categorized, there are six types of consciousness—from visual consciousness up to mental consciousness.

The definition of a mental factor is: a cognition that is concomitant with the main mind it accompanies in sharing the same five features. Mind and mental factors are concomitant in sharing the same five features: same basis, same object, same aspect, same time, same substance. They share the *same basis* in that a main mind and its accompanying mental factors are both based on the same sense faculty. For whatever sense faculty or dominant condition the main mind is based on, any mental factors accompanying it are also based on that. They share the *same object* in that they observe the same object. For whatever object the mind observes, any mental factors accompanying it also observe that. They share the *same aspect* in that their way of holding the object is the same. For whatever aspect of the object the mind arises having, any mental factors accompanying it also arise having that aspect of the object [66]—thus they have the same aspect or way of holding. They share the *same time* in that they arise at the same time. For a mind and any mental factors accompanying it both arise, abide, and cease together; thus they have the same duration. They share the *same substance* in that both the mind and any mental factors accompanying it arise only as a single substance. For in the case of one main mind, there is but one accompanying feeling, one discernment, and so on; thus a main mind and its accompanying mental factors are the same in being a single substance.

Consider, for example, an instance of visual consciousness and its accompanying factor of feeling. If the visual consciousness is based on its dominant condition of the visual sense faculty, then its accompanying feeling is also based on the visual sense faculty. If the visual consciousness observes an object that is a visible form, then its accompanying feeling also observes that form. If the visual consciousness arises having an appearance of blue, then its accompanying feeling also arises having an appearance of blue. When a visual consciousness arises, its accompanying feeling arises at the same time. And since a visual consciousness is a single substance, its accompanying feeling also arises just as a single substance. Pūrṇavardhana's *Investigating Characteristics: Explanatory Commentary on the Treasury of Knowledge* says:

> "Same basis, object, aspect, time, and substance" means that only the moments of a visual sense faculty and so on that act as the basis of such a consciousness can also act as the basis of the feelings and so on that are concomitant with it. Similarly, [67] since all their objects are not diverse, they have the same object; since they all arise with one aspect, they have the same aspect; since the main mind is present only when there is feeling, they have the same duration; and "same substance" means having an equal number.[106]

Asaṅga's *Compendium of Knowledge* has a somewhat different explanation of the five shared features. It says:

> Suppose someone asks, "What is it to be concomitant in the sense of cognizing in conjunction?" The phenomena of mind and mental factors cognize in conjunction with regard to the same object. And this concomitance by way of cognizing in conjunction involves what is different, not what is identical; and a concordant pair, not a discordant pair; two congruent times, not two incongruent times; and two concordant realms and levels, not two discordant realms and levels.[107]

Mind and mental factors are concomitant in sharing the same substance, same object, same nature, same time, and same realms and levels. So the way in which shared features are characterized here is slightly different from the way in which they were described earlier based on *Distinguishing the Middle from the Extremes* and its associated commentaries as well as Vasubandhu's *Treasury of Knowledge*.

The *sameness of substance* as well as the *sameness of time* as stated in this excerpt from the *Compendium of Knowledge* are just like the ones explained earlier. Furthermore, given that a mind and its accompanying mental factors apprehend by way of observing an object such as a visible form, they have the *same object*. [68] *Same nature* means that if the main mind is afflictive, then its accompanying mental factors will be afflictive, and if the main mind is a virtuous awareness, then its accompanying mental factors will be virtuous. *Same realm and level* means that there will never arise a mind of the desire realm accompanied by mental factors of the form realm, or a mind of the first absorption accompanied by mental factors of the second absorption, and such like; for whatever realm and level the main mind may be, the realm and level of its accompanying mental factors are the same. This text lists *same nature* and *same realm and level*, which are not listed among the five similar features in Vasubandhu's *Treasury of Knowledge*. In explaining these points, Yaśomitra's *Explanation of the Compendium of Knowledge* says:

> "What is it to engage by cognizing in conjunction?" The phenomena of mind and mental factors cognize in conjunction the same object. And this engagement by cognizing in conjunction involves what is different, not what is identical. Here, that mind is not in conjunction with another mind, and that feeling is not in conjunction with another feeling, and so on. Also, it involves a concordant pair—not a discordant pair like attachment and hatred, virtue and nonvirtue, and so on; and two congruent times—not two incongruent times [69] like the present and the future or the past and the present; and two concordant realms and levels—not two discordant realms and levels, like a combination of both the desire realm and the form realm or the first absorption and the second absorption.[108]

Now consider the following objection. Asaṅga's *Compendium of Ascertainments* says: "We say, 'Because they have the same substance, same object, same time, and same function.' If you ask, 'Why say they are endowed with aspects?' We say, 'They engage one object having many aspects.'"[109] This suggests that mind and mental factors have different aspects but the same substance and so on. Likewise, Sthiramati's *Explanatory Commentary on Distinguishing the Middle from the Extremes* says: "Since it would follow that they would be no different from primary consciousness, they do not have the same aspect."[110] Doesn't this indicate that mind and mental factors do not have the same aspect?

What these statements mean is as follows. For example, it is in terms of a visible form appearing that a visual consciousness apprehending a form satisfies the criterion of *mind*, merely because it perceives a form. Its accompanying intention or such like [70] does not satisfy the criterion of *mental factor* merely because of that. Rather, one must explain that it is in terms of its other attributes that a mind is moved toward an object or such like. That is what these treatises mean. They do not mean that a mind and its accompanying mental factors do not have the same aspect in general.

According to the Buddhist Sautrāntika school and above (i.e., the Cittamātra and Madhyamaka schools), a main mind and its accompanying mental factors are the same substance but conceptually distinct.[111] These schools also explain that the mental factors accompanying one main mind are reciprocally the same substance. While the mental factors accompanying one main mind are mutually concomitant in sharing the same five features, they differ in the relative strength of their function. For example, a person with a very strong mental factor of faith is known as someone with faith, a person with a very strong aspiration is known as someone with purpose, a person with very strong mindfulness is known as someone with mindful awareness, a person with very strong concentration is known as someone with mental stability, a person with very strong wisdom is known as someone with sharp intellectual faculties, and so on.

The mental factors (literally, "arisen from the mind" in Tibetan) are so called because they are transformations of the mind or they arise from the mind as individual parts, just as waves arise from a body of water.

If there were no mental factor of feeling accompanying the mind, it could not function to experience an object; and if there were no intention, it could not function to move the mind toward the object. The mind itself definitely has to arise simultaneously with its accompanying mental factors in every case and cannot do so without them. Thus the *Treasury of Knowledge* says:

> Mind and mental factors are definitely simultaneous.[112] [71]

The Fifty-One Mental Factors Presented in Asaṅga's *Compendium of Knowledge*

Presentations of the mental factors come from Abhidharma treatises such as the *Great Explanation*. These presentations are, in turn, based on earlier Abhidharma texts such as the *Attainment of Knowledge*, *Topic Divisions*, and *Compendium of Elements*, which are included among the seven canonical scriptures of the Abhidharma.[113] Here we will present the topic of mental factors based primarily on the two most famous treatises on the

Abhidharma—the *Compendium of Knowledge* by Asaṅga and the *Treasury of Knowledge* by Vasubandhu—which are grounded in those great earlier Abhidharma texts.

In the presentation of the five aggregates in Asaṅga's *Compendium of Knowledge*, the definition and function of the aggregate of *feeling* and of the aggregate of *discernment* are presented separately. Then the aggregate of conditioning factors is divided into two categories—the aggregate of conditioning factors that are concomitant with the mind and the aggregate of conditioning factors that are not concomitant with the mind. Then, in the context of explaining the aggregate of conditioning factors concomitant with the mind, this text presents the rest of the mental factors listed as fifty-one in number. We may wonder, "Do the fifty-one mental factors listed here in this text encompass all mental factors?" As will be explained in a later section, [72] there are many mental factors listed in *Finer Points of Discipline*, such as *offensive instigation*, which impels movements of body or speech as in scowling, *discouragement*, which is a deep despondency of mind, *thinking that one is deathless*, no doubt due to a fear of death, and *mental perturbation*. Furthermore, Asaṅga's *Compendium of Bases*, Nāgārjuna's *Precious Garland*, and many other texts present several that are not included among those enumerated in the list of fifty-one of the *Compendium of Knowledge*. This point is acknowledged in some of the classical commentaries on the *Compendium of Knowledge*. But since this presentation of fifty-one mental factors in the *Compendium of Knowledge* is quite well known, we will provide a fairly extensive explanation of it here.

The fifty-one mental factors are grouped in several different categories:

1. Five omnipresent mental factors
2. Five mental factors with a determinate object
3. Eleven virtuous mental factors
4. Six root mental afflictions
5. Twenty secondary mental afflictions
6. Four variable mental factors

6

Omnipresent Mental Factors

THE FIRST SET, the five omnipresent mental factors, contains: (1) feeling, (2) discernment, (3) intention, (4) attention, and (5) contact. Feeling experiences any of the three—pleasure, pain, or neutral feeling—as its object. Discernment distinguishes the attributes of the object, such as "This is blue and that is yellow." Intention moves the mind and mental factors toward the object, just as a magnet moves iron. Contact causes the mind [73] and object to meet. Attention directs the mind to a specific object. The *Verse Summary* reads:

> Feeling, discernment, intention,
> contact, and attention are called
> "omnipresent mental factors."[114]

First, the definition of *feeling* is: a mental factor that, by its own power, experiences any type of pleasure, pain, or neutral feeling. The *Compendium of Knowledge* says: "What is the definition of feeling? It has the characteristic of experience."[115] When categorized, there are three types of feeling: pleasure, pain, and equanimity. Or they can be divided into two types: physical feeling and mental feeling. According to the first way of categorizing, when a consciousness accompanied by feeling engages an object, it experiences pleasure, pain, or equanimity (which is neither pleasure nor pain). According to the second way of categorizing, feelings that accompany a sense consciousness are called *physical feelings*, whereas feelings that accompany a mental consciousness are called *mental feelings*. Āryadeva's *Four Hundred Stanzas*, for example, says: [74]

> Just as the tactile sense faculty pervades the whole body,
> delusion dwells within all [mental afflictions].[116]

The tactile sense faculty pervades all parts of the body, from the crown of the head to the soles of the feet, so any feelings accompanying the other four sense consciousnesses, such as the visual sense consciousness, are physical feelings. Additionally, there is a way of categorizing feeling as five: (1) pleasant physical feeling, (2) unpleasant physical feeling, (3) pleasant mental feeling, (4) unpleasant mental feeling, and (5) neutral feeling.

Second, the definition of *discernment* is: a mental factor that, by its own power, apprehends the distinguishing mark of its object. The phrase "apprehends the distinguishing mark" means to not mix up the distinguishing characteristics or fine distinctions of an object. Regarding this, the *Compendium of Knowledge* says: "What is the definition of discernment? That which has the essential nature of either apprehending a distinguishing mark or apprehending a mental image through which a convention is applied to the objects of seeing, hearing, differentiating, and understanding."[117] When categorized, discernment has two divisions: discernment that apprehends a distinguishing mark and discernment that apprehends images. The first is discernment accompanying any of the five sense consciousnesses that apprehends objects, such as blue and yellow, without mixing them up. The second is discernment accompanying a mental consciousness that apprehends thoughts, such as "This is blue" or "This is yellow," without mixing up these classificatory conventions. [75]

On what basis do these two types of discernment operate? Of the four of seeing and so on, *seeing* refers to applying a classificatory convention to things seen directly, *hearing* refers to applying a classificatory convention to trustworthy words heard, *differentiating* refers to applying a classificatory convention to things ascertained in dependence on evidence, and *understanding* refers to applying a classificatory convention (through discernment) to things ascertained directly. According to the explanatory commentaries on Abhidharma, this applies to what is seen by the eye, heard by the ear, understood by the mind, and differentiated by the three—nose, tongue, and bodily sensation.

Third, the definition of *intention* is: a mental factor that moves and incites the mind with which it is concomitant toward the object. Accordingly, the *Compendium of Knowledge* says: "What is intention? It is a mental action that actually conditions the mind. It has the function of making the mind engage in virtuous, nonvirtuous, or neutral actions."[118] The *Compendium of Ascertainments* says: "What function does intention have? It has the function of motivating actions of thought, body, and speech."[119] For example, just as a magnet moves iron, this mental factor's ability is to make the mind, or any of its mental factors, engage the object. [76]

When categorized in terms of the sense faculties, which are the bases, there are six, ranging from an intention that arises from contact involving the visual sense faculty to an intention that arises from contact involving the mental sense faculty. From among the two types of action—action that is intention and action that is intended—*action that is intention* refers to an intention that is concomitant with the mental consciousness it accompanies, and an action of body or speech motivated by that is called *action that is intended*. The *Treasury of Knowledge*, for example, says:

> Action is intention and what is done by it.
> Intention is an action of mind;
> actions of body and speech are produced by it.[120]

Candrakīrti too speaks about the distinction between the two along similar lines.

Fourth, the definition of *attention* is: a mental factor that by its own power directs the mind, with which it is concomitant, to a particular object. The difference between this and intention is that intention moves the mind with which it is concomitant toward a general object and attention directs the mind toward a particular object. The *Compendium of Knowledge* says: "What is attention? It is mental engagement; it has the function of mentally apprehending the object."[121] [77]

Fifth, the definition of *contact* is: a mental factor that—once the object, the sense faculty, and the sense consciousness have come together—by its own power selects an object in accord with whatever feeling of pleasure and so on is to be experienced. The phrase "selects an object" means to select a particular object as its own unique object—once the object, sense faculty, and sense consciousness have come together.

The phrase "the object, the sense faculty, and the sense consciousness have come together" does not indicate that they have come together at exactly the same time. For the sense faculty, which is the dominant condition, and the sense consciousness, which is based on it, are sequential not simultaneous. Thus the phrase "have come together" refers to the three conditions—the object, the sense faculty, and the sense consciousness—being complete. The *Compendium of Knowledge* says: "What is contact? It is what determines transformation of the sense faculty once the three have come together. It functions as the basis of feeling."[122]

The basis, when categorized, has six types—from contact when the visual sense faculty and so on have come together to contact when the mind sense faculty and so on have come together.

The reason that the five mental factors—feeling and so on—are called *omnipresent* is because they accompany every cognitive event. Indeed, if any of the five omnipresent mental factors were lacking for a mental state, then the object could not be fully experienced. Were there no feeling, there would simply be no experience. [78] Were there no discernment, it would be impossible to apprehend the unique distinguishing characteristic of the object or to differentiate the object's attributes. Were there no intention, there would be no moving toward an object. Were there no attention, there would be no directing toward a particular object. Were there no contact, there could be no meeting with the object. So for the mind to engage an object, the five omnipresent mental factors must be fully present. Accordingly, Jinamitra's *Commentary on the Levels of Yogic Deeds* says:

> Why are there exactly five? If any of them were lacking, the activity of experiencing an object would not be fully complete; thus they are not too few. And because there are cases of mistaken mindfulness, aspiration, and so on, they are not too many. Without attention, the collection would not be complete and there would be no arising of consciousness at all. The Buddha said, "If attention is present, visual consciousness will arise," and similar statements. Also, "If attention and intention were absent, not having been established or actually coming together, how would consciousness arise? If discernment and feeling were absent, there would be no apprehension of any sign of the object and no experience at all; therefore the arising of consciousness would be totally pointless. If contact is absent, then feeling, discernment, and intention would not arise at all because they would have no support."[123] [79]

7

Mental Factors with a Determinate Object

SECOND, the five mental factors with a determinate object are: (1) aspiration, (2) resolution, (3) mindfulness, (4) concentration, and (5) wisdom. The *Compendium of Ascertainments* says: "Do any that are not omnipresent arise? There are many that are not omnipresent, but the main ones are the following five: aspiration, resolution, mindfulness, concentration, and wisdom."[124] They are said to have determinate objects because they ascertain only particular objects, in this order: a desired object, a determined object, an object of acquaintance, and an analyzed object. The *Compendium of Ascertainments* says: "In relation to which determinate objects do any of the nonomnipresent mental factors arise? There are four types, in this order: desire, determination, acquaintance, and analysis. The last is for concentration and wisdom, whereas the rest are for the first three in that order."[125] The *Verse Summary* reads:

> The five with a determinate object are thus:
> aspiration, resolution, mindfulness,
> concentration, and wisdom. [80]

First, to explain *aspiration*, the cause of aspiration is to see the qualities of the object and so on. The definition of aspiration is: a mental factor that, upon observing a desired thing, seeks to attain the thing by its own power. An example is a mental factor that seeks to attain qualities that have not arisen in oneself in order to give rise to those qualities. As for the meaning of the term *aspiration*, it is so called because seeking the thing that is its object enables the mind and all the mental factors with which it is concomitant to *aspire* for its object. The *Compendium of Knowledge* says: "What is aspiration? It is the very wish to be endowed with this or that attribute of a desired thing. Its function is to act as a basis for generating diligence."[126] The function of aspiration is, directly, to produce diligence, and indirectly, to bring about an increase in all known good qualities. The *Compendium of Ascertainments* says: "What function does aspiration have? It has the function of giving rise to diligence."[127]

When categorized, there are three types of aspiration: aspiration wanting to meet with the object, aspiration wanting not to be separated from the object, and aspiration seeking to obtain the object. [81]

Second, to explain *resolution*, the cause of resolution is to see, hear, or remember and so on the qualities of the object. The definition of resolution is: a mental factor that by its own power, upon perceiving an object previously seen, heard, and so on, has the aspect of apprehending it to be just so. An example is an instance of a mental factor that, in relation to an object ascertained through analysis, thinks "It is just so" in accord with that ascertainment. Likewise, the *Compendium of Knowledge* says: "What is resolution? It apprehends the ascertained thing to be just as it is ascertained, so it functions to prevent diversion."[128]

According to the commentaries on the *Treasury of Knowledge*, resolution is not necessarily a mind that realizes its object, because there are nonvirtuous resolutions that apprehend their objects in accord with false ascertainments of those objects. To explain the etymology of the term, the Sanskrit word *adhimokṣa* has the connotation of determining its object, hence it is called *resolution*.

There are the following differences between resolution and discernment: resolution, by its own power, prevents diversion of the mind from its object, whereas discernment, by its own power, applies a classificatory convention to the object. Also, resolution, by its own power, apprehends its object to be just so, whereas discernment merely distinguishes its object.

The function of resolution is such that, in regard to an object's qualities or faults or neither, it resolves the object to be just so and prevents diversion to something else. The *Compendium of Ascertainments* says: "What function does resolution have? [82] It has the function of being resolved about good qualities, faults, or neither of those, regarding the object."[129] The *Compendium of Ascertainments* categorizes it into three divisions: resolution perceiving good qualities in its object, resolution perceiving faults in its object, and resolution perceiving neither good qualities nor faults in its object.

Third, to explain *mindfulness*, the cause of mindfulness is great appreciation, and only objects with which one has developed prior familiarity can induce mindfulness. Śāntideva's *Compendium of Training in Verses* says:

> Mindfulness arises from great appreciation.[130]

Also, Śāntideva's *Engaging in the Bodhisattva's Deeds* lists some other causes that produce mindfulness:

> Owing to accompanying their teacher,
> following their guru's instructions and out of fear,
> those fortunate ones who practice with respect
> easily generate mindfulness.[131]

Mindfulness arises from companionship with the virtuous friend, listening closely to the instructions taught by the learned master, having a sense of shame and fear of being reproached by others, [83] and respecting the holy person's advice.

The definition of mindfulness is: a mental factor that, focusing on an object to which there has been previous familiarization, prevents the forgetting of it by its own power. An example is mindfulness that recollects one's parents and siblings with whom one has been familiar since earliest childhood. The *Compendium of Knowledge* says: "What is mindfulness? It [is a mental factor that] prevents the mind from forgetting the familiar thing."[132]

Mindfulness has three distinguishing qualities. The distinguishing quality of its *object*—it is a familiar thing. The distinguishing quality of its *aspect*—not forgetting the object that it is focused on. The distinguishing quality of its *function*—it prevents the mind from being distracted from the object and serves as the basis for the mind to remain continuously on that object. Furthermore, since mindfulness does not arise if there is no prior familiarity, the distinguishing quality of the object is called *the familiar thing*. Since mindfulness does not arise if the familiar thing does not appear as an object to the mind right now, the distinguishing quality of the aspect is called the mind's *not forgetting*. And since the mind will become increasingly stable in dependence on mindfulness with these qualities, the distinguishing quality of the function is called the function of *nondistraction*.

Texts such as *Clarifying the Meaning of the Treasury of Knowledge* mention that there are nonvirtuous instances of mindfulness.[133] They say that nonvirtuous mindfulness, by its own power, [84] merely recollects an afflictive object but does not cause a virtuous object to be forgotten. On the other hand, the nonvirtuous factor of forgetfulness, by its own power, can either recollect an afflictive object or cause a virtuous object to be forgotten. To explain the etymology of the term, it is called *mindfulness* because it is mindful in not forgetting its object.

The function of mindfulness is to prevent movement of the mind away from the object and toward something else. However, the *Compendium of Ascertainments* says: "What function does mindfulness have? It functions to recall the memory of what one has thought, done, and said a long time ago."[134] From the point of view of recalling things thought of, actions done, and words spoken long ago, it functions to prevent movement of the mind away from the object.

When mindfulness is categorized according to the *Compendium of Ascertainments*, there are three divisions: being mindful of the mind's thoughts, the body's deeds, and the voice's utterances. The effects of mindfulness are: in reliance on mindfulness that does not forget the object, there is an ever-greater increase within one's own continuum of all the qualities of concentration single-pointedly placed on its object and of wisdom engaged in fine investigation. *Engaging in the Bodhisattva's Deeds* says:

> To those who wish to guard the mind,
> I join my palms together and pray: [85]
> "Make every effort to guard
> mindfulness and meta-awareness."[135]

Thus Śāntideva speaks of the great importance of applying mindfulness and meta-awareness.

Fourth, to explain *concentration*, the cause of concentration is twofold: (1) physical solitude away from social involvement, that is, away from gatherings of many people, and (2) mental solitude free from gross conceptualization, that is, free from running after various objects of thought. The *Questions of Gaganagañja Sūtra* says: "The accumulation of calm abiding unites physical solitude and mental solitude."[136] Furthermore, its causes are placing the mind on the object, mindfulness that does not forget its object, and diligence that places the mind again and again on the object. Maitreya's *Ornament for the Mahāyāna Sūtras* says:

> The mind remaining inwardly focused
> depends on mindfulness and diligence.[137]

The definition of concentration is: a mental factor that, by its own power, stabilizes the mind single-pointedly on its object and functions as the basis of wisdom. An example is a mental factor that, when fully focused on an attractive object, remains single-pointedly on that [86] without moving away to something else. The *Compendium of Knowledge* says: "What is concentration? It is the single-pointedness of mind placed on the thing investigated, and it functions as the basis of wisdom."[138]

To make the mind single-pointed necessarily means to focus intently on a single object without having a second object. Also, concentration is not the same as attention. Attention directs the mind with which it is concomitant to its object, but it does not, by its own power, give rise to wisdom; concentration does.

The etymology of the Sanskrit term *samādhi* for "concentration" has the connotation of "apprehending correctly," or that the mind continuously and firmly apprehends its object, hence the word *concentration*. The function of concentration is to serve as a basis for wisdom, which is its effect. The *Compendium of Ascertainments* says: "What function does concentration have? It functions as a basis for wisdom."[139]

The effect of concentration is as follows: in reliance on concentration, the wisdom of special insight arises. Also, it pacifies the mental disturbances and so on that come from attachment, excitation, scattering, and gross fluctuation of thought regarding the qualities of external sense objects, and it makes the mind and body serviceable and so on. The *Great Final Nirvāṇa Sūtra* says: "Śāriputra, [87] one who has engaged in concentration meditation attains correct understanding and correct insight."[140]

Fifth, to explain *wisdom*, the cause of wisdom is that it arises and increases through learning and through cultivating familiarity with what has been learned.[141] The *Sublime Continuum* says:

> Wisdom abandons all afflictive objects of knowledge;
> therefore
> it is utterly supreme. Its cause is learning.[142]

The definition of wisdom is: a mental factor that, focused on an object of inquiry, has the function of analyzing it by its own power. An example is a mental factor that analyzes properly the distinctions between gross and subtle atoms. The *Compendium of Knowledge* says: "What is wisdom? It fully differentiates the qualities of things to be investigated. It has the function of removing doubt."[143] So in terms of its function, it removes doubt and so on regarding its object.

"Differentiates" in the definition of the nature of wisdom means to differentiate individual objects without conflation. Wisdom and discernment are not the same: wisdom, by its own power, differentiates individual things, [88] whereas discernment, by its own power, applies a classificatory convention to them. Also, wisdom, by its own power, eliminates doubt, whereas discernment does not eliminate it. Inquiry and analysis are also not the same as wisdom because those two, by their own power, do not eliminate doubt. Furthermore, one could explain the term *wisdom*, or *prajñā* in Sanskrit, by noting that it is an awareness (*jñāna*) that is excellent (*pra*) for seeking its object's way of existing. Or it is called *wisdom* (*prajñā*) since it is an excellent (*pra*) form of knowing (*jñāna*).

When categorized in terms of its nature, wisdom has two types: innate wisdom and wisdom acquired through mental cultivation. When the latter is subdivided according to order of arising, there are three types: wisdom generated through learning, wisdom generated through critical reflection, and wisdom generated through meditative cultivation. The *Treasury of Knowledge Autocommentary*, for example, says:

> The phrase "also any" refers to any wisdom that has arisen from learning, critical reflection, and meditative cultivation, such as that belonging to the contaminated class, and that obtained since birth along with its attendant qualities.[144]

First is the wisdom that arises spontaneously within a person's continuum without having trained in learning and so on. Second is the wisdom that ascertains just the general meaning of the words through listening to any teachings not previously heard, memorizing them without forgetting, and reciting them, copying them, and so on in order to commit them to memory. Third is the wisdom that engages in inquiry and analysis of the meaning of the words listened to and eventually, through valid cognition, ascertains that meaning. Fourth is the wisdom that is a single-pointed mind, assisted by a special pliancy that makes the mind and body serviceable, [89] arisen from meditating again and again on the meaning ascertained by the wisdom arisen from critical reflection. Vasubandhu's *Explanation of the Ornament for the Mahāyāna Sūtras* says: "Learning involves placing latent potencies in the mind; reflection involves bringing about realization; meditative cultivation involves bringing about pacification through calm abiding and complete realization through special insight."[145]

Additionally, there is a categorization of wisdom into three types: virtuous, nonvirtuous, and neutral. The *Compendium of Ascertainments* says: "What is wisdom? It concerns this or that fact and what it entails and what it is to thoroughly analyze phenomena. It also concerns what arises through reasoning, what does not arise through reasoning, and what arises neither through reasoning nor not through reasoning."[146]

There are also these other types of wisdom: vast wisdom, swift wisdom, penetrating wisdom, clear wisdom, and so on.[147]

The fruits of wisdom are as follows: upon clearing away beings' confusion, it brings about an understanding of the object's way of existing. The *Range of the Bodhisattva Sūtra* says:

> This superior wisdom, which is like a light, [90]
> continuously clears away beings' darkness;
> like a lamp, it arises to illuminate;
> it acts as a light to the host of afflictions,
> for it clears away the root of craving and confusion.[148]

Special wisdom is like a light, continuously removing beings' darkness or distorted aspiration. It is like the illumination of a lamp, showing unerringly that which is to be adopted and to be abandoned. It is like a light, clearing away the darkness of attachment by removing the root of craving and confusion.

Thus the *Compendium of Knowledge* says that the five mental factors mentioned above—aspiration and so on—have a determinate object, and the reason why they are not listed as omnipresent is because they each ascertain their object as follows. Aspiration arises regarding a sought object but does not arise regarding anything else. Resolution arises regarding a previously ascertained object but does not arise regarding anything else. Mindfulness arises only regarding a familiar object. Both concentration and wisdom arise only regarding an object of investigation. However, in Vasubandhu's *Treasury of Knowledge*, a lower Abhidharma text, even these five are described as omnipresent. The reason is that even in, for example, a mind with doubt as a concomitant mental factor, a slight degree of wisdom exists. However, in such a mind, the wisdom aspect is weak; doubt is the dominant factor present. That this is so will be explained in the presentation of mental factors based on the *Treasury of Knowledge* below in chapter 12. [91]

8

Virtuous Mental Factors

THE THIRD SECTION presents the eleven virtuous mental factors: (1) faith, (2) shame, (3) embarrassment, (4) nonattachment, (5) nonhatred, (6) nondelusion, (7) diligence, (8) pliancy, (9) heedfulness, (10) equanimity, (11) nonviolence. Here is the *Verse Summary*:

> The eleven virtuous ones are as follows:
> faith, shame, embarrassment,
> nonattachment, nonhatred, nondelusion,
> diligence, pliancy, heedfulness,
> equanimity, and nonviolence.

The first among these is *faith*. This mental factor functions as the basis of aspiration aiming for a purpose. It has the aspect of trusting, admiring, or emulating a holy being in whom one has faith. The *Compendium of Knowledge* says: "What is faith? It acts as the basis of aspiration that is manifestly trusting, admiring, or emulating with regard to what actually exists, has qualities, or has ability."[149] Here the phrase "actually exists" indicates the object of *trusting faith*, [92] for it is faith that has confidence in what is true, existent, and nondeceptive. The phrase "has qualities" indicates the object of *admiring faith*, for upon seeing the qualities of a holy being in whom one has faith and so on, the mind becomes clear of defilements, like water becoming clear of mud, and this leads to faith that higher qualities can arise within one's own continuum. The phrase "has ability" indicates the object of *emulating faith*, the faith that thinks: "Knowing that unskillful conditioning can be abandoned and higher qualities can be attained, I must definitely attain them." Although in general the cause of all good qualities is diligence, in order to generate diligence, we need aspiration aiming for a purpose (emulating faith), and to generate aspiration we need to have trusting faith together with seeing good qualities (admiring faith). Therefore *Unraveling the Intention Sūtra* explains again and again that faith is the foundation of all good qualities.

The second of these mental factors is *shame*. This mental factor shuns wrongdoing out of consideration for oneself. It acts as a basis for refraining from bad conduct and for engaging in good conduct. The *Compendium of Knowledge* says: "What is shame? It avoids wrongdoing out of consideration for oneself. It has the function of acting as a basis for refraining from bad conduct."[150] [93]

The third mental factor is *embarrassment*. This mental factor shuns wrongdoing or bad conduct out of consideration for others. It functions to engage in good conduct and abandon bad conduct. The *Compendium of Knowledge* says: "What is embarrassment? It avoids wrongdoing out of consideration for others; that is its function."[151] Despite being the same in avoiding wrongdoing or bad conduct, there is a difference between shame and embarrassment. Shame shuns wrongdoing out of consideration for oneself, thinking: "If I engage in this bad conduct, not only will it not benefit me as a person, either temporarily or long term, but it will also bring harm, so I must not engage in it." Embarrassment shuns wrongdoing out of consideration for others, thinking: "If I engage in this bad conduct, the holy ones will be deeply concerned and others will disparage me, so it is not appropriate." These two mental factors are extremely important for ethical training. Beginners especially cannot do without a method for restraining nonvirtuous mental factors.

The fourth, *nonattachment*, is a mental factor that, having observed saṃsāric existence and what is needed to lead such existence, reverses attachment to them and functions as a basis for not engaging in bad conduct. [94] The *Compendium of Knowledge* says: "What is nonattachment? It reverses attachment to saṃsāric existence and what is needed to lead such existence. Thus it functions as a basis for not engaging in bad conduct."[152]

The fifth mental factor is *nonhatred*. This mental factor, upon perceiving any of the three objects that give rise to anger, destroys the arising of anger and reverses any wish to harm and so on. The three objects that give rise to anger are sentient beings, suffering, and circumstances that produce suffering. The *Compendium of Knowledge* says: "What is nonhatred? It reverses hostility toward sentient beings, suffering, or circumstances that produce suffering. Thus it functions as a basis for not engaging in bad conduct."[153]

The sixth mental factor is *nondelusion*. This mental factor refers to an aspect of fine investigative wisdom, and in dependence on its cause—whether innate or generated through training—it functions as the antidote to delusion. The *Compendium of Knowledge* says: "What is nondelusion? It is fine investigation and wisdom arisen from karmic ripening, from studying the scriptures, from contemplating their meaning, or from realizing their meaning. Thus it functions as a basis for not engaging in bad conduct."[154] Nondelusion, which is a root of virtue, has two possible causes: innate [95] and generated through training. *Innate* means generated through the ripening of powerful purification accomplished in an earlier life, so it is said to be "from karmic ripening." *Generated through training* refers to the three stages of training—learning, critical reflection, and meditative cultivation—stated in the *Compendium of Knowledge* as, respectively, wisdom generated through studying the scriptures, through contemplating their meaning, and through realizing that meaning. Now someone may ask, "What is the difference between the mental factors nondelusion and wisdom?" The mental factor nondelusion is just nonconfusion, and it functions as a basis for not engaging in bad conduct. Wisdom has the characteristic of finely analyzing things, and it functions to reverse afflictive doubt. The three mental factors—nonattachment, nonhatred, and nondelusion—are the roots of all virtuous qualities; they are the means of stopping all bad conduct.

The seventh is *diligence*. This mental factor consists of a completely joyous mental state that is focused on virtuous activity. The *Compendium of Knowledge* says: "What is diligence? It is a completely joyous mental state in the context of armoring, application, non-discouragement, non-reversal, and non-indolence. It functions to cause the virtuous side to become established and complete."[155] Diligence is necessarily delighting in virtue; therefore making effort in worldly activities is not diligence. [96] When categorized, diligence has five types: Delight that occurs prior to engaging in virtue is the *diligence of armoring*. Delight that occurs at the time of virtuous application, whether steady application or zealous application, is the *diligence of application*. Delight that occurs as non-discouragement of mind when generating virtue is the *diligence of non-discouragement*. Delight that occurs as not turning away from virtue in adverse conditions, such as other people, is the *diligence of non-reversal*. Delight that occurs as not being content with the virtue one has already developed is the *diligence of non-indolence*. These are outlined as "armoring" and so on in the *Compendium of Knowledge*.

The eighth mental factor is *pliancy*. This mental factor sets a tendency in the mindstream that enables the mind to become serviceable in focusing on a virtuous object exactly as desired and that interrupts the continuum of bodily and mental dysfunction. The *Compendium of Knowledge* says: "What is pliancy? It is a serviceability of body and mind that has the function of dispelling all hindrances, since it interrupts the continuum of bodily and mental dysfunction."[156] In general, pliancy has two divisions: bodily pliancy and mental pliancy. The pliancy presented here is mental pliancy. The difference between bodily and mental pliancy will be explained later in the context of calm abiding. [97]

The ninth mental factor is *heedfulness*. This mental factor is ascribed to a mind that lacks the three poisons and possesses diligence. Upon observing any type of virtuous factor, it has the aspect of protecting the mind from unfavorable factors. It functions to generate, maintain, or increase virtue. The *Compendium of Knowledge* says: "What is heedfulness? Based on non-attachment, nonhatred, and nondelusion, and accompanied by diligence, it is what habituates the mind to virtuous factors."[157] When categorized, heedfulness has three divisions: heedfulness regarding conduct belonging to the past, the future, or the present. Or we can count five divisions if we include conduct from an earlier time that continues to a later time. This order is expressed in the *Compendium of Knowledge*: "Purify past nonvirtues, refrain from them in the future, and do not engage in them in the present. Examine your motivation and continuously abide with heedfulness."[158]

The tenth mental factor is *equanimity*. This mental factor—having established the nine stages of mental abiding using methods that set the mind single-pointedly on an internal object—attains spontaneous mental abiding without needing to exert any effort in applying the antidotes to laxity and excitation, when the ninth stage of mental abiding is achieved. The *Compendium of Knowledge* says: "What is equanimity? It is based on nonattachment, nonhatred, and nondelusion, and it is accompanied by diligence. It functions to prevent an occasion for mental afflictions to arise. [98] It is evenness of mind, stillness of mind, and spontaneous mental abiding that counters the mental afflictions."[159] The functioning of equanimity does not allow an opportunity for mental afflictions such as laxity and excitation to arise.

In general, mere equanimity includes three types: equanimity that is a conditioning factor, equanimity that is a feeling, and immeasurable equanimity. In the present context, we are concerned with the first type of equanimity. *Śrāvaka Levels* says: "What is equanimity? It is a nonafflictive mind directed toward an object within the purview of calm abiding and special insight. It is a mind in equipoise without mental afflictions, flowing peacefully, and inwardly engaged, a mind that is blissfully balanced and serviceable, that is focused without needing to make any effort."[160]

The eleventh mental factor is *nonviolence*. This is a mental factor associated with nonhatred that, upon observing its object—sentient beings—thinks, with an aspect of loving kindness or being unable to bear others' suffering, "If only they were free from suffering." It functions to restrain oneself from beating, killing, and so on. The *Compendium of Knowledge* says: "What is nonviolence? It is the mind of compassion itself, which is associated with nonhatred. [99] Its function is to restrain from harming."[161] In brief, as the *Compendium of Knowledge* states, compassion has the nature of nonviolence.

9

Love and Compassion

HERE, TO EXPAND on the presentation of the eleven virtuous mental factors above, we now discuss love and compassion. As for the benefits of love and compassion, they accrue initially to the person in whom they arise. For instance, when kindhearted attitudes such as love, compassion, and forbearance arise and are sustained within one's mindstream, they reduce one's fear and boost one's confidence. In these and other ways, they increase one's inner strength. Love and compassion arouse a feeling of close connection with others as well as a sense of purpose and meaning in life; they also give one respite in difficult times.

It is true for all people, no matter who they are, that when they generate a kindhearted attitude toward others, their own life becomes happier. This attitude can also bring a greater sense of ease and more peace in the community within which they live. So if each individual is able to accord greater importance to his or her own ethical behavior and make qualities like love, compassion, forbearance, and so on an inseparable part of his or her own life, [100] then this will definitely produce the most wonderful results.

As human beings, we are born into and grow up under the loving care of our parents, and the powerful feeling of affectionate love during our youth remains in our lifeblood. Medical scientists have demonstrated with empirical evidence that when there is a strong feeling of loving kindness, happiness, and peace of mind in a person's life, it enhances physical health; and likewise a mind constantly agitated undermines physical health.

Love and compassion are values we all appreciate quite naturally without having to be told to by other people. Not only are they precious qualities of our minds, but they are also the basic sources of happiness for us as individual people and are the ground of social harmony. Therefore, whether one is seeking happiness for one's own sake or seeking happiness for others, one must practice love and compassion. Furthermore, having a motivation of kindness is the root of kind behavior; when one turns one's mind toward the well-being of others based on a pure altruistic motivation, one's own behavior naturally becomes flawless from the standpoint of ethical conduct. Therefore love and compassion definitely constitute the primary foundation that underpins all the paths of good ethical behavior. Also, what we call *peace* is not simply a matter of not harming other people but is also clearly an expression of loving kindness. [101]

Since the root of human happiness is loving kindness, all the major religious traditions in the world have teachings that focus on the practice of loving kindness. Just as people appreciate kindness, animals do as well. The most precious treasure a person can possess is a kind heart. To be helpful, good-natured, and kindhearted to others is the essence of human life, and when these are absent, life has no meaning.

Loving kindness is not something we have no means to generate. The potential for loving kindness exists within the continuum of every human being simply by virtue of the type of physical body we have. We need to nurture and enhance this potential using our intelligence. We need to develop loving kindness not only toward humans but also toward other living beings. Even if someone has no religious faith, he or she must be able to recognize that love and compassion are extremely important and profoundly helpful in one's life. Although we have strong feelings and experiences of loving kindness when we are very young, as we grow older and as time goes by, all sorts of internal factors and forces within our environment hinder the potential of loving kindness within us. Whether we are religious or not, we must surely understand that loving kindness is needed for our own happiness as well as for the happiness of our family and society, and that loving kindness is what gives rise to peace and happiness in the mind. [102]

In general, love and compassion are differentiated on two levels. First is the affectionate love that arises as a factor of physical development, like the love expressed by a mother carrying her infant child. The second level is when one has generated that natural affection and then enhanced it through extensive meditative cultivation so as to make one's love and compassion universal. To give rise to the latter, we must adopt an unbiased gaze, not paying attention to whether someone is master or servant, rich or poor, educated or uneducated, strong or weak, fair or dark skinned. It is most important to recognize that all of the nearly eight billion people in this world are the same in being human; everyone is the same in wanting happiness and not wanting suffering. When we have this type of understanding, notions of "us and them," and minds of attachment and hatred based on them, can diminish.

There are two ways of viewing others. One is to see that everyone is the same in wanting happiness and not wanting suffering and so to view everyone as one people. The other way to view others is in terms of their differences, such as of nationality, country, ethnicity, language, wealth, poverty, education, religious tradition, and so on. The first way of viewing others does not give rise to attachment and hatred in our minds, whereas the second way of viewing gives rise to biased attitudes of "us versus them," which provide the basis for notions of friend and foe. Although love may arise (for some) on the basis of discriminatory feelings of "me versus you," [103] it would be a biased type of affectionate love. And where affectionate love is biased, hatred can arise as well. It is on the basis of such biased attitudes that attachment toward one's own side and suspicion toward the other side, as well as resentment and ill will, can arise. In contrast, love that arises out of simply considering someone to be just another human being is an unbiased type of loving kindness. So the love that we need to develop is the loving kindness that arises in dependence on recognizing someone to be a human being just like ourselves. Accordingly, Maitreya's *Ornament for Clear Knowledge* says: "Adopt an attitude of sameness toward sentient beings."[162] Likewise, Haribhadra's *Short Commentary on the Ornament for Clear Knowledge* says: "Adopt an attitude of sameness toward all sentient beings."[163]

Śāntideva's *Engaging in the Bodhisattva's Deeds* says:

> In that both myself and others
> are the same in wanting happiness,
> what is so special about me
> that I strive for my happiness alone?
>
> In that both myself and others
> are the same in not wanting suffering,
> what is so special about me
> that I protect myself but not others?[164]

In that there is no difference between oneself and others—since both are alike in wanting to be happy—it becomes unreasonable to strive for one's own happiness alone and not to strive for others' happiness. [104] Similarly, in that there is no difference between oneself and others—since both are alike in not wanting to suffer—it is unreasonable to guard one's own happiness alone and not to guard the happiness of others. This is what Śāntideva declares in the above verses.

Furthermore, when we understand the way in which human beings live in mutual dependence on one another, this too can give rise to unbiased loving kindness toward others. Human society survives through community and not through each individual living in isolation. Even if we just look at the constitution of the physical body, an individual's happiness necessarily arises in dependence on others. Thus throughout all of human society, from large units like nations to smaller units like single households, among all our needs—education, health, commerce and wealth, food, drink, clothing, utensils, and so on—not a single thing is not dependent on others. Given this fact of our existence, if we continue to despise or mistrust one another, deceive one another, and hurt one another, then there is no possibility for us to attain happiness.

Nowadays, all over the world, one finds the belief that happiness is increased through mere material progress and that this is the root cause of a happy human life. But even with the best possible material conditions, if everyone's mind is filled with self-centeredness, attachment, anger, pride, ill will, jealousy, rivalry, expectation, fear, prejudice, and so on, [105] then no one can have a happy life. Conversely, even without the best material conditions, if everyone has a subdued, peaceful mind—an attitude that cherishes others and is helpful, content, loving, tolerant, and so on—then everyone will have a happy life. Thus there is no denying that a happy or unhappy life depends mainly on one's attitude. The causes of happiness are produced within each person's mind. If people in individual households have no mental happiness, then it will be difficult for those households, and the societies that they comprise, to find a path to happiness.

The conditions that produce so many of the human-made problems in our world these days are greed, competitiveness, resentment, ill will, pride, jealousy, attachment, hatred, and so on, and the root of these is none other than the self-grasping thought of "me" and self-centeredness. *Exposition of Valid Cognition* says:

> When there is "self," there is a notion of "other";
> from the distinction between self and other comes attachment
> and hatred;
> fully linked with these two,
> all faults arise.[165]

When there is self-grasping, an exaggerated notion of the "other" occurs. From this arises a division into essentially distinct partitions or categories of "self" and "others." This gives rise to grasping and attachment to one's own side and hatred toward the others' side. When one's mindstream is linked with these mental factors of attachment and hatred, [106] all the faults such as killing and stealing occur. Similarly, *Engaging in the Bodhisattva's Deeds* says:

> All those who are happy in the world
> are so from wanting others to be happy.
> All those who suffer in the world
> are so from wanting themselves to be happy.[166]

All the various types of happiness in the world come from benefiting others and wanting others to be happy. All the various types of suffering in the world come from wanting oneself to be happy, which is self-centeredness and self-grasping. Therefore one must counteract self-centeredness and self-grasping from a variety of angles.

In brief, people's verbal and physical conduct is classified as good or bad from the point of view of motivation, even down to the smallest deed. The crucial point is that when engaging in any action, whether large or small, one must first generate a good motivation. The way for all people born in this world to attain happiness is to turn our backs on all disharmony based on race, philosophical view, religious tradition, and so on. This is a crucial cause of enhancing good conduct among all races and nations regardless of their religious traditions. We should understand "good conduct" to mean not harming others through one's own body, speech, or mind. [107]

Since this worldwide human society must live in mutual dependence, it is extremely important to create a society that lives with love and affection, like a family sharing food and drink in equal measure. Mentally, we need to take up the responsibility of abandoning hatred and ill will toward the human race on which each of us depends.

LOVE

In the Buddhist texts, love and forbearance are recognized as the two antidotes against hatred. Of these, we will first explore love more specifically. Among the fifty-one mental factors, both compassion and love are in the nature of the mental factor referred to as *nonhatred*. Yet there is a difference between them. Compassion arises from observing sentient beings to be suffering and has the aspect of wishing them to be free of that. Love, on the other hand, arises upon observing sentient beings from the perspective of their well-being and has the aspect of wishing them to be happy.

The definition of love is: a mental factor that, having observed sentient beings, thinks how wonderful it would be if they had happiness and wishes for them to have it. The *Concentration Combining All Merit Sūtra* says: "*Love* is to think 'May all sentient beings be happy.'"[167] [108] The function of love is to help pacify resentment, rage, and ill will toward sentient beings. Furthermore, the *Teachings of Akṣayamati Sūtra* says: "Love protects oneself and consistently benefits others, for it is supremely nonargumentative and thoroughly destroys all the severe faults of ill will, rage, and resentment."[168]

Moreover, it acts as an antidote to hatred and functions as a basis for not engaging in bad conduct. Pṛthivībandhu's *Explanation about the Five Aggregates* says:

> Love that engages in benefiting sentient beings is indeed the antidote to hatred. It also functions as a basis for not engaging in bad conduct, for when nonhatred is present, there is no engaging in evil actions such as killing and so on.[169]

The results of love are pacification of jealousy, of rivalry, and of ill will. Also, one has mental happiness, courage, great inner strength and confidence, and enduring tolerance. And with mental peace and happiness, one's blood circulation and respiratory flow become even—based on which one has a healthy body, a long life, and so on. The *Sūtra on the Application of Mindfulness* says:

> Wishing to benefit all sentient beings, [109] one's blood becomes very clear; owing to one's blood becoming very clear, the color of one's face becomes clear; owing to the color of one's face becoming clear, one becomes fair to behold. Owing to this, all sentient beings become joyful in this present life. This can be experienced directly.[170]

FORBEARANCE

To explain what forbearance is, the *Play of Mañjuśrī Sūtra* says:

> Mañjuśrī asked, "Daughter, how do you explain *nonhatred*?"
> His spiritual daughter replied, "O Mañjuśrī, it is that which stops animosity arising in the mind and prevents the harming of any object; this I understand to be *forbearance*."[171]

Candrakīrti's *Entering the Middle Way* says:

> Giving is the essence of the perfection of generosity;
> absence of anguish is the essence of ethical discipline;
> nonhatred is the essence of forbearance.[172]

In defining the essence of generosity and so on, Candrakīrti identifies forbearance in terms of nonhatred, or a state of mind that is not perturbed in the face of suffering or harm.

Three kinds of forbearance are outlined in [110] *Commentary on Difficult Points in Engaging in the Bodhisattva's Deeds.* "Forbearance consists of three kinds, which are presented as follows in the *Compendium of Teachings Sūtra*: the forbearance of consciously accepting suffering, the forbearance of certitude in contemplating the nature of reality, and the forbearance of disregarding harm done by others."[173]

(1) The forbearance of consciously accepting suffering refers to accepting with equanimity painful circumstances from which one cannot escape, such as great difficulty and hardship in one's home life. If you cannot access resilient tolerance when experiencing your own suffering, then consider what *Engaging in the Bodhisattva's Deeds* says:

If there is a remedy,
what use is it to feel upset?
If there is no remedy,
what use is it to feel upset?[174]

If there is a remedy for something that causes suffering, then what reason is there to be unhappy? By remedying its immediate cause, that suffering will no longer exist. Conversely, if there is no remedy for it, then what benefit is there in being unhappy? It is as useless as being unhappy with space for being unobstructive.

(2) The forbearance of certitude in contemplating the nature of reality refers to analyzing and contemplating the meaning of the object to be meditated on. [111] For example, while debating, a student of Buddhist philosophy contemplates the meaning of what he or she is debating on.[175]

(3) The forbearance of disregarding harm done by others refers to an attitude of restraint and tolerance when, for example, your enemy and others cause you harm; instead of being angry, you practice forbearance. Anger immediately destroys one's peace of mind, causes psychological imbalance, and results in damage to our immediate environment. By recognizing these faults of anger, we adopt forbearance. We do this by contemplating the faults of anger so that before it has arisen when it arises, we can resist the conditions that fuel the anger.

Of the three types of forbearance, the last occurs in the context of being harmed, whereas the other two may occur in any context. The first two types of forbearance are both present, for example, when we are studying for a long time, in that we are listening single-pointedly to the teachings and contemplating the meaning, and even when we are hiking along a trail, in that we are paying attention to our physical conduct and tolerating hunger and thirst.

To contemplate the benefits of adopting forbearance, the *Sūtra Teaching the Great Compassion of the Tathāgata* says: "Bodhisattvas do not have hatred in their minds; they see all sentient beings as dear and are fond of them."[176] Similarly, the *Sūtra on the Application of Mindfulness* says: "Those with forbearance, having abandoned rage, [112] are in harmony with all living beings; they see those beings as dear and grow fond of them. Those with forbearance become a basis of supreme trust and develop a very radiant appearance and a very radiant mind."[177]

Also, *Engaging in the Bodhisattva's Deeds* says:

> There is no evil like hatred;
> there is no resilience like forbearance.[178]

Śāntideva states that there is no evil like hatred for disturbing the mind and destroying virtue; and there is no resilience like forbearance for destroying the feverish torment of the afflictions.

Candrakīrti's *Entering the Middle Way* says:

> Forbearance brings qualities opposite to those just described.
> Forbearance makes one beautiful, dear to holy beings,
> and skillful in knowing right from wrong.[179]

Here Candrakīrti says that through practicing forbearance, one will attain a beautiful body, be dear to and cherished by holy beings, and become skillful in knowing what is and is not correct, which in this tradition means to become versed in sound moral reasoning. [113]

The practice of forbearance itself is a function of one's own state of mind, as is explained in *Engaging in the Bodhisattva's Deeds*:

> Wicked people are as limitless as space;
> there is no possibility of conquering them.
> But if I conquer my mind of anger alone,
> this is like conquering all of my enemies.
>
> Where would enough leather be found
> to cover the entire surface of the earth?
> But if I put leather just on the soles of my shoes,
> this is like covering the whole surface of the earth.[180]

Since wicked people are as limitless as space, it is impossible to overcome them all. Yet overcoming and subduing your mind of anger alone would be like conquering all of your external enemies. As an analogy, if you had to cover the entire surface of the earth with soft leather to protect your feet from injury by thorns and so on, you would never find enough leather; but if you protected your feet by covering just the soles of your shoes with leather, it would be like covering the entire surface of the earth. The way to cultivate forbearance and so on as an antidote to the fault of hatred is demonstrated superbly in Śāntideva's *Engaging in the Bodhisattva's Deeds*, so one can learn much about mastering this skill from that text. [114]

The way to cultivate forbearance, the antidote to hatred, is as follows. Contemplate that anger is inappropriate because it yields no benefits and brings serious problems. Say someone harms you, and you become enraged and retaliate. This will not undo the harm they have already done, so what point is there in seeking revenge? Candrakīrti says in *Entering the Middle Way*:

> If you respond with vengeance when someone harms you,
> does your vengeance reverse what was inflicted?
> Vengeance surely serves no purpose in this life.[181]

Furthermore, not wanting to experience future suffering while simultaneously seeking revenge for the harm done to you by others is a contradiction. Responding with harm—through engaging in quarrels, disputes, and so on—damages not only yourself but also your relatives and friends. Also, grave consequences occur in the present and will continue to occur, even so far as losing your life. Therefore, just as you tolerate the pain when a doctor pierces you with a sharp instrument as part of a treatment, so you should cultivate firm tolerance in the face of slight temporary suffering so as to avoid endless long-term pain. Thus by thinking "It is unsuitable to respond with harm," you put a stop to hatred. [115]

Also, some people overwhelmed with mental sickness even harm their doctor, yet their doctor thinks "This is beyond their control," and without getting angry, he tries various methods to cure their mental sickness. Similarly, when an abuser harms you, consider, "He is behaving in this way because he is impelled by mental afflictions beyond his control," and make a distinction between the mental affliction and the person in whose continuum it arises. So without getting angry at the person, you think "May he be free from mental affliction," and this is how you put a stop to hatred. Āryadeva says in *Four Hundred Stanzas*:

> Just as a doctor does not get angry with
> an enraged person seized by demonic forces,

the buddhas see mental afflictions as the enemy,
not the person who has them.[182]

Also, Candrakīrti says in his *Commentary on the Four Hundred Stanzas*:

Thinking "The fault here is not sentient beings,
the fault is the mental afflictions,"
wise ones who have thoroughly analyzed this
do not get angry with sentient beings.[183]

When someone strikes you, if the appropriate response is to get angry with the one *directly* inflicting the harm, then you should get angry with the stick or other weapon. And if the appropriate response is to get angry with the one *indirectly* inflicting the harm, [116] then since it is anger that incites the attacker, you should get angry with the anger. Either way, it is not appropriate to get angry with the person. *Engaging in the Bodhisattva's Deeds* says:

If I become angry with the wielder
who directly employs the stick and so on,
then since he is impelled only by hatred,
at worst I should get angry with hatred instead.[184]

If it is the nature of unwise beings to harm others, then it is just as inappropriate to get angry with them as it is with fire for being hot and burning in nature. And if it is but an adventitious fault, then it would be like blaming the sky for being filled with smoke. As *Engaging in the Bodhisattva's Deeds* says:

If it is the nature of childish beings
to act hurtfully toward others,
then it is as unreasonable to get angry with them
as it is to begrudge a fire for burning.

And if this is an adventitious fault
in beings whose nature is agreeable,
then it is as unreasonable to get angry with them
as it is to blame the sky for bearing smoke.[185]

There are many techniques such as these to destroy the supreme enemy, our anger. So if we use several kinds of reasoning, based on thorough analysis with fine investigative awareness, to prevent the arising of anger, many types of anger will cease, and forbearance will arise in manifold ways. [117]

COMPASSION

The definition of compassion is: a mental factor that thinks, upon observing other sentient beings, "How wonderful it would be if they were free from the causes of suffering; may they be free from suffering." It functions to counteract violence. Examples are loving kindness that thinks, upon seeing stricken sentient beings tortured by suffering, "How wonderful it would be if they were free from that suffering," and loving kindness that thinks, upon observing sentient beings creating the cause of suffering, "May they be free from that cause." Based on that, the scriptures also explain a special type of compassion that wishes to protect sentient beings by thinking "I myself will free sentient beings from suffering." The *Concentration Combining All Merit Sūtra* says: "That which completely frees all sentient beings from every suffering is great compassion."[186]

The Sanskrit word for compassion is *karuṇā*, and here the syllable *kaṃ* means "happiness" and the syllable *ruṇa* means "block." This indicates that when one sees the specific sufferings of others, it is unbearable and can block one's own happiness. Therefore *karuṇā* means "blocking happiness."[187] [118] The function of compassion is to counteract violence toward sentient beings. The *Explanation of the Compendium of Knowledge* says: "What is nonviolence? Associated with nonhatred, it is the mind of compassion itself; it functions to counteract violence."[188] Then the *Treasury of Knowledge Autocommentary* says: "Love is in the nature of nonhatred" and "Compassion too is like that."[189] As stated above, among the fifty-one mental factors, both love and compassion are in the nature of the mental factor nonhatred. Yet there is a difference between them. Compassion arises upon observing sentient beings to be suffering and has the aspect of wishing them to be free of that. Love arises upon observing sentient beings in terms of their happiness and has the aspect of wishing them to possess that.

As for the cause of generating compassion, this includes: recognizing that one must not abandon other sentient beings, since oneself and other sentient beings are equal in wanting happiness and not wanting suffering; recognizing how other sentient beings have been one's relatives from the distant past; contemplating how they are tortured by various sufferings; reflecting deeply on how one's own happiness and suffering are the results of benefiting or harming other sentient beings; and so on.

The *Perfection of Wisdom Sūtra in Twenty-Five Thousand Lines* says, "Regarding all sentient beings, one must think of them as one's mother, father, brother, sister, friend, relative, family member."[190] [119] Many sūtras and commentaries say that such a teaching—that one must think of all sentient beings as one's relatives and friends—is the cause of compassion. Thinking of sentient beings as our relatives is intended to train the mind in compassion, as in the case of seeing a neutral person with whom we have no special relationship or seeing someone who harms us, such as an enemy. In general we do not need to train in compassion toward our dear relatives or toward a child tormented by grief at the loss of his or her mother. Compassion arises naturally in such contexts. Say we are training in a compassionate attitude toward neutral persons. We cultivate the thought of them as family members to help generate the feeling of finding their suffering unbearable when we see them in that state. This is because without finding their suffering unbearable, compassion for them will not arise.

Śāntideva's *Engaging in the Bodhisattva's Deeds* does not stress the cultivation of recognizing other sentient beings as one's relatives and instead emphasizes training the mind in equanimity—that oneself and others are the same in wanting happiness and in not wanting suffering. It is on such a basis that the contemplation of exchanging the self and others is generated. This is a unique instruction for training the mind in compassion, and one should develop an understanding of it from this source. [120]

The results of compassion are as mentioned in the *Sūtra on the Application of Mindfulness*:

> That which is called *compassion* sees all beings as dear, it enables all beings to trust, it is the foundation of faith for those who fear the terrors of saṃsāra, it places beyond sorrow those who have good mental discipline, and it is the basis of refuge for those who lack protection.[191]

The *King of Concentrations Sūtra* says:

> It makes one kind, excellent in demeanor, and always serene;
> it benefits oneself as well as other sentient beings;
> so one should cultivate love and compassion.[192]

In general the scriptures speak of five types of virtue: (1) ultimate virtue, (2) virtue by its essential nature, (3) virtue through concomitance, (4) virtue owing to motivation, and (5) virtue through relationship. The eleven virtuous mental factors listed above are characterized as being *virtue by their essential nature*, in that their being virtuous is not contingent on some other factor, such as an underlying motivation or some concomitant factor. They are virtuous by their very existence. *Ultimate virtue* is identified in terms of ultimate reality, and it is so called because when one meditates by taking ultimate reality as one's focus, it gradually leads to the removal of all the mental obscurations. It is therefore not an actual virtue; it is just given the name *virtue*. [121] *Virtue through concomitance* refers to any minds or mental factors that have arisen as concomitant factors to any of the eleven virtuous mental factors of faith and so on. *Virtue owing to motivation* refers to being motivated by compassion and so forth in one's physical and verbal actions. *Virtue through relationship* refers to virtuous latent potencies and so on.

10

Mental Afflictions

The Six Root Mental Afflictions

The fifth section presents the six root mental afflictions: (1) attachment, (2) anger, (3) pride, (4) afflictive ignorance, (5) afflictive doubt, (6) afflictive view.

Here is the *Verse Summary*:

> The six root afflictions are thus:
> attachment, anger, pride,
> ignorance, doubt, and afflictive view.

The definition of a mental affliction in general is: a mental factor that functions to disturb the mindstream of the person in whose continuum it occurs. The *Compendium of Knowledge* says: "The definition of a mental affliction is: any factor that, upon arising, arises with the characteristic of being thoroughly disturbing, so that through its arising, the mindstream arises as thoroughly disturbed."[193] Yaśomitra's *Explanation of the Compendium of Knowledge* says: "The characteristic of being thoroughly disturbing should be understood as the definition of mental afflictions in general. [122] There are six types of disturbance: (1) disturbance due to distraction, (2) disturbance due to distortion, (3) disturbance due to excitation, (4) disturbance due to dullness, (5) disturbance due to heedlessness, and (6) disturbance due to non-refraining."[194]

The first of the six root mental afflictions is *attachment*. This mental factor keenly seeks to acquire something contaminated on account of conceiving it to be intrinsically attractive. Also, upon seeing a physical body, food, clothing, jewelry, and so on as something attractive, it fixates on what was seen and does not want to be separated from it. Afflictions other than attachment are relatively easier to remove from the mind, just as, by analogy, it is relatively easy to remove dirt from a dry cloth. Attachment, in contrast, is as difficult to remove as dirt from an oil-soaked cloth. It clings to its object and gives rise to other afflictions. Thus attachment is a mental factor that is very difficult to remove on account of increasing one's fixation on objects of desire, intensifying longing to see, touch, and do such things with its chosen objects.

The function of attachment is to produce its resultant suffering. The *Compendium of Knowledge* says: "What is attachment? It is desire pertaining to the three realms. It functions to give rise to suffering."[195] [123]

There is, in general, no difference between attachment and *craving*. There are three types of craving: craving for the desirable, craving for dissolution, and craving for continued existence. The first is attachment that longs for happiness and wishes not to be separated from it. The second is craving for dissolution out of fear of pain, thinking "Oh, how I wish I would die" or "How I wish this did not exist or that would not happen," and such like. The third is craving for continued existence, such as attachment to one's own aggregates of body and mind. When that very craving becomes more intense and has the power to proliferate saṃsāra, then it is called *grasping* (or *appropriation*). As a provisional antidote against attachment, one needs to meditate on ugliness, and the way to do this will be discussed later in the section on how to engage in the meditative application of mindfulness.

The second root mental affliction is *anger*. This is a mental factor that, upon perceiving any of the three objects of anger, arises as intolerance and hostility wishing to cause harm. It has the aspect of a very harsh mind that perceives the enemy as repellent and so on.[196] For example, when someone speaks to us with harsh words, intolerance immediately arises in the mind. The *Compendium of Knowledge* says: "What is anger? It is hostility toward sentient beings, manifest suffering, or things that cause suffering. It functions as a basis for unhappy states and for bad conduct."[197] There are three types of objects of anger: sentient beings, one's own suffering, and circumstances that produce suffering, such as thorns and weapons. [124]

As for objects that induce anger, the sūtras teach nine basic causes of hostility. These are stated in Nāgārjuna's *Precious Garland*:

> Ill will arises from nine causes:
> it is the intent to harm others on the part of one

> who is worried about some misfortune in the three times
> involving past harm, present harm, or possible future harm
> in regard to oneself, one's friends, or one's enemies.[198]

This passage explains that there are nine types of worry in this context: worry about whether one has been harmed by someone in the past, is being harmed by them in the present, or will be harmed by them in the future; worry about whether one's relatives have been harmed by someone in the past, are being harmed by them in the present, or will be harmed by them in the future; and worry about whether one's enemies have been benefited by someone in the past, are being benefited by them in the present, or will be benefited by them in the future.

The function and faults of anger are explained in *Sūtra on the Application of Mindfulness*:

> When anger arises, and for as long as it lasts, it makes a person's heart-mind burn; it alters his appearance; it distorts the blood vessels in his face; it makes him knowingly heedless toward others; it generates fear in worldly beings; it continuously creates defilement, unpleasantness, and disharmony again and again in all places and lands.[199] [125]

Many evil consequences are explained in scriptures such as this. A person under the influence of anger in this life becomes an unsuitable basis of virtue; the color of his face becomes unappealing; others will see him as reckless; it creates disharmony among all peoples; it makes everyone appear ugly; and it causes the three gateways of behavior—body, speech, and mind—to become defiled. *Engaging in the Bodhisattva's Deeds* says:

> When pierced by painful thoughts of hatred,
> the mind experiences no peace.
> Finding neither joy nor happiness,
> it becomes unsettled and cannot sleep.
>
> Those reliant upon a master,
> though honored by him with wealth and status,

may nevertheless wish to kill him
if that master is repugnantly filled with anger.

Even his friends are disenchanted;
though he gives gifts, he is not approached.
In brief, there is nobody at all
who lives happily with anger.

Sufferings such as these are
created by the enemy: anger.[200]

In that it produces vehement suffering, hatred is like a stabbing pain. When in the grip of thoughts of hatred, neither physical pleasure nor mental joy can pacify such suffering. Sleep will not come, and the mind refuses to relax and settle. When overpowered by anger, [126] even those masters who kindly sustain one with wealth and status might get killed. Even their friends become wearied and disenchanted. Other people may be attracted by gifts, but they are not happy to stay. In brief, there is no opportunity for someone under the sway of anger to live happily. Therefore it is absolutely necessary to put a stop to anger by thinking that anger must not be given an opening. This has already been explained extensively in the section on love.

The third root mental affliction is *pride*. This is a mental factor that has the aspect of a grandiose mind caused by perceiving all kinds of excellent personal attributes, such as one's good qualities, wealth, and so on. It is an inflation of the mind that arises upon perceiving anything such as one's own power, acquisitions, social class, family lineage, and good qualities—or even just a pleasant voice or great strength. Just as when seen from a high mountain peak, other people below appear very small, this mental factor has an aspect of loftiness holding oneself to be superior and others to be inferior.

Pride functions as a basis for the arising of suffering in that it causes one to disrespect others and prevents one from developing higher qualities. We are advised to meditate on the groups of elements as an antidote to intellectual pride, to reflect on the good qualities of exceptional beings superior to oneself, and to contemplate the vast number of things one does not know. The *Compendium of Knowledge* says: [127] "What is pride? It is a grandiose mind based on the view of the perishable collection (Skt. *satkāyadṛṣṭi*). It functions as a basis for being disrespectful and for the arising of suffering."[201] Here "based on the view of the perishable collection" indicates that every time pride arises, it arises in dependence on an innate self-grasping attitude thinking "me" in the mindstream.

When categorized, pride has seven divisions: pride, pride of superiority, excessive pride, pride of thinking "me," pretentious pride, emulating pride, distorted pride. *Pride* is a grandiose mind thinking "I am better than this one who is considered inferior to me." *Pride of superiority* is a grandiose mind thinking "I am better than this one who is considered my equal." *Excessive pride* is a grandiose mind thinking "I am better than this one who is considered superior to me." *Pride of thinking "me"* is a grandiose mind that, based on perceiving one's own aggregates, thinks "me." *Pretentious pride* is a grandiose mind that believes oneself to have attained good qualities that one has not attained. *Emulating pride* is a grandiose mind that, having perceived someone who is very much superior to oneself, thinks "I am only slightly inferior to this person." *Distorted pride* is a grandiose mind that believes what is not a good quality within oneself to be a good quality. Alternatively, *Precious Garland* says:

> Any deriding of oneself, [128]
> thinking "I am useless,"
> is the pride of inferiority.[202]

Deriding oneself, thinking "There is no purpose in my being alive," is the pride of inferiority.

The fourth root mental affliction is *ignorance*. In general, just as in expressions like "not seeing," "not knowing," the Sanskrit word *avidyā* (for "ignorance") is a term wherein the word *vidyā* ("knowing") is conjoined with a negative particle. So the negative particle in the word for ignorance (*avidyā*) functions in exactly the same manner as those in "not knowing," "not seeing," "not understanding," and being "unclear." For example, when one closes one's eyes and everything becomes black or covered in darkness, this prevents one from seeing any external forms and so on. In the same manner, ignorance prevents one from understanding the way things exist. So the definition of ignorance is: a mental factor that engages its object in a confused manner. However, the *Compendium of Knowledge* says: "What is ignorance? It is an absence of knowing within the three realms.

It functions as the basis for the distorted ascertainment of things, doubt, and all that is thoroughly afflictive."[203]

Ignorance can be understood in two different forms: as a mental factor that is the confusion consisting in the absence of knowing or as a cognition that apprehends in a distorted manner. Asaṅga and Vasubandhu consider ignorance to be a mental factor that is primarily the absence of knowing. Dharmakīrti and others, in contrast, consider ignorance to be a confusion that involves distorted apprehension. Thus the first interpretation understands ignorance to be [129] a form of non-cognition, whereas the latter understands it to be a form of mis-cognition. That said, both interpretations converge on identifying the wisdom realizing the ultimate nature of reality to be the opposite of ignorance as well as the principal antidote that counters ignorance. The *Compendium of Ascertainments* says: "Ignorance is defined as not cognizing reality as it is. Given that afflictive views do not cognize reality as it is, they are defined as adhering to a distorted nature."[204] *Exposition of Valid Cognition* says:

> Ignorance is distorted cognition
> because it opposes wisdom,
> because it is cognitive due to being a mental factor,
> and because [the Buddha] said [that this was the case].
> Any other explanation is not correct.[205]

The function of ignorance is as follows. In dependence on ignorance, other mental afflictions arise; in dependence on those, karma is created; and in dependence on karma, suffering arises. Thus ignorance functions as the basis for the arising of all mental afflictions and faults. *Exposition of Valid Cognition* says:

> Every type of fault arises from
> the view of ["me" and "mine" in] the perishable collection.
> That view is ignorance; when it is present, there is attachment
> to that ["me" and "mine"],
> and from that attachment hatred and so on arise. [130]
> Due to this, the cause of faults is said to be ignorance.[206]

Candrakīrti's *Entering the Middle Way* says:

> Seeing with wisdom that all afflictions and faults
> arise from the view of the perishable collection . . .[207]

A detailed explanation of this will be given below when presenting the causes and conditions of the mental afflictions.

The fifth root mental affliction is *afflictive doubt*. This is a mental factor of doubt that, upon considering any of the four truths, cause and effect, and so on, wavers between two standpoints. For example, one may have wavering doubt, thinking, "Is the self impermanent or not?" Now suppose you want to travel along a road. If you doubt whether it is the right road, this creates an obstacle to your following it. Similarly, afflictive doubt creates an obstacle to seeing the way in which things exist and so on. The *Compendium of Knowledge* says: "What is doubt? It is to be of two minds about the truths and so on. It functions as a basis for not engaging in virtue."[208]

Afflictive doubt functions to prevent engaging in virtue and abandoning nonvirtue as appropriate. In general, there are many types of doubt, not all of them afflictive: we may wonder if this man is Tashi or not, [131] or what the weather will be like tomorrow, or if this is Tsering's house.

The sixth root mental affliction is *afflictive view*, of which there are five types: (1) view of the perishable collection, (2) view holding to an extreme, (3) wrong view, (4) holding views to be supreme, and (5) holding ethics and vows to be supreme. The *view of the perishable collection* is an afflictive intelligence that, focused on "me" or "mine" within one's own continuum, thinks of "me" or "mine" as autonomously "me" or "mine." For example, it is an apprehension of an exaggerated sense of "me" that arises in the depths of your heart when someone praises you, or criticizes you, and so on, and you think "Why me?" This mind is called "view of the perishable collection" because it views "me" or "mine" on the basis of the aggregates that are assembling and disintegrating.[209] The *Compendium of Knowledge* says: "What is the view of the perishable collection? It is any acquiescing, desiring, discriminating, conceiving, or viewing that views the five aggregates of appropriation as the self or as belonging to the self. It functions as the basis of all views."[210] Here where the *Compendium of Knowledge* explains the definition of the view of the perishable collection, the meaning of acquiescing and so on is as follows. *Acquiescing* means not being wary of a distorted meaning, *desiring* means engaging its object distortedly, *discriminating* means fully differentiating its object, *conceiving* means actively adhering to its object, *viewing* means perceiving its object. [132]

A *view holding to an extreme* is an afflictive intelligence that grasps the focal object of the view of the perishable collection as either permanent or annihilated. The *Compendium of Knowledge* says: "What is the view holding to an extreme? It is any acquiescing and so on that views the five aggregates of appropriation as either permanent or annihilated."[211] This falling to an extreme that views something as permanent or as annihilated is the main obstacle to making progress on the Middle Way, which is free from the extremes of viewing as permanent and viewing as annihilated. When categorized, the view holding to an extreme has two types: the view of permanence and the view of annihilation.

Wrong view is an afflictive intelligence that, upon considering something that exists—such as karmic cause and effect, action and agent, and so on—views it to be nonexistent. It functions to make one behave perversely regarding what to take up or cast aside, such as avoiding virtue and severing the roots of virtue as well as engaging in nonvirtue and welcoming evil intention. The *Compendium of Knowledge* says: "What is wrong view? To denigrate the functioning of causes, effects, and agents and to deny things that actually exist is an acquiescence and so on that conceives distortedly."[212]

Holding views to be supreme is an afflictive intelligence that, focused on other pernicious views or the aggregates based on which pernicious views arise, [133] holds either of these to be supreme. The *Compendium of Knowledge* says: "What is holding views to be supreme? It is any acquiescing and so on that views those views, or indeed the five aggregates of appropriation that are the basis of those views, to be supreme, principal, superior, and sacred."[213] Here the meaning of listing "supreme" and so on is as follows. Holding something as *supreme* means believing it to be the most excellent, holding it as *principal* means believing it to be unsurpassed by others, holding it as *superior* means believing it to be superior to others, viewing it as *sacred* means believing it to be unmatched by others.

Holding ethics and vows to be supreme is an afflictive intelligence that is focused on defective ethics motivated by a wrong view, on a vow involving improper ethics, on defective austerities of body and speech, or on the aggregates they are based upon. This view, taking one of these as its object, views it to be the cause of purification and liberation—that is, to be supreme. The *Compendium of Knowledge* says: "What is holding ethics and vows to be supreme? It is any acquiescing and so on that views ethics and vows, or the five aggregates of appropriation that are the basis of ethics and vows, to be purifying, liberating, and definitively releasing."[214] [134]

These five views and the five nonview afflictions—attachment, anger, pride, ignorance, doubt—are called *root mental afflictions* because there are no secondary mental afflictions that do not arise from one of these ten. Also, secondary mental afflictions are secondary to one of these ten, and given that these are the primary causes of the mind being afflictive, they are called *root mental afflictions*.

The Twenty Secondary Mental Afflictions

The sixth section presents the twenty secondary mental afflictions as follows: (1) rage, (2) resentment, (3) concealment, (4) spite, (5) jealousy, (6) avarice, (7) pretense, (8) guile, (9) arrogance, (10) violence, (11) shamelessness, (12) nonembarrassment, (13) dullness, (14) excitation, (15) faithlessness, (16) laziness, (17) heedlessness, (18) forgetfulness, (19) lack of meta-awareness, and (20) distraction. Here is the *Verse Summary*:

> The twenty secondary ones are thus:
> Rage, resentment, concealment,
> spite, jealousy, avarice, pretense,
> guile, arrogance, violence, shamelessness,
> nonembarrassment, dullness, excitation,
> faithlessness, laziness, heedlessness,
> forgetfulness, lack of meta-awareness,
> and distraction, making twenty.

The first is *rage*. This is a mental factor associated with anger that is the wish to inflict harm when any of the nine basic causes of hostility are present. [135] The *Compendium of Knowledge* says: "What is rage? It is an attitude that is hostile when a cause of ill will is present, since it is a mind that is associated with anger. It functions as a basis for taking up weapons, punishing, and so on, and preparing to harm."[215] What is the difference between anger and rage? Anger is a disturbance within the depths of the heart that is intolerant of the object of anger even when the object merely appears in the mind without being actually present; it is like igniting a fire. Rage is an amplification of anger, wanting to physically beat and so on when a basic cause of rage is present; it is like a fire when butter is poured onto it, blazing with tongues of flames.

The second is *resentment*. This is a mental factor associated with anger that does not let go of the wish to respond with harm. Since it firmly holds a continuous grudge, it is called *resentment*, or literally, *holding a grudge*. The *Compendium of Knowledge* says: "What is resentment? Associated with anger and following in its wake, it does not let go of an intention to retaliate. It functions as a basis for intolerance."[216]

The third is *concealment*. This is a mental factor associated with delusion that wants to hide one's misdeeds or keep them secret when others mention them out of a wish to help. It has the function of directly causing regret and indirectly [136] causing unhappiness or a wretched state. The *Compendium of Knowledge* says: "What is concealment? Associated with delusion, it keeps secret one's misdeeds when one is rightly confronted. It functions as a basis for regret and a wretched state."[217]

The fourth is *spite*. This is a mental factor that has no intention of regretting and confessing one's own faults when they are mentioned by others but, with animosity empowered by rage and resentment, wants to utter harsh words. It functions to generate misery by doing many unsuitable things, such as uttering harsh words. The *Compendium of Knowledge* says: "What is spite? Associated with anger, it is a thoroughly hostile attitude that is preceded by a mind of anger and resentment. It functions as a basis of wrath, as well as harsh, insulting words, and to increase demerit and a state of wretchedness."[218]

The fifth is *jealousy*. This is a mental factor associated with hatred that, out of attachment to gain and respect, is an inner disturbance of mind unable to bear the success of others. The *Compendium of Knowledge* says: "What is jealousy? [137] Associated with hatred, it is an inner disturbance of mind that, out of excessive attachment to gain and respect, cannot bear the notable success of others. It functions to cause unhappiness and wretched states."[219] When jealousy is categorized there are two types: jealousy of someone perceived to be equal to oneself and jealousy of someone perceived to be superior to oneself. It is called *jealousy* (*phrag dog*) because jealousy is a narrowing and contraction of the mind—literally, in Tibetan, a narrowing (*dog*) of the in-between space (*phrag*).

The sixth is *avarice*. Associated with attachment, this is a mental factor that, out of attachment to gain and respect, wants to hold on to things and cannot let go of them. It functions to accumulate unnecessary things without allowing them to decrease. The phrase "not allowing unnecessary things to decrease" means to accumulate unneeded things without reducing them. The *Compendium of Knowledge* says: "What is avarice? Associated with attachment, it is a mind that firmly holds on to things out of excessive attachment to gain and respect. It functions as a basis for not allowing unnecessary things to decrease."[220]

The seventh is *pretense*. Associated with attachment or delusion, this is a mental factor that, out of attachment to gain and respect, wants to show, with an intention to deceive others, that one has good qualities that one does not actually have. It functions to establish wrong livelihood. [138] The *Compendium of Knowledge* says: "What is pretense? Associated with attachment and delusion, this is a mental factor that, out of excessive attachment to gain and respect, displays what is not a genuine quality. It functions as a basis of wrong livelihood."[221]

The eighth is *guile*. Associated with attachment or delusion, this is a mental factor that, out of attachment to gain and respect, wants to mislead others and make them unaware of one's faults. It functions to prevent one from obtaining correct advice. The *Compendium of Knowledge* says: "What is guile? Associated with attachment and delusion, it is a mental factor that, out of excessive attachment to gain and respect, treats faults as good qualities. It functions to prevent one from obtaining correct advice."[222]

The ninth is *arrogance*. Associated with attachment, it is a mental factor that, upon seeing any signs of contaminated good fortune in oneself, such as good health, inflates the mind with joy and mental bliss. The *Compendium of Knowledge* says: "What is arrogance? It is joy and mental bliss associated with attachment that arises [139] upon seeing any signs of long life and contaminated good fortune based on good health and youth. It functions to assist all the root and secondary mental afflictions."[223] Regarding the function of arrogance, the *Sūtra Invoking the Supreme Intention* says:

> Arrogance is the root of all heedlessness.[224]

Arrogance functions as the basis of all root and secondary mental afflictions, such as heedlessness and so on.

The tenth is *violence*. Associated with anger, this is a mental factor that, in opposition to loving kindness and so on, wants to inflict injury on others. The *Compendium of Knowledge* says: "What is violence? Associated with anger, it is to be unloving, uncompassionate, and unempathetic. It functions to be thoroughly injurious."[225] Here "unloving" is when one wants to inflict injury, "uncompassionate" is when one wants another to be injured, and "unempathetic" is when one relishes hearing or seeing someone else inflict injury. The function of violence is to injure or to harm. [140]

The eleventh is *shamelessness*. Associated with any of the three poisons, it is a mental factor that does not shun wrongdoing, whether for the sake of oneself or the Dharma. The mental factor shamelessness is the opposite of a sense of shame. The *Compendium of Knowledge* says: "What is shamelessness? Associated with attachment, hatred, or delusion, it is to not shun wrongdoing for the sake of oneself. It functions to assist all the root and secondary mental afflictions."[226]

The twelfth is *nonembarrassment*. Associated with any of the three poisons, it is a mental factor that does not shun wrongdoing for the sake of others. Nonembarrassment is the opposite of embarrassment. The *Compendium of Knowledge* says: "What is nonembarrassment? Associated with attachment, hatred, and delusion, it is to not shun wrongdoing for the sake of others. It functions to assist all the root and secondary mental afflictions."[227]

The thirteenth is *dullness*. Associated with delusion, it is a mental factor that causes physical and mental heaviness and makes the mind hazy and unserviceable in apprehending the aspects of its object. It functions to increase all the mental afflictions. [141] The *Compendium of Knowledge* says: "What is dullness? Associated with delusion, it is the unserviceability of the mind."[228]

The fourteenth is *excitation*. Associated with attachment, it is a mental factor that, upon seeing attractive characteristics such as those in desirable sense objects, scatters the mind so that it becomes unpeaceful. It functions to prevent the mind from abiding on its object. The *Compendium of Knowledge* says: "What is excitation? Associated with attachment, it is a mind that is disquieted upon beholding attractive characteristics. It functions as a hindrance to calm abiding."[229]

The fifteenth is *faithlessness*. Associated with delusion, it is a mental factor that does not trust, admire, or emulate an object worthy of faith. It functions as a basis for laziness. The *Compendium of Knowledge* says: "What is faithlessness? Associated with delusion, it is a mind that does not trust, admire, or emulate virtuous dharmas. It functions as a basis for laziness."[230] [142]

The sixteenth is *laziness*. Associated with delusion, it is a mental factor that, in dependence on sleep and so on, causes the mind to lack enthusiasm for virtue. It functions to reduce the side of virtue. The *Sūtra on the Application of Mindfulness* says:

> The single basis of mental afflictions
> is laziness, for it is present in all afflictions;
> where there is laziness,
> there is no Dharma there.[231]

The *Compendium of Knowledge* says: "What is laziness? Associated with delusion, it is a mind's lack of enthusiasm, based on the pleasures of sleeping, resting, and lying down. It functions to hinder virtuous practice."[232]

When laziness is categorized, there are three types: the laziness of indolence, the laziness of adherence to unwholesome activities, and the laziness of self-disparagement. The laziness of indolence is having no wish to engage in any virtue and allowing it to slip away every moment of the day under the influence of procrastination. The laziness of adherence to unwholesome activities is to cling to worldly activities and so on and not delight in virtue. The laziness of self-disparagement is to deride oneself in thinking, out of indolence, that someone like me cannot accomplish virtue. An example is to think with discouragement, "How can someone like me accomplish the purpose of sentient beings?" *Engaging in the Bodhisattva's Deeds* says: [143]

> Laziness: adherence to unwholesome activities,
> indolence, and self-disparagement.[233]

The seventeenth is *heedlessness*. This is a mental factor that, based on any of the three poisons in association with laziness, behaves carelessly without guarding the mind against accumulating mental afflictions and faults. The *Compendium of Knowledge* says: "What is heedlessness? Based on attachment, hatred, and delusion in association with laziness, it is to not cultivate virtuous dharmas and not protect the mind from impure dharmas. It functions as a basis for increasing nonvirtue and decreasing virtue."[234]

The eighteenth is *forgetfulness*. This is a mental factor that makes the mind unclear and forgetful of virtue as a result of recollecting the objects of focus of the mental afflictions. It functions to distract the mind toward such objects of focus or toward the ways that mental afflictions apprehend their objects of focus. The *Compendium of Knowledge* says: "What is forgetfulness? It is recollection concomitant with mental afflictions. It functions as a basis for distraction."[235]

The nineteenth is *lack of meta-awareness*. This mental factor is an afflictive intelligence that engages in activities of body, speech, or mind without meta-awareness. [144] It thus functions as a basis for evil conduct and moral downfalls. The *Compendium of Knowledge* says: "What is lack of meta-awareness? It is an intelligence concomitant with mental afflictions; it engages in activities of body, speech, and mind without meta-awareness. It functions as a basis for moral downfall."[236]

The twentieth is *distraction*. Associated with any of the three poisons, this is a mental factor that causes the mind to become scattered and distracted from its object of focus. The *Compendium of Knowledge* says:

> What is distraction? Associated with attachment, hatred, or delusion, it is a scattering of the mind. It may be natural distraction, external distraction, internal distraction, distraction involving distinguishing marks, distraction involving nonvirtuous tendencies, or distraction involving attention.[237]

These six of natural distraction and so on are identified as follows: (1) *Natural distraction* refers to the five sense consciousnesses. (2) *External distraction* refers for the most part to mental scattering and excitation. (3) *Internal distraction* refers to gross and subtle laxity and to craving for the delicious experience of concentration. (4) *Distraction involving distinguishing marks* is like the virtuous mental activity of thinking "It would be wonderful if others considered me to be a great meditator" or such like. [145] (5) *Distraction involving nonvirtuous tendencies* refers to the pride of thinking "me." (6) *Distraction involving attention* refers to attention that thinks "Perhaps I should give up the supreme path, or the higher concentrations, and practice the lower ones," or such like. Although these are all called "distraction," they need not be distraction that is a secondary mental affliction. The attributes of gross and subtle laxity and excitation, as well as distraction and so on, are explained in detail in the section on calm abiding (see part 6). The reason that these twenty—from rage to distraction—are called *secondary mental afflictions* is because each of them causes the mind to be afflicted according to whichever root mental affliction it is associated with.

The way in which specific root and specific secondary mental afflictions are or are not concomitant is explained in the *Compendium of Knowledge*:

> Which are the concomitant factors? Attachment is not concomitant with anger; and just as it is not with anger, likewise it is not with doubt. However, it is concomitant with the remaining ones. And just as attachment is not with anger, likewise anger is not concomitant with attachment, or pride, or view. Pride is not concomitant with anger or doubt. Ignorance is of two types: that which is concomitant with all the other mental afflictions and that which is unmixed. Unmixed ignorance is an unknowing of the truths. View is not concomitant with anger or doubt. [146] Doubt is not concomitant with attachment, pride, or view. The secondary mental afflictions of rage and so on are not mutually concomitant. Shamelessness and nonembarrassment are concomitant with all nonvirtuous minds. Dullness, excitation, faithlessness, laziness, and heedlessness are concomitant with all the mental afflictions.[238]

Why are attachment and anger not concomitant? Because two contradictory ways of apprehending the object do not accompany one mind simultaneously. Why is attachment not concomitant with doubt? When a mind becomes doubtful, it does not maintain a single position, in which case it cannot be attached to its object. And in the case of the remaining root mental afflictions—pride, ignorance, and view—since attachment does not have a way of apprehending its object that is incompatible with those, they can be concomitant. Why is anger not concomitant with pride and view? Respectively, whatever object animosity may arise toward, the mind does not become inflated on account of that and it does not mistakenly conceive the object in the sense of viewing it.

Why is pride not concomitant with anger? Because two contradictory ways of apprehending the object do not arise simultaneously accompanying one mind. [147] Why is it not concomitant with doubt? Because the mind does not become inflated on account of whatever ambivalent mental experience doubt has.

According to those who assert the foundation consciousness, ignorance is divided into two types: ignorance that is concomitant with all the other mental afflictions and ignorance that is unmixed. [The latter is twofold: ignorance that is unmixed from the point of view of concomitance, and ignorance that is unmixed from the point of view of the basis.] Ignorance that is unmixed from the point of view of concomitance is ignorance that is confused about how the truth exists; it accompanies the sixth mental consciousness (Skt., *manovijñāna*) but is not concomitant with the other root mental afflictions. Ignorance that is unmixed from the point of view of the basis is said to accompany only the afflictive mental consciousness (*kliṣṭamanas*). Most Buddhist masters uphold the standpoint that says this ignorance is concomitant with all the other five root mental afflictions.

Why is view not concomitant with either anger or doubt? This can be understood by means of the above explanation as to why doubt is not concomitant with attachment, pride, or view.

Why are the secondary mental afflictions associated with anger, such as rage, not concomitant with the secondary mental afflictions associated with attachment, such as avarice? Because they have incompatible ways of apprehending their objects, just as anger and attachment do. Why must both shamelessness and nonembarrassment be concomitant with all nonvirtuous minds? Because one cannot create nonvirtue unless one avoids shunning wrongdoing out of consideration for oneself or others. [148] Why are dullness, excitation, faithlessness, laziness, and heedlessness concomitant with all the other mental afflictions? Because something cannot be a mental affliction unless it makes the mind unclear, scatter outward, and become turbid, unless it does not delight in virtue and is bereft of protecting the mind from nonvirtue. It is explained in the commentaries on the *Compendium of Knowledge* that the statement about dullness and excitation being concomitant with all mental afflictions is to be understood as being made by Asaṅga in order to conform to the tenets of the Hīnayāna schools. The point is that some form of mental unserviceability and lack of serenity are present in all mental afflictions.

11

Variable Mental Factors

Now we present the four variable mental factors: (1) sleep, (2) regret, (3) inquiry, (4) analysis.

Here is the *Verse Summary*:

> The four changeable ones are sleep,
> regret, inquiry, and analysis.

The first is *sleep*. This is a mental factor that draws the object-perceiving sense consciousnesses inward as a result of its causes, such as the body becoming heavy, weak, and weary, the mental attention attending to a dark image, and so forth. It may be virtuous, nonvirtuous, or neutral. [149] The *Compendium of Knowledge* says: "What is sleep? Associated with delusion, it draws the mind within, and in dependence on the cause of the sleep, it may be virtuous, nonvirtuous, neutral, timely, untimely, suitable, or unsuitable; it functions as a basis for loss of activity."[239] "Timely" indicates that the time for sleep is the middle part of the night, whereas all else is "untimely." "Suitable" indicates a wish to make an effort in virtue by increasing the energy of the body. "Unsuitable" indicates prohibited or not permitted sleep (according to the monastic rule in certain circumstances), even if timely. "Associated with delusion" and "a basis for loss of activity" indicates any sleep that is a lack of activity in terms of being afflictive sleep. Yaśomitra's *Explanation of the Compendium of Knowledge* says: "'Associated with delusion' indicates instances that are distinguished from concentration. Such qualifications as 'virtuous' indicate instances that are definitely not in the nature of delusion." And "'A basis for loss of activity' indicates instances that are in the nature of the secondary mental afflictions. This is how it should be understood."[240]

The second variable mental factor is *regret*. This is a mental factor that is a mental repentance of suitable or unsuitable actions done according to either one's own intention or others' insistence. The *Compendium of Knowledge* says: [150] "What is regret? Associated with delusion, it is mental repentance of intended or unintended actions that are virtuous, nonvirtuous, neutral, timely, untimely, suitable, or unsuitable. It functions to hinder stability of the mind."[241]

When regret is categorized, there are three types: virtuous, nonvirtuous, and neutral. However, in stating that it is "associated with delusion," the *Compendium of Knowledge* presents regret in terms of being afflictive. Yaśomitra's *Explanation of the Compendium of Knowledge* says: "'Associated with delusion' means that it belongs to the secondary mental afflictions."[242]

The third variable mental factor is *inquiry*, and the fourth is *analysis*. Within the twofold categorization of inquiry and analysis, inquiry is a mental factor that, in dependence on intention or wisdom, examines any object in only a rough manner; and analysis is a mental factor that, in dependence on intention or wisdom, analyzes its object finely. The *Compendium of Knowledge* says:

> What is inquiry? Based on intention or wisdom, it is a deeply searching mental discourse. [151] It is a coarse aspect of mind. What is analysis? Based on intention or wisdom, it is a finely investigating mental discourse. It is a refined aspect of mind. They both function as a support for pleasant or unpleasant states.[243]

The function of inquiry and analysis is to act as a basis for pleasant or unpleasant states because each of those two include virtuous, nonvirtuous, and neutral instances. Inquiry and analysis on the side of virtue produce pleasing results, so are said to act as a support for pleasant states, whereas inquiry and analysis on the side of nonvirtue produce unpleasing results, so are said to act as a support for unpleasant states.

These four mental factors (sleep, regret, inquiry, analysis) can become virtuous, nonvirtuous, or neutral owing to the force of motivation and other accompanying factors. Thus they are called *variable*.

The fifty-one mental factors presented above found in the *Compendium of Knowledge* also appear in another text by Asaṅga, *Levels of Yogic Deeds*, which says:

> What are the accompanying ones? They are as follows: attention, contact, feeling, [152] discernment, intention, aspiration, resolution, mindfulness, concentration, wisdom, faith, shame, embarrassment, nonattachment, nonhatred, nondelusion, diligence, pliancy, heedfulness, equanimity, nonviolence, attachment, anger, ignorance, pride, view, doubt, rage, resentment, concealment, spite, jealousy, avarice, pretense, guile, arrogance, violence, shamelessness, nonembarrassment, dullness, excitation, faithlessness, laziness, heedlessness, forgetfulness, distraction, lack of meta-awareness, regret, sleep, inquiry, and analysis. In this way, dharmas that are mental factors concomitant with and occurring simultaneously with the mind are called "concomitant complements." Given that one object has many aspects, they arise simultaneously with one another, and each definitely arises from its own seed, having concomitance, having an aspect, having an object, and having a basis of arising.[244] [153]

Also, in Vasubandhu's *Treatise on the Five Aggregates*, fifty-one mental factors are enumerated, just as in Asaṅga's *Compendium of Knowledge* and *Levels of Yogic Deeds*. After listing each of the fifty-one mental factors, Vasubandhu concludes: "Among these, five are omnipresent, five have a determinate object, eleven are virtuous, six are root afflictions, the rest are secondary afflictions, and four are variable ones."[245] These presentations occur, therefore, as a type of synopsis in several texts of Asaṅga and his brother Vasubandhu. The text mentioned here, *Treatise on the Five Aggregates*, represents a summary of the extensive explanation of the five aggregates in the first chapter of Asaṅga's *Compendium of Knowledge*. Similarly, the Tengyur contains an extensive text called *Explanation about the Five Aggregates*, which is composed by Pṛthivībandhu as an introduction to the *Compendium of Knowledge*. Although in the opening section of the text there is the phrase "Introduction to the treasury of Abhidharma (or knowledge)," this "treasury" must be understood to refer to the *Compendium of Knowledge*. In conclusion, to acquire a detailed understanding of the explanation of the five aggregates as presented in Asaṅga's *Compendium of Knowledge*, both Vasubandhu's *Treatise on the Five Aggregates* and Pṛthivībandhu's *Explanation about the Five Aggregates* are key supporting sources. [154]

12

Mental Factors in Other Works

THE PRESENTATION IN VASUBANDHU'S *TREASURY OF KNOWLEDGE*

VASUBANDHU'S *Treasury of Knowledge* presents forty-six mental factors. It says:

> Mental factors are of five types by virtue of
> the division into the extensive factors (*mahābhūmika*) and so
> on.[246]

Such a passage explicitly indicates five categories: (1) extensive mental factors, (2) extensive virtuous factors, (3) extensive afflictive factors, (4) extensive nonvirtuous factors, and (5) narrower afflictive factors. Also, if we include the category of indeterminate factors, six categories are taught.

The Ten Extensive Mental Factors

The first category contains the ten extensive mental factors: (1) feeling, (2) intention, (3) discernment, (4) aspiration, (5) contact, (6) intelligence or wisdom, (7) mindfulness, (8) attention, (9) resolution, and (10) concentration. The *Treasury of Knowledge* says: [155]

> Feeling, intention, and discernment,
> aspiration, contact, and intelligence,
> mindfulness, attention, resolution,
> and concentration accompany all minds.[247]

In Asaṅga's *Compendium of Knowledge* and its commentarial texts, as well as Vasubandhu's own *Treatise on the Five Aggregates*, aspiration, wisdom, mindfulness, resolution, and concentration are listed not as extensive mental factors but as belonging to the group of mental factors with a determinate object. Here in the *Treasury of Knowledge* tradition, however, these five are asserted as accompanying every mental state. Vasubandhu's *Treasury of Knowledge Autocommentary* presents their definitions as follows:

> Intelligence, or wisdom, is to thoroughly analyze the dharmas, mindfulness is to not forget the object, attention is to engage the mind, resolution is to wish for, concentration is to single-pointedly focus the mind.[248]

One might think that since wisdom must accompany a mind that is concomitant with doubt, it would be incorrect to claim that wisdom has the function of eliminating doubt. However, even though a minimal degree of wisdom is present in the mind that is concomitant with doubt, it is doubt that remains predominant in that mind. [156] It is like, for example, a stream of fresh water mixed into the salty ocean. This tradition similarly explains, in terms of their greater or lesser dominance, the presence of aspiration in a mind concomitant with rage, concentration in a mind concomitant with distraction, mindfulness in a mind concomitant with forgetfulness, and resolution in a mind concomitant with faithlessness. Now why are these called *extensive mental factors*? They are called extensive mental factors because they accompany all states of mind whatsoever.

The Ten Extensive Virtuous Factors

The second category contains the ten extensive virtuous factors: (1) faith, (2) heedfulness, (3) pliancy, (4) equanimity, (5) shame, (6) embarrassment, (7) nonattachment, (8) nonhatred, (9) nonviolence, and (10) diligence. The *Treasury of Knowledge* says:

> Faith, heedfulness, pliancy,
> equanimity, shame, embarrassment,
> the two roots, nonviolence, and diligence
> always arise in those that are virtuous.[249]

While the rest are similar to the ones found in the *Compendium of Knowledge*, here in the *Treasury of Knowledge* system, *shame* and *embarrassment* are described as the opposites of shamelessness, defined as disrespect toward virtue and the virtuous, and nonembarrassment, defined as a failure to see evil deeds such as killing as morally reprehensible (see category 4 below). [157] Just as the *Compendium of Knowledge* does, other Buddhist systems also identify shamelessness and nonembarrassment in terms of engaging in bad behavior and not shunning wrongdoing for the sake of *oneself* and for the sake of *others*, respectively. The "two roots" are *nonattachment* and *nonhatred*. *Nonviolence* is defined as not doing violence or harm toward others, and *diligence* is defined as delighting in virtue. The reason why the Vaibhāṣika system refers to these as *extensive virtuous factors* is that, according to them, these factors accompany all virtuous states of mind.

The Six Extensive Afflictive Factors

The third category contains the six extensive afflictive factors: (1) confusion, (2) heedlessness, (3) laziness, (4) faithlessness, (5) dullness, (6) excitation. These six are called *extensive afflictive factors* because they arise accompanying all afflictive states of mind. The *Treasury of Knowledge* says:

> Confusion, heedlessness, laziness,
> faithlessness, dullness, and excitation
> always arise in those that are afflictive.[250] [158]

The Two Extensive Nonvirtuous Factors

The fourth category contains the two extensive nonvirtuous factors. The *Treasury of Knowledge* says:

> The nonvirtuous are nonembarrassment
> and shamelessness.[251]

Here the definitions of nonembarrassment and shamelessness are, as briefly noted above, slightly different than what is presented in the *Compendium of Knowledge*. The *Treasury of Knowledge* says:

> Nonembarrassment is disrespect, and shamelessness
> is the failure to see evil deeds as morally reprehensible.[252]

This stanza defines the mental factor shamelessness in terms of disrespecting virtue and the virtuous, and the mental factor nonembarrassment in terms of failing to see evil deeds such as killing as morally reprehensible. These two are called *extensive nonvirtuous factors* because they arise accompanying all nonvirtuous states of mind.

The Ten Narrower Afflictive Factors

The fifth category contains the ten narrower afflictive factors: (1) rage, (2) resentment, (3) guile, (4) jealousy, (5) spite, (6) concealment, (7) avarice, (8) pretense, [159] (9) arrogance, and (10) violence. The *Treasury of Knowledge* says:

> Rage, resentment, guile,
> jealousy, spite, concealment, avarice,
> pretense, arrogance, and violence
> are the narrower afflictive ones.[253]

These are said to be "narrower" afflictive mental factors because they are concomitant with ignorance alone and not with other root afflictions; hence, they are "narrower" with respect to their concomitant complements. Also, because they do not distortedly conceive the truth and are abandoned only by the path of meditation, they are "narrower" with respect to their category as an object to be abandoned. And because they are present only within the mental consciousness and are not concomitant with any sense consciousness, they are "narrower" with respect to their basis of arising.

The Eight Indeterminate Mental Factors

The sixth category contains the eight indeterminate mental factors: (1) inquiry, (2) analysis, (3) regret, (4) sleep, (5) anger, (6) attachment, (7) pride, (8) doubt. *Investigating Characteristics* says:

Inquiry, analysis, and regret,
sleep, anger, and attachment, [160]
pride, and doubt are said to be
the eight indeterminate ones.[254]

The *Treasury of Knowledge* posits the difference between inquiry and analysis as follows:

Inquiry and analysis are coarse and fine [respectively].[255]

According to this, inquiry engages its object in a coarse way, whereas analysis engages it in a refined way. The Vaibhāṣikas say that when these two accompany one mind, inquiry does not render it especially coarse and analysis does not render it especially subtle, just as when butter is placed in cold water illuminated by sunlight, the sunlight cannot make it especially hot and the cold water cannot make it especially cold. The Sautrāntikas say that inquiry yields judgments about the object itself, whereas analysis yields judgments about its attributes. The Cittamātras say that inquiry searches for the object, whereas analysis ascertains the object.

These eight mental factors are called "indeterminate" within the system of the *Treasury of Knowledge* because they are not identified as any of the first five categories. Since the definitions, subdivisions, and functions of the specific factors within the list of forty-six mental factors presented from the *Treasury of Knowledge* are very similar to those explained in the *Compendium of Knowledge* (which was covered above), we will content ourselves here with this brief outline. [161]

Five mental factors mentioned in the *Compendium of Knowledge* do not appear among those listed in the *Treasury of Knowledge*: (1) the root virtue that is nondelusion, (2) view that is a root affliction, (3) lack of meta-awareness, (4) forgetfulness, and (5) distraction. The classification of mental factors into the list of forty-six in the *Treasury of Knowledge* seems to be based primarily on identifying those factors that are distinct with respect to their definitions. It turns out that three factors—the root virtue that is nondelusion, view that is a root affliction, and lack of meta-awareness—have the nature of wisdom. Similarly, forgetfulness and distraction have the nature of mindfulness and concentration, respectively. So the understanding seems to be that these factors do not exist as separate from those forty-six mental factors. Also, it is stated here (in the *Treasury of Knowledge* as well as its associated texts) that *nondelusion* refers to virtuous wisdom, *view* refers to afflictive wisdom, *lack of meta-awareness* refers to wisdom that is concomitant with afflictive minds and mental factors, and *forgetfulness* and *distraction* refer to mindfulness and concentration. [162]

How Many Mental Factors Can Accompany One Mind?

How many of the forty-six mental factors listed above arise accompanying one mind? The most that can accompany one mind is twenty-three. In the case of accompanying a virtuous mind concomitant with regret, there arise the ten extensive mental factors, the ten extensive virtuous factors, and both inquiry and analysis, which along with regret add up to twenty-three. The least that can arise is ten, since there are just ten extensive mental factors that arise accompanying an unobstructed neutral mind of the second absorption.[256] By comparison, the mind of the special actual basis of the first absorption has one more, making it eleven, owing to the addition of analysis. In the case of an unobstructed neutral mind of the desire-realm, there are twelve because it is accompanied by both inquiry and analysis.

THE PRESENTATION IN *FINER POINTS OF DISCIPLINE*

Supplementing their explanations of the secondary afflictions, the commentaries on the *Treasury of Knowledge* enumerate a list of twenty-four secondary mental afflictions found in *Finer Points of Discipline*. These are discussed in Yaśomitra's *Clarifying the Meaning of the Treasury of Knowledge* as follows: [163]

> The statement "These are what appear in *Finer Points of Discipline*" indicates that they are what appear in *Finer Points of Discipline*, which is part of the canon, and that intention and so on are not the only mental factors. They are listed as follows: (1) dislike, (2) yawning, (3) discouragement, (4) nausea, (5) not comprehending the right measure of food, (6) discrimination, (7) inattentiveness, (8) bodily dysfunction, (9) malice, (10) insult, (11) stench, (12) nonstench, (13) restlessness, (14) thinking of the desirable, (15) thinking of hurting, (16) thinking about one's rel-

atives, (17) thinking about one's homeland, (18) thinking that one is deathless, (19) thinking with contempt, (20) thinking with caste arrogance, (21) sorrow, (22) pain, (23) grief, and (24) despair.[257] [164]

THE PRESENTATION IN THE *GREAT EXPLANATION*

The *Great Explanation* (*Mahāvibhāṣa*), which is the famed treatise of the Vaibhāṣika tradition and the source based on which Vasubandhu composed his *Treasury of Knowledge*, enumerates mental factors slightly differently from the *Treasury of Knowledge*. It is said that this text lists fifty-seven or fifty-eight mental factors in seven major groups.[258] However, in the Tibetan translation of the *Great Explanation* that is now available, the part pertaining to the presentation of concomitant mental factors is missing, so we cannot draw any conclusion with confidence. However, one of the Abhidharma treatises, the *Compendium of Elements* (*Dhātukāya*), classifies the mental factors as follows: (1) ten omnipresent mental factors accompanying every mind (as adopted in the *Treasury of Knowledge*), (2) ten extensive nonvirtuous factors, (3) ten extensive afflictive factors accompanying every afflictive mind, (4) five hindrances—sensory desire, attachment to the form[-realm absorption], attachment to the formless[-realm absorption], ill will, and doubt, (5) five wrong views, (6) five qualities related to contact and five qualities related to faculties—mental satisfaction, mental discomfort, contentment, discouragement, and equanimity, and (7) five qualities related to conditioning factors—inquiry, analysis, consciousness, shamelessness, and nonembarrassment—and the qualities related to each of the following—feeling, discernment, conditioning action, and craving.[259] [165]

THE PRESENTATION IN ASAṄGA'S *COMPENDIUM OF BASES*

There are also many secondary mental afflictions mentioned in *Finer Points of Discipline* on which Asaṅga offers some explanation in his *Compendium of Bases*. Among these mental factors, several are not found in the lists of mental factors enumerated either in Asaṅga's *Compendium of Knowledge* or Vasubandhu's *Treasury of Knowledge*.[260]

The secondary afflictions enumerated in Asaṅga's *Compendium of Bases* are: (1) attachment, (2) hatred, (3) delusion, (4) rage, (5) resentment, (6) concealment, (7) mental torment due to the force of affliction, (8) jealousy, (9) avarice, (10) pretense, (11) guile, (12) shamelessness, (13) nonembarrassment, (14) pride, (15) pride of superiority, (16) excessive pride, (17) pride of thinking "me," (18) pretentious pride, (19) emulating pride, (20) distorted pride, [166] (21) arrogance, (22) heedlessness, (23) haughtiness, (24) offensive instigation, which is inciting fights and so on, (25) hypocrisy, which is, owing to the force of affliction, giving an impression of excellent conduct and so on so that others will believe that one has good qualities, (26) flattery, which is saying sweet words owing to the force of affliction, (27) hinting, which is revealing one's desirous thoughts owing to the force of affliction, (28) soliciting, which is begging insistently for requisites, (29) pursuing gain with gain, which is wishing excessively for goods, (30) disrespect, (31) not appreciating instruction, which is not accepting another's helpful advice, (32) evil companionship, which makes one engage in activities that harm others, (33) evil desire, which is owing to the force of attachment to material things, (34) great desire, which is wanting more honor and gain than those who are superior, (35) possessing desire, which is wanting to show ascetic qualities and arousing belief owing to the force of affliction, and the four types of intolerance: (36) wanting to retaliate with scolding upon another's scolding, (37) wanting to retaliate with rage upon another's rage, (38) wanting to retaliate with beating upon another's beating, and (39) wanting to retaliate with criticism upon another's criticism, (40) adherence, which is attachment to one's own resources, (41) thorough adherence, which is hankering after another's resources, (42) attachment, which is excessive adherence to objects, (43) improper attachment, which is excessive adherence to unethical conduct, (44) intolerable attachment, which is tightly grasping [167] at one's own and others' resources, (45) view of the perishable collection, (46) view of extreme existence, which is viewing conditioning factors as permanent, (47) view of extreme nonexistence, which is viewing them as annihilated, (48–52) the five hindrances, (53) apathy, which is being overwhelmed by sleep at unsuitable times, (54) dislike, which is wishing for a situation to be otherwise, (55) stretching, which is elongating and bending the body when the mind has become unserviceable, (56) not knowing the right measure of food, which is not knowing whether one's food is too much or too little, (57) inattentiveness, which is not engaging in learning, reflection, and meditation due to heedlessness, (58) engaging inefficiently, which is to be overwhelmed by waves of sleep while the mind is trying to focus on its object, (59) discouragement, which is contempt toward oneself, (60) malice, which is harming others, (61) slander, which is blaming others, (62) dishonesty, which is deceiving one's teacher and friends, (63) ungentleness, which is acting roughly in the three gateways of conduct (body, speech, and mind), (64) applying discordantly, which is to engage in discordant practice, view, and so on, (65) thinking of the desirable, which is placing attention again and again on an object of desire owing to the force of attachment, (66) thinking about harm, which is placing attention again and again on harm done by another, (67) thinking of hurting, which is placing attention again and again on wanting to harm another, (68) thinking about one's friends, which is placing attention again and again on one's friends owing to an afflictive mind, (69) thinking about one's people at home, which is placing attention again and again on one's people at home owing to an afflictive mind, (70) [168] thinking that one is deathless, which is placing attention on living a long time owing to an afflictive mind, (71) thinking that inclines toward contempt for others, (72) thinking that inclines toward class wealth, which places attention again and again on the wealth of sponsors and so on owing to an afflictive mind, (73) sorrow, (74) lamentation, (75) pain, (76) grief, and (77) despair. The *Compendium of Bases* says:

> As for the secondary afflictions, attachment is the root of nonvirtue, and hatred, delusion, rage, and resentment likewise increase it, as explained in the *Finer Points of Discipline*, which says: "With regard to the motivation of all nonvirtue, attachment is the root of nonvirtue." One should understand hatred and delusion to be like that also. *Rage* is to be completely ensnared in anger that is enveloped in evil. *Resentment* exists as an inner intention. *Concealment* is the covering up of faults. *Torment* is complete despair due to the force of affliction. *Jealousy* is an utter disliking of others' achievements owing to an afflictive mind. *Avarice* is to hold on tightly to possessions. *Pretense* is to give an impression to others with the intention to mislead so that they think differently about oneself. *Guile* is insincerity, lack of clarity, and lack of frankness in speech

because of a dishonest mind. [169] *Shamelessness* is not shunning wrongdoing for the sake of oneself. *Nonembarrassment* is not shunning wrongdoing for the sake of others.

Pride is an attitude of mind thinking "I am greater than my inferiors" or "I am equal to my peers." *Pride of superiority* is an attitude thinking "I am greater than my equals" or "I am equal to my superiors." *Excessive pride* is an attitude thinking "I am much greater even than my superiors." *Pride of thinking "me"* is an attitude viewing conditioning factors as "me" or "mine." *Pretentious pride* is an attitude thinking, with regard to special realizations that one has not attained, "I have attained it." *Emulating pride* is an attitude thinking "I am only slightly inferior to those especially great noble ones." *Distorted pride* is an attitude thinking "I have good qualities" while having no such qualities.

Arrogance is the conceit of an afflictive mind on account of any fortune at all. *Heedlessness* is not guarding the mind from evil nonvirtuous dharmas owing to not cultivating the side of virtue. *Haughtiness* is not honoring teachers and holy ones. *Offensive instigation* is being completely ensnared with afflictions so as to bear weapons, carry cudgels, fight, criticize, cause divisions, and incite disputes. [170]

Hypocrisy is altering one's usual conduct, owing to an afflictive mind, in order to make others believe that one has good qualities. *Flattery* is using sweet words and employing ways of speech to show one's own good qualities owing to an afflictive mind. *Hinting* is showing one's desirous thoughts owing to the characteristics of an afflictive mind. *Soliciting* is begging insistently. *Pursuing gain with gain* is excessively wishing for gains by pointing out another's gains without considering sufficient what one already has.

Disrespect is not honoring one's spiritual teachers by way of great love and faith in their good qualities. *Not appreciating instruction* is not accepting advice. *Evil companionship* is a friendship that leads to harm. *Evil desire* is wanting to show ascetic qualities and arouse belief owing to longing for material things. *Great desire* is wanting more honor and gain than those who are superior. *Possessing desire* is wanting to show ascetic qualities and arouse belief owing to the force of affliction. *Intolerance* is retaliating with scolding upon another's scolding, rage upon another's rage, beating upon another's beating, and criticism upon another's criticism. *Adherence* is attachment to one's own resources. *Thorough adherence* is hankering after another's resources. One must understand these two to apply, respectively, to those who are kind and to those who are unkind. [171] *Attachment* is excessive adherence to objects. *Improper attachment* is excessive adherence to unethical conduct. *Intolerable attachment* is tightly grasping at the resources of oneself and one's family, at suitable resources, and at the belongings of others.

The *view of the perishable collection* consists of a conceptual and an innate view seeing the conditioning factors as the self or as belonging to the self. Based on the view of the perishable collection, the *view of extreme existence* views the conditioning factors as permanent, and the *view of extreme nonexistence* views them as annihilated. The *five hindrances* are as described earlier; accordingly, they must be understood in terms of how they occur in *Levels of Meditative Equipoise*.[261] *Apathy* is to be enveloped and overwhelmed by sleep at unsuitable times, even while not wanting to. *Dislike* is wishing for a situation to be otherwise. *Stretching* is elongating and bending the body when the mind has become unserviceable owing to gross bodily dysfunction. *Not knowing the right measure of food* is not knowing whether one's food is too much or too little. *Inattentiveness* is not engaging in the Dharma through learning, reflection, and meditation on the way to engage in correct conduct and to not engage in wrong conduct owing to the motivating force of heedlessness. *Engaging inefficiently* is to be overwhelmed by waves of sleep while the mind tries to focus on its object. *Discouragement* is to have contempt toward oneself.

Malice has the nature of harming another. *Slander* has the nature of backbiting. [172] *Dishonesty* is to deceive the masters, the teachers, the holy ones, and those in harmony with the Dharma. *Ungentleness* is a harsh unrefined mind intent on rough behavior of body and speech. *Applying discordantly* is to

> engage in morality, view, conduct, and livelihood that do not accord with the Dharma. *Thinking of the desirable* is to focus on desirable things that one has perceived and to mentally become accustomed to this again and again owing to a mind of attachment. *Thinking about harm* is to focus on a cause of harm done by another that one has perceived and to mentally become accustomed to this again and again owing to a mind of hatred. *Thinking of hurting* is to focus on the harm done to another that one has perceived and to mentally become accustomed to this owing to a mind of violence. *Thinking about one's friends* is to focus on the friends that one has perceived and to mentally become accustomed to this owing to an afflictive mind. *Thinking about one's people at home* is to focus on one's people at home and tobmentally become accustomed to this owing to an afflictive mind. *Thinking that one is deathless* is to focus on viewing another time in order to attain one's goal and to mentally become accustomed to this owing to an afflictive mind. *Thinking that inclines toward contempt for others* is to focus on the superiority or inferiority of oneself and others and to mentally become accustomed to this owing to an afflictive mind. *Thinking that inclines toward class wealth* is to focus on the wealth of sponsors and householder acquaintances [173] and to mentally become accustomed to this again and again owing to an afflictive mind. *Sorrow* and so on are to be understood as presented above.[262]

The final reference here is to earlier in the same source, where it says:

> *Sorrow* is mental anguish, in the sense of sorrow about something painful; *lamentation* is crying out from that; *pain* is bodily torment; *grief* is inner mental torment; and *despair* is complete confusion arising from that. Alternatively, *sorrow* is any grief based on pain newly arisen due to the pain of either loss of one's resources, health, relatives, or friends. *Lamentation* and *pain* are, respectively, any crying out and the bodily torment that follows on from it. *Grief* is any unhappiness beyond that of lamentation and bodily torment, which becomes slightly

> pacified after the day it first arose. *Despair* is unhappiness that extends beyond the first day, the second or third day, the fifth or tenth day, half a month or one month, or any longer period.[263] [174]

These mental factors include many, such as offensive instigation, that are not in the lists of mental factors in either the *Treasury of Knowledge* or the *Compendium of Knowledge*.

13

Substantial and Imputed Mental Factors

AMONG THESE MENTAL FACTORS, depending on whether they are posited in terms of their own distinct nature or imputed on the basis of an aspect or function of other mental factors, the texts distinguish between "substantial mental factors" versus "imputed mental factors." There are various explanations of the distinction between substantially existent and imputedly existent, but we will follow the one presented in Asaṅga's *Compendium of Ascertainments*, which is as follows. Something is imputedly existent if it can be identified only in relation to identifying something else, and something is substantially existent if it can be identified without having to identify something else. The *Compendium of Ascertainments* says:

> In brief, what should one understand to be substantially existent? And what should one understand to be imputedly existent? Whatever it may be, anything whose own characteristic is posited without dependence on some other thing and not in relation to something other than itself, one should understand this to be substantially existent. [175] Whatever it may be, anything whose own characteristic is posited in dependence on some other thing and in relation to something other than itself, one should understand this to be imputedly existent.[264]

Generally, from the point of view of its usage, there are four different senses in which the term *substantial existence* is applied: (1) substantial existence in the sense of being established by reasoning, (2) substantial existence in the sense of being capable of functioning, (3) substantial existence in the sense of being stable and not subject to change, and (4) substantial existence in the sense of being self-sufficient. Among these, the first refers to all objects of knowledge, the second refers to all functioning things, the third refers to all permanent phenomena, and the fourth refers to substantially existent things within the aforementioned binary distinction of substantially existent versus imputedly existent.

Likewise, *imputed existence* also has four senses: existing as (1) imputed on parts, (2) imputed on circumstances, (3) imputed despite being nonexistent, (4) imputed in some other fashion. The first of these refers to mental factors that are imputedly existent within the aforementioned binary distinction of substantially existent versus imputedly existent. The second refers to what is imputed on changing circumstances, such as an impermanent vase. The third refers to what is imputed by mind even though it is nonexistent, such as a rabbit's horn. The fourth refers to something such as a "person" that must be cognized in relation to positing something else.

Something is said to be substantially existent by reason of being self-sufficient in that it can appear to the mind without needing to rely on anything else as a basis of imputation. [176] Something is said to be imputedly existent because it becomes an object of the mind in dependence on something else that is the basis of imputation appearing to the mind. Higher and lower Abhidharma each have their own explanations as to which mental factors are substantially existent and which are imputed. First, in higher Abhidharma, Asaṅga's *Compendium of Knowledge* explains that out of the fifty-one mental factors, twenty-two are substantially existent: the five omnipresent ones, the five with a determinate object, seven of the eleven virtuous ones—faith, pliancy, shame, embarrassment, nonhatred, nonattachment, and diligence—and five of the six root afflictions, leaving out view. In contrast, twenty-nine are characterized as imputed: four of the eleven virtuous ones—nondelusion, heedfulness, equanimity, and nonviolence—as well as view that is a root affliction, the twenty secondary afflictions, and the four changeable factors.

Although texts of the higher Abhidharma system explain the twenty secondary afflictions to be imputed mental factors, some followers of the Śrāvaka path explain these to be substantially existent mental factors, as mentioned in Pṛthivībandhu's *Explanation about the Five Aggregates*:

> In this regard, although the proponents of the Śrāvakas' system say that the secondary afflictions too are substantially existent,

> according to the proponents of the Mind Only system, who view all as consciousness only, since the secondary afflictions arise from the six root afflictions, one should understand them to be only imputed, not substantially existent factors.[265]

As to which one is imputed on the basis of which, this is as follows. Nondelusion, afflictive view, and lack of meta-awareness are imputed only on wisdom. [177] Heedfulness and equanimity are imputed on the three root virtues and diligence. Nonviolence is imputed on nonhatred. Rage, resentment, spite, jealousy, and violence are imputed on anger. Avarice, arrogance, and excitation are imputed on attachment. Concealment, dullness, faithlessness, and laziness are imputed on delusion. Guile and pretense are imputed on attachment and delusion. Shamelessness, nonembarrassment, and distraction are imputed on all three poisons (attachment, hatred, and delusion). Heedlessness is imputed on the three poisons together with laziness. Forgetfulness is imputed on afflictive mindfulness. Sleep and afflictive regret that are included in the secondary afflictions are imputed on delusion. The virtuous mental factors and the nonafflictive neutral ones are imputed on wisdom. Inquiry and analysis are imputed on intention and wisdom. This manner of imputing mental factors can be understood from the way these mental factors are defined in the *Compendium of Knowledge*, as cited above.

In the *Compendium of Ascertainments*, however, the imputedly existent mental factors are referred to as "conventionally existent" mental factors. As this text says:

> Suppose someone asks, "Among the virtuous factors, how many are conventionally existent and how many are substantially existent?" We say that three are conventionally existent: heedfulness, equanimity, and nonviolence. Heedfulness and equanimity are associated with nonattachment, nonhatred, nondelusion, and diligence. [178] This is because those factors exemplify equanimity, which is the absence of manifest afflictions, and they exemplify heedfulness, which is the antidote to manifest afflictions. Nonviolence is associated with nonhatred.[266]

This same text says:

> Suppose someone asks, "Among the root afflictions, how many are conventionally existent and how many are substantially existent?" We say that view is conventionally existent because it is associated with wisdom. The rest are substantially existent, they exist as mind and mental factors.[267]

And:

> Suppose someone asks, "Among the secondary afflictions, how many are conventionally existent and how many are substantially existent?" We say, since rage, resentment, spite, jealousy, and violence are associated with anger, they are imputedly existent; since avarice, arrogance, and excitation are associated with attachment, they are imputedly existent; and since concealment, pretense, guile, dullness, sleep, and regret are associated with delusion, they are also imputedly existent. Shamelessness, nonembarrassment, faithlessness, and laziness are substantially existent. [179] Heedlessness, like the previous ones, is imputedly existent. Since forgetfulness, distraction, and faulty wisdom are only associated with delusion, they are all imputedly existent. Since inquiry and analysis are associated with wisdom and associated with the conditioning factors of mind that motivates speech, they are imputedly existent.[268]

However, the explanation in Asaṅga's *Compendium of Bases* differs slightly from the *Compendium of Knowledge*. The text states that nondelusion within the category of eleven virtuous mental factors and shamelessness, nonembarrassment, faithlessness, and laziness within the category of secondary afflictions are substantially existent. The reason given is because they have a stronger potency than other secondary afflictive mental factors.

Second, according to the texts of the lower Abhidharma system, the explanation of which mental factors are substantially existent and which ones are imputed is as follows: the forty-six mental factors enumerated in the *Treasury of Knowledge*—five categories of determinate ones, and one category of indeterminate ones that includes a subset of eight mental factors—plus the twenty-four mental factors listed in the *Finer Points of Discipline*, making a total of seventy mental factors, are substantially existent. In contrast, the seven—the root virtue that is nondelusion, view that is a root affliction, as well as the five extensive afflictive mental factors listed in the [seven] Abhidharma treatises [such as the *Compendium of Elements*]—lack of meta-awareness, forgetfulness, distraction, inappropriate attention, and wrong resolution—are explained as imputed mental factors. Nondelusion is imputed only on virtuous wisdom, [180] view only on afflictive wisdom, lack of meta-awareness only on wisdom that is concomitant with afflictive minds and mental factors, forgetfulness only on afflictive mindfulness, distraction only on afflictive concentration, inappropriate attention only on afflictive attention, and wrong resolution only on afflictive resolution.

When analyzed more precisely, however, it appears that the two—violence and nonviolence—must be accepted as being imputed mental factors, in that nonviolence is imputed on nonhatred and violence is imputed on hatred. For example, the *Treasury of Knowledge* states:

> Nonhatred is love and compassion.[269]

Here nonhatred is explained as compassion, nonviolence as heartfelt love, and violence as wanting sentient beings to suffer.

Similarly, *Finer Points of Discipline* explains that the four—dislike, sorrow, pain, and grief—are imputed on feeling, discouragement on laziness, inattentiveness on afflictive wisdom, disturbance on anger, and so forth. So, when collated, it seems that out of the total of seventy-seven mental factors, we need to posit sixty-one mental factors as substantially existent and sixteen as imputed.[270] [181]

14

Alternate Presentations of Mental Factors

The Presentation in the *Sūtra on the Application of Mindfulness*

The mental factors appear in the *Sūtra on the Application of Mindfulness* in two ways: subsumed into categories and scattered throughout the text. Where categorized, the mental factors are included within four sets: (1) ten that arise simultaneously with the mind,[271] (2) ten extensive mental afflictions, (3) ten narrower mental afflictions, and (4) ten virtuous mental factors.

The first category, the ten that arise simultaneously with the mind, lists the following: (1) feeling, (2) discernment, (3) intention, (4) contact, (5) attention, (6) aspiration, (7) resolution, (8) mindfulness, (9) concentration, and (10) wisdom.

As for their enumeration and definitions, the *Sūtra on the Application of Mindfulness* says: "Thus the ten universal factors are feeling, discernment, intention, contact, attention, aspiration, resolution, mindfulness, concentration, and wisdom."[272] This sūtra also states: "Feeling is to experience."[273] [182] It further tells us:

> Those factors arise at the same time, just like the sun and its light. The defining characteristic of those factors is that they arise simultaneously with the mind, and their number is neither too many nor too few. *Discernment* has the characteristic of bringing about right understanding; for discernment is that which rightly knows the characteristic nature of the individual dharmas. *Intention* moves the mind toward any of the three aspects—virtuous, nonvirtuous, and neutral—and the three aspects are characterized on the basis of body, speech,

> or mind. *Contact*, which has three kinds of experience, occurs when the three [object, sense faculty, and sense consciousness] meet. What do the three kinds of experience primarily refer to? They are pleasant experience, painful experience, and neither pleasant nor painful experience. *Attention* is primarily placement of the mind on the dharmas. *Aspiration* is contemplation of the object. *Resolution* is primarily confidence, owing to whatever gives rise to confidence. *Faith* is faithfulness. *Diligence* is primarily delight. *Mindfulness* is simply not forgetting the object, where the mind is not confused. *Concentration* is primarily single-pointedness of mind. *Wisdom* is differentiating the dharmas, [183] or what causes the dharmas to be fully differentiated.[274]

This text also presents the defining characteristics of faith and diligence as an aside.

The second category, the ten extensive mental afflictions, contains the following: (1) faithlessness, (2) laziness, (3) forgetfulness, (4) distraction, (5) flawed wisdom, (6) inappropriate attention, (7) wrong resolution, (8) excitation, (9) ignorance, and (10) heedlessness. In the concise list in the sūtra, *faithlessness* is followed by *regret*, but in the extended presentation elaborating on it, *faithlessness* is followed by *laziness* [omitting any mention of *regret*]. Here our explanation accords with the extended presentation.

As for their enumeration and definitions, the *Sūtra on the Application of Mindfulness* says: "Faithlessness, regret, forgetfulness, mental distraction, flawed wisdom, inappropriate attention, wrong resolution, excitation, ignorance, and heedlessness. These ten dharmas are the extensive mental afflictions."[275] And:

> Their definitions are to be explained as follows. *Faithlessness* is a lack of resolution, or what causes a lack of resolution toward the dharmas. [184] *Laziness* is simply laziness that abandons diligence. *Forgetfulness* is any non-mindfulness of the dharmas. *Mental distraction* is having many objects of focus. *Flawed wisdom* is a mind that is unskillful. *Inappropriate attention*, since it is not a path, does not think in accordance with the dharmas and includes thinking that something impure is pure. *Wrong*

> *resolution* is primarily wrong apprehending, which takes what has been distortedly apprehended to be correct. *Excitation* is primarily a totally unpeaceful mind. *Ignorance* is not knowing the dharmas of the three realms. *Heedlessness* is that due to which the dharmas are not accomplished, for those mental afflictions undermine the extensive dharmas.[276]

The third category, the ten narrower mental afflictions, lists the following: (1) rage, (2) resentment, (3) nonrevelation,[277] (4) spite, (5) pretense, (6) guile, (7) jealousy, (8) avarice, (9) pride, and (10) pride of superiority. [185]

As for their enumeration and definitions, the *Sūtra on the Application of Mindfulness* says: "There are another ten factors that arise, known as the narrower ones. What are they? They are listed as: rage, resentment, nonrevelation, spite, pretense, guile, jealousy, avarice, pride, and pride of superiority."[278] Also, apart from *pride* and *pride of superiority*, this text clearly presents the definitions of the remaining eight, for it says: "*Rage* is mental disturbance and enragement. *Resentment* is a firm intention toward one's enemy. *Nonrevelation* is to die of the three evils [of body, speech, and mind]. *Spite* firmly holds on to what is improper. *Pretense* misleads or is that which causes deception."[279] Also, it says: "The mental factor *guile* is primarily dishonesty; it is an evil that causes one to be tightly bound in saṃsāra. *Jealousy* is in the nature of causing harm to others. *Avarice* is primarily fear dreading that one's material wealth will run out and the mental factor greed."[280] [186]

The fourth category, the ten virtuous mental factors, includes the following: (1) nonattachment, (2) nonhatred, (3) shame, (4) embarrassment, (5) faith, (6) pliancy, (7) heedfulness, (8) diligence, (9) equanimity, and (10) nonviolence. As for their enumeration and definitions, the *Sūtra on the Application of Mindfulness* says:

> Also, there are ten main virtuous factors that arise universally. What are they? They are the following: nonattachment, nonhatred, shame, embarrassment, faith, pliancy, heedfulness, diligence, equanimity, and nonviolence. These virtues arise universally. Their definitions are as follows. All these virtuous factors arise from the basis of *nonattachment* because that is the root of all of them, as is *nonhatred*. *Shame* is fear on account of

> oneself, out of consideration for oneself only. *Embarrassment* is out of consideration for others. *Faith* is admiration within a being's mental continuum. *Pliancy* is a virtuous mind and a physical state that eliminates the development of bodily and mental dysfunction for it brings complete bliss. *Heedfulness* accomplishes the virtuous dharmas. *Equanimity* is any meditative absorption focused on causes, conditions, and what is to be done or not to be done. [187] *Nonviolence* is primarily the nonharming of sentient beings.[281]

As for the definition of diligence, the same sūtra states: "*Diligence* is delight."[282] And as the definition of pliancy, it states: "As a quality of the body and a quality of the mind itself, as much as bodily serviceability, softness, and lightness arises, so bliss gathers within the yogin; thus correct pliancy is a factor conducive to enlightenment."[283] Apart from *pride* and *pride of superiority*, this sūtra clearly explains all forty mental factors in the above categories in terms of their definitions and functions.

Furthermore, the following mental factors, although not included in the above categories, are mentioned and scattered throughout the *Sūtra on the Application of Mindfulness*: (1) attachment, (2) hatred, (3) inquiry, (4) analysis, (5) afflictive doubt, (6) arrogance, (7) shamelessness, (8) nonembarrassment, (9) dullness, (10) sleep, (11) concealment, (12) wrong view, (13) regret, (14) terror, (15) fear, (16) conceit, (17) disheartenment, (18) discontent, [188] (19) discouragement, (20) disenchantment, (21) dissatisfaction, (22) intolerance, (23) animosity, (24) mental agitation, (25) sadness, (26) unruliness, (27) satisfaction, (28) sorrow, (29) not wanting, (30) posing, (31) stupefaction, (32) distorted pride, (33) pride of thinking "me," (34) pretentious pride, (35) embellishment, and so on. Furthermore, there is (36) courage, (37) tolerance, (38) lesser desire, (39) courtesy, (40) contentment, (41) respect, (42) wishing, (43) love, (44) compassion, (45) joy, (46) trust, (47) gentle attitude, (48) generosity, (49) ethical conduct, (50) honesty, (51) right view, (52) rejoicing, (53) meta-awareness, and so on. Thus there are many mental factors mentioned in that sūtra that are not enumerated in the above four categories. When the forty mental factors that are part of the determinate categories in *Sūtra on the Application of Mindfulness* are added to the fifty-three factors we have identified as scattered, we arrive at a total of ninety-three mental factors mentioned in that sūtra. [189]

The Presentation in the Theravāda *Compendium of Abhidhamma*

The second chapter of the Theravāda work *Compendium of Abhidhamma*, composed in the Pali language by the master Anuruddha, contains the following summary of the presentation of fifty-two mental factors:

> Arising and ceasing together with it,
> having the same object and base,
> are fifty-two factors associated with the mind;
> these are known as mental factors.[284]

A *mental factor* is identified as an awareness that is concomitant with a main mind in terms of being the same in four respects: (1) arising, (2) ceasing, (3) object, and (4) base. Here "same base" means that the basis or dominant condition is the same. This text presents the divisions of mental factors as follows:

> Thirteen are common,
> fourteen are nonvirtuous,
> and twenty-five are virtuous;
> thus fifty-two are stated.[285] [190]

According to this presentation, there are thirteen mental factors that are common to both virtuous and nonvirtuous minds, fourteen that are nonvirtuous, and twenty-five that are virtuous, making a total of fifty-two mental factors subsumed into three categories. Within the first category, seven mental factors occur with all minds universally, and six occur with particular ones. This text states the first seven to be: "(1) contact, (2) feeling, (3) discernment, (4) intention, (5) single-pointedness, (6) life-force, and (7) attention. These seven mental factors are termed *universal to all minds*."[286] Among them, *life-force* is explained to be a mental factor that engages and empowers the sustaining of the dhammas that arise together with itself.

The six particular mental factors are said to be: "(1) inquiry, (2) analysis, (3) resolution, (4) diligence, (5) joy, and (6) aspiration. These six are termed *particular mental factors*."[287] Here, *joy* is characterized as a mental factor that makes the mind that it accompanies glad and pleased to focus on its object. It functions to cause the mind and body to flourish and be satisfied. There are five types described, such as *minor joy* and so on. [191]

The second category contains fourteen nonvirtuous mental factors, stated to be: "(1) confusion, (2) shamelessness, (3) nonembarrassment, (4) excitation, (5) desire, (6) view, (7) pride, (8) hatred, (9) jealousy, (10) avarice, (11) regret, (12) laxity, (13) sleep, and (14) doubt. These fourteen are termed *nonvirtuous mental factors*."[288]

The third category containing the virtuous mental factors includes nineteen common to all virtuous minds, three abstinences,[289] two immeasurables, and the faculty of wisdom, which together make twenty-five mental factors. This text says:

> (1) Faith, (2) mindfulness, (3) shame, (4) embarrassment, (5) nonattachment, (6) nonhatred, (7) balance, (8) bodily pliancy, (9) mental pliancy, (10) bodily lightness, (11) mental lightness, (12) bodily softness, (13) mental softness, (14) bodily serviceability, (15) mental serviceability, (16) bodily proficiency, (17) mental proficiency, (18) bodily straightness, (19) mental straightness. These nineteen mental factors are termed *universal to virtuous minds*.[290]

Here *balance* (Pali: *tatramajjhattatā*) means equanimity. As for their definitions, *bodily lightness* and *mental lightness* [192] are mental factors that, respectively, have the nature of pacifying bodily and mental heaviness, and thereby bodily and mental unserviceability. *Bodily softness* and *mental softness* are mental factors that, respectively, have the nature of pacifying bodily and mental hardness and inflexibility. The pair of mental factors that make the body and mind serviceable are mental factors that, respectively, pacify bodily and mental unserviceability. The pair of mental factors that make the body and mind proficient are mental factors that, respectively, have the nature of pacifying bodily and mental sickness and dysfunction. The pair of mental factors that make the body and mind straight are mental factors that, respectively, pacify bodily and mental crookedness—that is, pretense, guile, and so on.

Regarding the second set, the three abstinences, this text says: "(1) right speech, (2) right action, and (3) right livelihood. These three are called *abstinences*."[291] These three are defined, respectively, as mental factors that prevent engagement in faulty verbal and physical conduct and wrong livelihood. They function to give rise to shame and the fear of behaving badly. [193]

Regarding the third set, the two immeasurables, and the fourth, the faculty of wisdom, this text says: "Both compassion and joy are called *immeasurable.* Along with the faculty of wisdom, these twenty-five mental factors are together called the *thoroughly wholesome.*"[292] The definition of *compassion* is the wish that others may be free from suffering. It functions to make one unable to bear the suffering of others. In terms of its definition, *joy* is a mental factor that produces mental clarity and delight upon seeing the happiness of others. Its function is to clear the mind of jealousy and so on when seeing other people's excellent attributes, such as their wealth. The *faculty of wisdom* is defined as a mental factor that understands the way things exist in terms of their ultimate reality and manifold nature. It functions to distinguish individually appearing objects.

A more extensive enumeration of the definitions, functions, and causes of each of the fifty-two mental factors of the noble Theravāda tradition must be understood as presented in the *Compendium of Abhidhamma* and its commentaries. [194]

THE PRESENTATION IN NĀGĀRJUNA'S *PRECIOUS GARLAND*

Nāgārjuna's text lists fifty-seven sources of faults mentioned in the sūtras, which both monastic and lay disciples are advised to abandon. Within this group are some that appear on the lists of mental factors in the *Treasury of Knowledge* and *Compendium of Knowledge* and some that are not on either of these two lists. Nāgārjuna's *Precious Garland* says:

> Even the least of faults must be known,
> and the sources of faults abandoned;
> make effort to clearly ascertain
> those widely known as the fifty-seven.[293]

The fifty-seven sources of faults are identified as: (1) rage, (2) resentment, (3) concealment, (4) spite, (5) guile, (6) pretense, (7) jealousy, (8) avarice, (9) shamelessness, (10) nonembarrassment, (11) haughtiness, which is not honoring one's teacher and so on, (12) offensive instigation, which is engaging in verbal and physical actions such as scowling and so on under the influence of rage, (13) arrogance, (14) heedlessness, (15) pride, which is of seven types, here counted as one, (16) [195] hypocrisy, which is restraining the doors of the sense faculties out of a desire for honor and gain, (17) flattery, which is speaking pleasing words out of a desire for honor and gain, (18) hinting, which is praising something belonging to someone else in order to acquire it, (19) soliciting, which is, for the sake of gain, openly deriding others as being very avaricious and so on, (20) pursuing gain with gain, which is extolling something one previously acquired, (21) carping, which is stating again and again the faults of others, (22) discomposure, which is mental irritation toward others, or (23) discomposure having laziness, which is attachment toward lowly things, (24) discrimination between self and others, which is differentiating between oneself and others in a biased manner when obstructed by hatred and attachment, (25) not watching one's mind, which is not analyzing whether this or that mind is virtuous or nonvirtuous, (26) weakened esteem for what is to be done in accordance with the Dharma, which occurs owing to laziness, (27) being a vile person, which is to take another person as a spiritual guide without treating them in the manner of a buddha, (28) adherence, which arises from attachment and is a minor entanglement, (29) thorough adherence, which arises from desire for the five objects of the senses—form, sound, and so on—and is a major entanglement, (30) attachment toward one's belongings, (31) inappropriate attachment toward others' belongings, (32) improper attachment, which is wanting to praise a woman or a man who is an improper object of desire out of desire for him or her, (33) hypocrisy, which is giving an impression of having qualities that one does not have,[294] (34) [196] great desire, which is an extreme greed for things that extends far beyond contentment, (35) desire for attainment, which is wanting others to believe that one possesses great qualities, no matter what, (36) intolerance, which is being unable to bear suffering when it arises within oneself or when one is harmed by others, (37) impropriety, which is a lack of respect for the activities of preceptors and gurus, (38) not appreciating instruction, which is to think "It is all right for me to engage in virtue or vice" and so on when one has been given beneficial advice by others, (39) thinking about one's relatives, which is attachment to one's family members, (40) craving an object, which is to extol even a lowly object's good qualities for the purpose of obtaining that object, (41) thinking that one is deathless, which is being untroubled by a fear of death, (42) thinking with concern for recognition, which is determining how to behave so that others will see oneself as having excellent qualities, thus gaining their respect, (43) thinking with attachment, which is lusting after other men or women, (44) ascertaining ill will, which is to consider whether someone intends to benefit or harm oneself, (45) mental instability, which is a mind of dislike, (46) mental perturbation, which is owing to desire, (47) apathy, which is slothfulness of the body with no energy for virtue, (48) alteration of color and body, which is owing to the force of the mental afflictions, (49) not wanting food, which is explained as physical discomfort from overeating, (50) deep despondency of mind, which is explained as discouragement of the mind, (51) sense desire, which is wanting and seeking the five desirable sensory qualities, (52) ill will, (53) dullness, (54) sleep, [197] (55) excitation, (56) regret, and (57) doubt.

Here, in number (15) above, the seven types of pride are included within pride, so they are counted as one; from (16) to (20), the five wrong livelihoods are counted separately; and from (51) to (57), the five hindrances—sense desire, ill will, dullness and sleep, excitation and regret, and doubt—are counted separately, thus making a total of fifty-seven. Among these fifty-seven, (12) offensive instigation, (16–20) the five wrong livelihoods, (21) carping, (22) discomposure, (24) discrimination between self and other, (25) not watching one's mind, and so forth are not found among the mental factors listed in both Asaṅga's *Compendium of Knowledge* and Vasubandhu's *Treasury of Knowledge*. [198]

The Presentation in the *Compendium of Teachings*

Also, the *Compendium of Teachings*, which is attributed to Nāgārjuna, lists forty mental factors:

> The forty conditioning factors concomitant with the mind are as follows: (1) feeling, (2) discernment, (3) intention, (4) aspiration, (5) contact, (6) intelligence, (7) mindfulness, (8) attention, (9) resolution, (10) concentration, (11) faith, (12) heedfulness, (13) pliancy, (14) equanimity, (15) shame, (16) embarrassment,

> (17) nonattachment, (18) nonhatred, (19) nonviolence, (20) diligence, (21) confusion, (22) heedlessness, (23) laziness, (24) faithlessness, (25) dullness, (26) excitation, (27) shamelessness, (28) nonembarrassment, (29) rage, (30) resentment, (31) guile, (32) jealousy, (33) spite, (34) concealment, (35) avarice, (36) pretense, (37) arrogance, (38) violence, (39) inquiry, and (40) analysis.[295]

On this list, apart from absence of regret, sleep, anger, attachment, pride, and doubt, the enumeration of mental factors seems to be exactly the same as the one in Vasubandhu's *Treasury of Knowledge*. [199]

THE PRESENTATION IN CANDRAKĪRTI'S *CLASSIFICATION OF THE FIVE AGGREGATES*

According to *Classification of the Five Aggregates*, which is attributed to Candrakīrti, the mental factors are explained as being concomitant with the mind in terms of bearing five similarities—same basis, same object, same aspect, same time, and same substance—just as explained in Vasubandhu's *Treasury of Knowledge*.

The subdivisions of mental factors listed here are: (1) intention, (2) contact, (3) attention, (4) aspiration, (5) resolution, (6) faith, (7) diligence, (8) mindfulness, (9) concentration, (10) wisdom, (11) inquiry, (12) analysis, (13) heedlessness, (14) heedfulness, (15) disenchantment, (16) joy, (17) pliancy, (18) nonpliancy, (19) violence, (20) nonviolence, (21) shame, (22) embarrassment, (23) equanimity, (24) liberation, (25) roots of virtue, (26) roots of nonvirtue, (27) roots of neutral karma, (28) fetters, (29) bindings, (30) proclivities, (31) secondary mental afflictions, (32) entanglements, (33) contaminants, (34) streams, (35) [200] bonds, (36) graspings, (37) ties, (38) hindrances, (39) knowledge, and (40) acquiescence. Although this text lists forty sets altogether, those from *roots of virtue* up to *acquiescence* have subdivisions, making the total over a hundred. *Classification of the Five Aggregates* lists these as follows:

> Regarding this, those that are concomitant with the mind are the following: intention, contact, attention, aspiration, resolution, faith, diligence, mindfulness, concentration, wisdom, inquiry, analysis, heedlessness, heedfulness, disenchantment,

> joy, pliancy, nonpliancy, violence, nonviolence, shame, embarrassment, equanimity, liberation, roots of virtue, roots of nonvirtue, roots of neutral karma, fetters, bindings, proclivities, secondary mental afflictions, entanglements, contaminants, floods, bonds, graspings, ties, hindrances, knowledge, and acquiescence. The [meaning of the] term *concomitant with the mind* was already explained above.[296]

Of these mental factors, the following—intention, contact, attention, aspiration, [201] resolution, faith, diligence, mindfulness, concentration, wisdom, inquiry, analysis, heedlessness, heedfulness, pliancy, violence, nonviolence, shame, embarrassment, and equanimity—can be understood on the basis of the explanations provided above in the presentation of mental factors in the texts of the upper and lower Abhidharma systems.

Disenchantment is a mental factor identified as being fed up with saṃsāra once one has seen its faults, such as the suffering of birth, aging, and death. It functions to cause one to abandon the mental afflictions. This text says: "Disenchantment is so called because of being a mental factor that is concomitant with a mind of disenchantment with saṃsāra upon seeing the faults of saṃsāra, leading to a gradual abandonment of the mental afflictions."[297]

Complete joy, according to the Vaibhāṣika system, is a mental factor having the aspect of bliss that is a mental feeling, which is considered to be a different substance from the bliss of pliancy that is a mental factor. This text says: "Complete joy of the mind is a gladness of the mind that is different from mental bliss; it is complete joy."[298] [202] According to the higher Abhidharma system, however, bliss is a mental factor that is an experience of pleasure accompanying a primary mental consciousness, thereby enhancing the sense faculty that is its basis; and joy is pleasure accompanying a mental consciousness. So bliss and joy are said to be the same substance.

Nonpliancy is a mental factor that is included in all types of laxity and dullness where the object is unclear; it hinders the ability to employ as one wishes any type of virtuous mind. This text says: "Nonpliancy is a heaviness of body and mind; and since to be heavy is to be unclear, it has the nature of laxity and dullness."[299]

Liberation is a mental factor that abandons the mental afflictions. This text says, "Liberation is a freeing of the mind from stains. Whatever mental factor arises that frees the mind from stains is called *liberation* because it abandons the mental afflictions completely."[300]

Roots of virtue are mental factors that are virtuous by nature without depending on motivation or some concomitant factor; they function as the basis of all virtues, which are their results. When categorized, there are three: the root of virtue that is nonattachment, the root of virtue that is nonhatred, and the root of virtue that is [203] nondelusion. This text says: "The three roots of virtue are nonattachment, nonhatred, and nondelusion."[301]

Roots of nonvirtue are nonvirtuous mental factors that have the power to sever the roots of virtue; they motivate nonvirtuous actions of body and speech. When categorized there are three: nonvirtuous attachment, hatred, and nonvirtuous delusion. *Classification of the Five Aggregates* says: "Opposite to the three roots of virtue are the three roots of nonvirtue: attachment, hatred, and delusion."[302]

Roots of neutral karma are mental factors that are neither virtuous nor nonvirtuous; they function as the basis of all neutral actions, which are their results. When categorized, there are three: neutral craving, neutral ignorance, and neutral view. This text says: "The three roots of neutral karma are: craving, ignorance, and intelligence."[303]

Fetters are mental factors that, when present in the mindstream, yoke it to the arising of craving for an object or yoke it to many instances of suffering. When categorized there are nine: the fetter of attachment, the fetter of anger, [204] the fetter of pride, the fetter of ignorance, the fetter of view, the fetter of holding unsuitable ethics and vows to be supreme, the fetter of doubt, the fetter of jealousy, and the fetter of avarice. *Classification of the Five Aggregates* says: "Fetters are ninefold: the fetter of attachment, the fetter of anger, the fetter of pride, the fetter of ignorance, the fetter of view, the fetter of holding [unsuitable ethics and vows] to be supreme, the fetter of doubt, the fetter of jealousy, and the fetter of avarice."[304]

Bindings are mental factors that cause ordinary beings in the three realms to be bound without choice to the suffering of suffering and so on.[305] When categorized, there are three: the bindings of attachment, the bindings of hatred, and the bindings of delusion. This text says: "Bindings are of three types: the bindings of attachment, the bindings of hatred, and the bindings of delusion. These bindings tightly bind one to the three realms, thus they are called *bindings*."[306] [205]

Proclivities are mental factors that are the roots of both projecting and actualizing causes of rebirth in saṃsāra. Their nature is hard to realize, and they proliferate owing to their object and their concomitant factors. They can be classified in two ways. They may be counted as six: (1) the proclivity that is attachment, (2) the proclivity that is anger, (3) the proclivity that is pride, (4) the proclivity that is ignorance, (5) the proclivity that is view, and (6) the proclivity that is doubt. Or they may be counted as seven, where the proclivity of attachment is further divided into the following two classes: the proclivity of attachment belonging to the desire realm and the proclivity of attachment belonging to the higher realms of saṃsāric existence. This text says:

> In this context there are six proclivities, which are the root proclivities: the proclivity that is attachment, the proclivity that is anger, the proclivity that is pride, the proclivity that is ignorance, the proclivity that is view, and the proclivity that is doubt. Within these six, the proclivity that is attachment may be divided into two—the proclivity that is desire for the desirable and the proclivity that is desire for saṃsāric existence—thus making seven.[307]

Secondary mental afflictions are mental factors that defile the mindstream and are close to the six proclivities, the mental afflictions that are the roots of both projecting and actualizing causes of rebirth in saṃsāra. They are listed as the *six stains*: pretense, arrogance, violence, spite, [206] resentment, and guile. The "such as" [in the citation below] includes the ten entanglements—shamelessness, nonembarrassment, jealousy, avarice, excitation, afflictive regret, dullness, afflictive sleep, rage, and concealment—which brings the list up to sixteen, and it also includes the twenty-four secondary mental afflictions listed in *Finer Points of Discipline*. On this point, *Classification of the Five Aggregates* says:

> The expression *secondary mental afflictions* can be explained as follows. Since the six proclivities function as the causes of afflictions of body, speech, and mind, they are called *afflictions*; and

> any that are afflictive temporarily are called *secondary* mental afflictions. Alternatively, any afflictive mental factors included within the aggregate of conditioning factors are secondary mental afflictions because they cause the mind to become afflictive. Which ones are they? Many types are mentioned in the Abhidharma, such as pretense, arrogance, violence, spite, resentment, and guile.[308]

Entanglements are mental factors that remain entwined in the minds of ordinary beings at all times. They function to obstruct engaging in virtue. When categorized, there are ten entanglements: dullness, afflictive sleep, excitation, regret, jealousy, avarice, shamelessness, nonembarrassment, rage, and concealment. This text says: [207] "Entanglements are tenfold: dullness, afflictive sleep, excitation, regret, jealousy, avarice, shamelessness, nonembarrassment, rage, and concealment," and "Through ensnaring and entangling sentient beings, entanglements obstruct engagement in the side of virtue."[309]

Contaminants are afflictive mental factors that overflow from the wounds of the six sources of saṃsāra. When categorized, they are: contaminants of desire, contaminants of saṃsāric existence, and contaminants of ignorance. This text says: "Contaminants are threefold: contaminants of desire, contaminants of [saṃsāric] existence, and contaminants of ignorance."[310]

Floods are afflictive mental factors that carry one away to other realms, migrations, and places of birth in the ocean of saṃsāra. When categorized, there are four: the flood of desire, the flood of saṃsāric existence, the flood of views, and the flood of ignorance. This text says: "Floods are fourfold: the flood of desire, the flood of [saṃsāric] existence, the flood of views, and the flood of ignorance."[311] [208]

Bonds are afflictive mental factors that cause the consciousnesses of ordinary beings to reincarnate in saṃsāra, in its realms, its migrations, and in places of birth, or they are what cause attachment to a place. When categorized, there are four: the bond of desire, the bond of saṃsāric existence, the bond of views, and the bond of ignorance. *Classification of the Five Aggregates* says: "These four should be known as the four bonds."[312]

Graspings are afflictive mental factors that cause the consciousnesses of ordinary beings to firmly grasp on to saṃsāra. When categorized, there are four: grasping at objects of desire, grasping at views, grasping at ethics and vows, and grasping at a doctrine of self. *Classification of the Five Aggregates* says: "Graspings are fourfold: grasping at [objects of] desire, grasping at views, grasping at ethics and vows, and grasping at a doctrine of self."[313]

Ties are afflictive mental factors that obstruct the development of an unfluctuating mental body. When categorized, there are four: the bodily tie of covetousness toward enjoyable things, the bodily tie of ill will toward people, [209] the bodily tie of holding ethics and vows to be supreme, and the bodily tie of holding bad views to be supreme, thinking "This is the truth" and clinging to these views. *Classification of the Five Aggregates* says: "Ties are fourfold: the bodily tie of covetousness, the bodily tie of ill will, the bodily tie of holding ethics and vows to be supreme, and the bodily tie of clinging to [the view] 'This is the truth.'"[314] For example, just as tightly knotted ties obstruct one's freedom, so in making the mind distracted, those mental factors are like ties that obstruct single-pointed meditative equipoise.

Hindrances are nonvirtuous mental factors of desire that obstruct meditative absorption, liberation, concentration, and equipoise. When categorized, there are five: sensory desire, ill will, dullness and sleep, excitation and regret, and doubt. *Classification of the Five Aggregates* says: "Hindrances are fivefold: sensory desire, ill will, dullness and sleep, excitation and regret, and doubt. Hindrances that are associated with the desire realm are only completely nonvirtuous; therefore, since they obstruct the side of virtue, they are hindrances."[315] [210] *Dullness and sleep* are presented together because they have the same remedy (the same discernment of an appearance), the same nourishment or causal basis (the same stupefaction, dislike, stretching, unsuitable measure of food, and discouragement), and the same function (making the mind discouraged). Also, *excitation and regret* are presented together because they have the same remedy (calm abiding), the same nourishment (thinking about one's relatives, thinking about one's country, thinking that one is deathless, and recollecting earlier hilarity, fun, and enjoyment), and the same function (making the mind unpeaceful). Furthermore, sensory desire and ill will are hindrances to ethical conduct, dullness and sleep are hindrances to wisdom, excitation and regret are hindrances to concentration, and doubt is a hindrance to concentration and wisdom.

Knowledge is either contaminated knowledge or knowledge having the nature of definitely abandoning the doubt that is its object to be abandoned. When categorized, there are ten: knowledge of the Dharma teaching, subsequent knowledge, knowledge of others' minds, knowledge of conventionality, knowledge of suffering, knowledge of its origin, knowledge of its cessation, knowledge of the path, knowledge of extinction, and knowledge of nonarising. *Classification of the Five Aggregates* says: "Knowledge is tenfold: [211] knowledge of the Dharma teaching, subsequent knowledge, knowledge of others' minds, knowledge of conventionality, knowledge of suffering, knowledge of its origin, knowledge of its cessation, knowledge of the path, knowledge of extinction, and knowledge of nonarising."[316]

Acquiescence is an uncontaminated mental factor that acts as the direct antidote to its specific object to be abandoned. When categorized, there are eight: acquiescence in the knowledge of the teaching on suffering, acquiescence in the subsequent knowledge of suffering, acquiescence in the knowledge of the teaching on its origin, acquiescence in the subsequent knowledge of its origin, acquiescence in the knowledge of the teaching on its cessation, acquiescence in the subsequent knowledge of its cessation, acquiescence in the knowledge of the teaching on the path, and acquiescence in the subsequent knowledge of the path. *Classification of the Five Aggregates* says: "What is acquiescence? There are eight types of acquiescence that are clear realizations: acquiescence in the knowledge of the teaching on suffering, acquiescence in the subsequent knowledge of suffering, acquiescence in the knowledge of the teaching on its origin, acquiescence in the subsequent knowledge of its origin, acquiescence in the knowledge of the teaching on its cessation, acquiescence in the subsequent knowledge of its cessation, acquiescence in the knowledge of the teaching on the path, and acquiescence in the subsequent knowledge of the path."[317] [212]

An Estimate of Mental Factors Collated from Various Enumerations

Although in general it may be difficult to speak in terms of a definitive list that includes all the mental factors, if we combine the various lists of mental factors enumerated in Asaṅga's *Compendium of Knowledge*, Vasubandhu's *Treasury of Knowledge*, and so on as presented above, the following seems to emerge. To begin with, there are the forty-six mental factors such as feeling and so on that are on the lists of both the *Treasury of Knowledge* and *Compendium of Knowledge.* On top of that, there are an additional five listed in the *Compendium of Knowledge*—the root virtue that is nondelusion, view that is a root affliction, and included among the twenty secondary afflictions, forgetfulness, lack of meta-awareness, and distraction—bringing the total to 51. Then there is an additional list of twenty-four secondary afflictions presented in *Finer Points of Discipline,* such as dislike and so on, bringing the total to 75. On top of that, there are an additional three found in *Sūtra on the Application of Mindfulness*—inappropriate attention and wrong resolution, which are among the ten universal mental afflictions, and nonrevelation,[318] which is included in the ten narrower mental afflictions—bringing the total to 78. Then there are sixteen additional factors enumerated in the Pali text *Compendium of Abhidhamma*—life-force that is among the seven universals, joy that is among the six particulars, laxity that is among the fourteen nonvirtues, [213] then bodily lightness, mental lightness, bodily softness, mental softness, bodily serviceability, mental serviceability, bodily proficiency, mental proficiency, bodily straightness, right speech, right livelihood, right action, and immeasurable joy that are among the twenty-five virtues—which brings the total to 94. On top of that, there are another eighteen factors to be added from the *Precious Garland*—offensive instigation, soliciting, hypocrisy causing one to restrain the doors of the sense faculties out of a desire for gain and so on, flattery, pursuing gain with gain, hinting, carping, discomposure, not watching, weakened esteem, being a vile person, hypocrisy giving an impression of having qualities that one does not have, intolerance, not appreciating instruction, ill will, mental perturbation, alteration of color, and not wanting food—bringing the total to 112. There are still an additional five found in the *Compendium of Bases*—evil companionship, engaging inefficiently, ungentleness, applying discordantly, and lamentation—which brings the total to 117. In addition to these, there are two more from Candrakīrti's *Classification of the Five Aggregates*—disenchantment that is fed up with saṃsāra, and ties. Thus when we add up the various unique mental factors found in the texts identified in this compendium of Buddhist sciences, we see that altogether the total number of mental factors comes to 119.

However, the question concerning the total number of mental factors still requires further research. [214] More specifically, there are some factors enumerated in Candrakīrti's *Classification of the Five Aggregates*, such as liberation, knowledge, and acquiescence, that are just different names for mental factors already listed in the Abhidharma texts presented earlier. So since, in terms of their definitions, they are included in feeling, discernment, wisdom, aspiration, resolution, and so on, we have not counted them as separate factors in the above list of the aggregated total. Likewise, roots of virtue, roots of nonvirtue, roots of neutral karma, fetters, bindings, proclivities, secondary mental afflictions, entanglements, contaminants, floods, bonds, graspings, hindrances, and so on occur in both the upper and lower Abhidharma texts; and since they too do not occur separately from the mental factors found in the upper and lower Abhidharma texts as presented earlier, we also have not counted these separately in the aggregated list. By means of this illustration, we can understand which mental factors are not counted here—such as those in the *Sūtra on the Application of Mindfulness*, as well as Anuruddha's *Compendium of Abhidhamma*, Nāgārjuna's *Precious Garland*, Asaṅga's *Compendium of Bases*, and Candrakīrti's *Classification of the Five Aggregates*—by comparing their definitions with those of the fifty-one mental factors found in the Abhidharma texts and the twenty-four secondary mental afflictions found in *Finer Points of Discipline*. [215]

Why do Buddhist texts present taxonomies of the mental factors as well as detailed analyses of individual mental factors in terms of their definitions, their functions, and their internal causal relationships? In general our external behavior of body and speech flows primarily from the motivating forces operating within our minds; and our minds, in their turn, follow principally from the activity of their concomitant mental factors. It therefore becomes extremely important for those of us who wish to transform our minds to have some understanding, even in terms of a rough outline, of the definitions of the most significant mental factors and their functions. It is with this recognition that extensive explanations of the mental factors are presented in the Buddhist texts. [217]

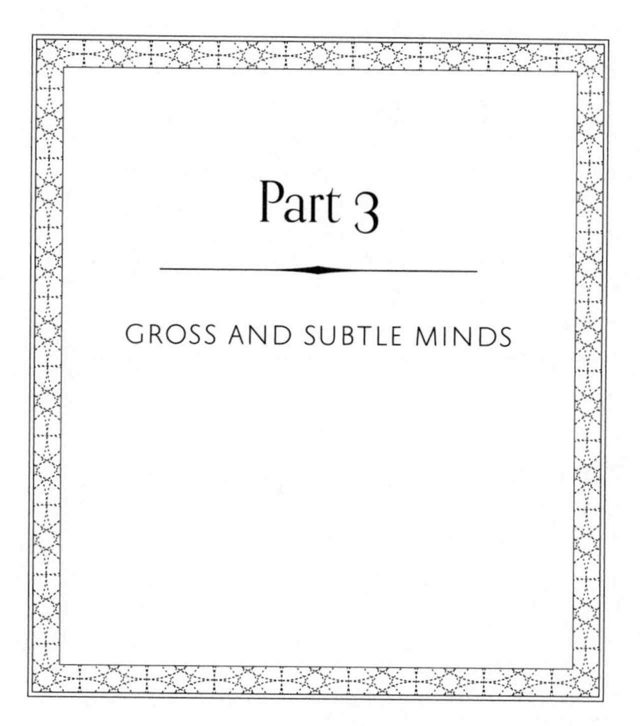

Part 3

GROSS AND SUBTLE MINDS

A Tantric Perspective

The notion that consciousness occurs at various degrees of subtlety is found in many Indian traditions, and Buddhism is no exception. According to Buddhist theorists, subtle states can occur both naturally and as a result of contemplative practice, and while they can be harnessed to various purposes, they especially offer the opportunity to explore the nature of consciousness in forms that are not cluttered by the hubbub—or even the chaos—of a mind in the usual "gross" consciousness of everyday life.

Our authors begin with a well-known typology of consciousness in three contexts: awake, dreaming, and dreamless sleep, with dreamless sleep being the subtlest. They further note that, during the waking state, mental consciousness is subtler than sense consciousness: thinking of an apple is subtler than seeing one. These notions of the subtlety of consciousness are formulated from the theoretical standpoint assumed by most of this volume. But our authors begin to leave that standpoint behind when they move on to a discussion of the subtle levels of consciousness that transpire in the death process. This in turn quickly leads to this part's main theme: the levels of consciousness as depicted in Buddhist tantric texts. To introduce this account of "gross and subtle minds," I will thus focus on the tantric perspective, which shifts our exploration of the mind into a context quite unlike the other parts of this compendium.

Although the precise history of tantric practice in India is obscure, by the time Buddhism began to be established in Tibet in the eighth century, the dominant style of Buddhist practice in India was tantric. The term *tantra* refers to a genre of Buddhist literature that was said to be transmitted in secret to a select audience. These texts stand in distinction from the *sūtras*, discourses taught to a wide audience by Śākyamuni, the historical Buddha. Traditional accounts maintain that the tantras were also taught by the historical Buddha, although not in his ordinary form. Whatever one might say about the difficult question of authorship, from the time Buddhism first reached Tibet until the end of the thirteenth century, when the historically attested interchange between Tibetan and Indian Buddhists dwindled to the point of disappearance, tantric practice was indisputably at the center of all the Tibetan traditions, where the texts and practices fall under the rubric of the Vajrayāna, or "Diamond Vehicle."

The Vajrayāna includes a strikingly wide array of texts and methods, but most are not relevant to understanding our authors' discussion of gross and subtle minds. Instead, only two aspects of Vajrayāna theory require our attention. First, in relation to the subtlest level of consciousness, Vajrayāna theory proposes a nondual interpretation of the relationship between mind and body, such that the substance dualism assumed in other sections of this compendium does not apply. Second, Vajrayāna offers a set of techniques—perhaps even a "technology"—that manipulates the mind-body system in ways that are said to be especially effective in producing the transformations that are central to Buddhist practice. A brief look at this nondual account and the methods or technology that flow from it will help to clarify our authors' discussion of gross and subtle minds.

"Wind" and Mind: A Nondual Perspective

In other parts of this volume, the general level of analysis adopted by our authors assumes what is often called *substance dualism*. From this perspective the mental and the physical are distinct because they are made of different kinds of "substances"—namely, material stuff and mental stuff. Importantly, even this dualistic account does not assume that mind can simply exist in complete independence of some type of embodiment. Instead, the mind stands in an interdependent relationship with its "basis" (Tib., *rten*), the physical body, which acts as a container or substratum within and through which mind functions. In the form of Buddhist tantra most relevant to our authors' analysis, this relationship between mind and its basis moves beyond substance dualism by articulating a fundamental energy called *wind* (Tib., *rlung*; Skt., *vāyu*). In the first volume of this series, wind figures prominently as one of the four primary elements (Skt., *mahābhūta*) that constitute the physical world, where wind accounts for the basic lightness, mobility, and motility of physical entities. As tantric traditions emerge in India, however, *wind* takes on meanings that, connecting to developments in Indian medical theories, attribute to it a central role in the embodied mind, where this fundamental energy facilitates many capacities, including physical movement, cognitive processes, and consciousness itself.

Chapter 30 of the first volume in this series discusses in detail the most typical Tibetan account of the Vajrayāna model that presents wind as playing a key role in the functioning of the embodied mind. This Vajrayāna model involves an elaborate model of a "subtle body" featuring 72,000 channels—branching from three main channels—in which ten types of wind energy flow and where drops (Skt., *bindu*) or vital essences occur at key locations. Our authors' detailed account of this complex, subtle physiology need not be reiterated here, except for one important point: in its subtlest state, known as the extremely subtle mind (Tib., *shin tu phra ba'i sems*), consciousness is indistinguishable from its basis, the extremely subtle wind (Tib., *shin tu phra ba'i rlung*). In other words, to adopt a typical metaphor, the subtlest level of consciousness is nothing other than the extremely subtle energy on which it is "mounted," and that extremely subtle energy is nothing other than the extremely subtle consciousness that is its "rider." Thus, while the mind-body distinction can be maintained at a coarse level, that distinction falls away at the subtlest level from the perspective of Vajrayāna theory. In their own discussion of this issue in the previous volume, our authors thus say that "at the subtlest level, given that wind and consciousness exist as a single entity, no differentiation can be made between the two in terms of their reality" (p. 425). Clearly, we have arrived at a nondual account of the relation between mind and body, albeit one that presumes that the only "body" left is an extremely subtle form of energy. As it turns out, that particular mind-body configuration occurs reliably only at death, and that is precisely the main target of what we might call the technology of tantra.

THE TECHNOLOGY OF TANTRA

Most of our authors' discussion of gross and subtle minds focuses on a Vajrayāna tantric model that is formulated from the perspective of what the Tibetan traditions call highest yoga tantra (Tib., *bla na med pa'i rnal 'byor gyi rgyud*). For this account, the main purpose of tantric practice is one that we have discussed before: the complete eradication of the fundamental ignorance that, by virtue of distorting our awareness of reality, causes suffering and prevents us from achieving the full awakening (Skt., *samyaksaṃbodhi*) of a buddha. Recall also that removing that fundamental cognitive distortion involves cultivating the wisdom that uproots ignorance precisely because it directly cognizes the nature of reality without any such distortion. While the various Tibetan traditions differ somewhat on the details, the Vajrayāna tantric insight is that our gross level of experience contains so many distortions that systematically undoing them can take an extraordinarily long time, on the order of "three incalculable eons," according to a traditional estimate. Yet if we can somehow experience wisdom at a subtler level of experience that acts as a foundation for the gross level of experience, that subtle level of wisdom can uproot all the gross confusions subserved by that level of consciousness. According to Vajrayāna theory, that kind of experience, which requires diving down to the subtlest levels of consciousness, enables us to achieve full awakening in "a single body, a single lifetime" (Tib., *lus gcig tshe gcig*).

That subtlest level of consciousness occurs reliably at the moment of death, so tantric technology involves reproducing the experience of death, without actually dying. From the Vajrayāna perspective, the various winds in the body must be manipulated to achieve this feat, but to achieve that level of control, practitioners must first move beyond their "ordinary perceptions and conceptions" (Tib., *tha mal kyi snang zhen*). This is necessary because the processes of perceiving and thinking are constituted by the fluctuations of these winds in various channels, and those fluctuations follow the deeply habituated patterns that manifest as our ordinary identities. Hence, before attempting to manipulate the winds, practitioners must first disrupt those deeply ingrained patterns, and to do so, they engage in the *generation stage* (Skt., *utpattikrama*), the first phase of practice in the Vajrayāna style discussed by our authors. By transforming their sense of identity through elaborate visualizations, recitations, and rituals, tantric practitioners transform their identities in ways that make the energy winds available for manipulation. When they reach that point, practitioners are ready for the next phase of practice, the *completion stage* (*sampannakrama*).

When tantric practitioners engage in the completion stage, they employ techniques to actually manipulate the winds, with the eventual goal of inducing a "dissolution" sequence said to closely resemble the process of dying. In actual death, the physical processes associated with the gross elements "dissolve" or cease to operate, and although the physical body is still present, its coarse functions such as digestion have ceased. As this process continues, only the energy winds remain functional, and as these also dissipate, conceptual and affective processes also shut down. The final stages of death involve just three progressively subtler forms of wind that are inseparable from three progressively subtler levels of consciousness. If awareness can be sustained through this process, various phenomenological appearances are said to occur, and in the final step, all that remains is an extremely subtle form of wind-mind called the *clear light* (Tib., *'od gsal*, Skt., *prabhāsvara*).

In actual death, it is said that all of the winds dissipate, and the clear-light wind-mind that marks the end of this process must also decay; by that point, the process is long-since irreversible, and death is inevitable. In Vajrayāna practice, practitioners use techniques that induce a simulation of this process, without actually allowing it to end in death. In so doing, these practitioners learn to access what is said to be the subtlest level of conscious awareness. Ordinary persons, if they somehow could sustain awareness into the clear light, would still experience that state with the distortions that come from ignorance, but tantric practitioners are said to have the training to see that clear-light wind-mind in its true nature, thus counteracting ignorance at the very subtlest level of consciousness. In a sense, the clear-light wind-mind is the foundation for all other levels of consciousness, so counteracting ignorance at that level has a tremendous impact, potentially leading directly to buddhahood itself.

The clear-light state induced through tantric practice is said to be both extremely subtle and also extremely intense or even "blissful," and these account for its tremendous potential for transformation. Nevertheless, the practice-induced clear-light state is said to be "metaphorical" (Tib., *dpe'i 'od gsal*) in relation to the actual clear-light wind-mind that occurs in true death. For Vajrayāna practitioners, harnessing the extreme subtlety and power of the metaphorical clear-light mind to the task of uprooting ignorance is certainly part of the goal while they are alive, but inducing that state has another purpose: it prepares practitioners for death and the opportunity to bring their contemplative practice into the actual clear-light mind itself.

John Dunne

Further Reading in English

An account of the basic features of Buddhist tantra is presented by Jeffrey Hopkins in *The Tantric Distinction: A Buddhist's Reflections on Compassion and Emptiness* (Boston: Wisdom Publications, 1999).

For an in-depth presentation of the key features of the tantric path, including the stages of dissolution, see Daniel Cozort, *Highest Yoga Tantra: An Introduction to the Esoteric Buddhism of Tibet* (Ithaca, NY: Snow Lion Publications, 2005).

One of the key sources for this part's material is translated and discussed by Christian Wedemeyer in *Āryadeva's Lamp That Integrates the Practices (Caryāmelāpakapradīpa): The Gradual Path of Vajrayāna Buddhism According to the Esoteric Community Noble Tradition* (New York: American Institute of Buddhist Studies, 2008).

15

Gross and Subtle Minds in the Shared Traditions

HAVING SURVEYED THE MENTAL states in the various presentations of minds and mental factors, we turn now to examining degrees of coarseness and subtlety among mental states. If we look at human consciousness broadly, consciousness during life is grosser and consciousness at the time of death is subtle. Even while alive, within a twenty-four-hour cycle, consciousness while awake is gross, consciousness while dreaming is subtler, and consciousness during deep dreamless sleep is even subtler. Also, during the waking state, sense consciousness is grosser and mental consciousness is subtler.

Buddhist texts speak of the way in which, having accomplished single-pointed concentration, one makes the mind progressively subtler and thereby attains higher levels of mind. These consist in mental states of the three realms—the mental states of the desire realm, of the form realm or absorption, [218] and of the formless realm. From among these three, a formless-realm mind is said to be the subtlest—and within this, the mind of the *peak of cyclic existence* is the subtlest—whereas the lower minds are progressively grosser.

The tradition of highest yoga tantra distinguishes between gross and subtle minds on the basis of whether the wind[319] that is the mount of a given consciousness is gross or subtle. According to the texts of the Perfection Vehicle, however, such differentiation between gross and subtle levels is made on the basis of whether a mental state is easy or difficult to identify, whether or not it has a physical sense faculty as its uncommon dominant condition, whether a mind's object is gross or subtle, and whether the mind's object and aspect are clear or not. We will examine each of these in turn.

Among states of mind, those that are difficult to identify are *subtle* and those that are easy to identify are *gross*. We can illustrate this distinction between grossness and subtlety with respect to the four nonphysical aggregates—feeling, discernment, conditioning factors, and consciousness. In the case of feeling, since "the body experiencing pleasure" easily appears to the mind, it is grosser than the subsequent aggregates. Since in thinking "This person has excellent qualities" discernment apprehends a mental image, it is grosser than conditioning factors and consciousness. Conditioning factors concomitant with the mind, which encompasses all the remaining mental factors besides feeling and discernment, are grosser than the primary consciousness because they apprehend an object's attributes, as in the thought "May I be happy." Primary consciousness, since it apprehends just the nature of the object, and since this is slightly more difficult to bring to mind than the previous ones, is considered to be subtler than the previous three of feeling and so on. *Investigating Characteristics* says: [219]

> With regard to the statement that begins "Among the nonmaterial [aggregates]," it says, "Feeling [is more gross] because its way of apprehending is gross," for although it is not physical, it is very easy to realize; thus it adds, "such as in my hand." It says, "Discernment is grosser than the next two" because its functioning is much easier to realize than both conditioning factors and consciousness; discernment apprehends the signs of a man, a woman, and so on. What about conditioning factors in comparison with consciousness? This is gross, because it realizes very clearly the functioning of attachment and so on. Consciousness is the final one because it is the subtlest of all. The reason for this order is that it accords with guiding the trainees' understanding.[320]

In the above manner, the texts explain the relative grossness and subtlety among the five aggregates.

Further, this way of distinguishing the difference between gross and subtle from the viewpoint of being easy or difficult to identify can be applied to the three types of feeling: pleasant, painful, and neutral. Among these three, both pleasant and painful feelings are grosser, whereas neutral feeling is subtler. Since pleasant and painful feelings are associated with benefit and harm for persons who have them in their minds, they are easier to notice, and therefore such feelings are grosser. Neutral feelings, not being associated with benefit and harm for persons who have them in their minds, cannot be noticed clearly as in the previous two cases, and therefore neutral feelings are subtler. [220] *Investigating Characteristics* says:

> The mental consciousness can be concomitant with all three types of feeling—the capacities for mental comfort, mental discomfort, and equanimity. Since the state of feeling in death, transmigration, and rebirth is unknown, it says that the feelings "at death, transmigration, and rebirth are neutral." Mental comfort and mental discomfort, insofar as they immediately help and hinder, are apparent when analyzed. Thus "the other feelings are clearly evident" refers to mental comfort and mental discomfort.[321]

Now, to present the second way of distinguishing degrees of subtlety, we can again use the example of feelings. While the feelings that depend on the five physical sense faculties such as the eye are not physical themselves, they rely on a physical sense faculty as a basis; therefore they are grosser. Feelings that accompany a mental consciousness are also not physical, but since they rely on a mental sense faculty as a basis, they are subtler. *Clarifying the Meaning of the Treasury of Knowledge* says: "The gross level of feelings and so on are those that are included among the four aggregates—feelings and so forth—that are based on the five sense-faculties, but because they are nonphysical, they have no grossness within them as such. The subtle level are those that arise from the mental faculty because even their basis is nonphysical."[322] This shows the difference between gross and subtle in terms of whether or not a given mental state depends on a physical sense faculty as its dominant condition. [221]

The texts also speak of a way of distinguishing a difference between gross and subtle in terms of grossness or subtlety of the object of mind. The *Compendium of Ascertainments* says: "An awareness knowing reality is subtle indeed because it realizes the subtle dimension of the object."[323] This explicitly states that in the case of an object difficult to realize (the subtle object reality), its cognizing subject (the consciousness knowing reality) is also subtle. It thereby implicitly indicates that, in the case of an object easy to realize, the subjective mind is also grosser.

There is also, in the case of the mind during life, a further way of distinguishing between gross and subtle in terms of the clarity of the object and its mental image. Take, for example, the process of falling asleep. Through the force of an inward-turning mental factor, the mind is deprived of control and loses its ability to act as the immediately preceding condition for sense consciousness. As a result, sense consciousness ceases. Then the gross mental consciousnesses cease in stages. Finally, from within a mental consciousness with an unclear object and mental image, the mind of sleep occurs. This mind is subtler than the mind when it is awake. Furthermore, even within the mind of sleep, the mind when dreaming is grosser, and the mind during deep sleep is subtler, since the object and the image are even more unclear.

Moreover, in the case of ordinary beings, the potency of the four elements of the body decreases during the time of dying owing to the force of either wind, bile, or phlegm. The heat gradually gathers so that the sense consciousnesses cease, and finally only the mental consciousness remains. The gross mental consciousnesses too gradually cease. [222] And when the subtle mind of death finally manifests, this is the actual death. The subtle mind of death is considered to be the subtlest mind of the basic state [i.e., the uncultivated state of existence].

In general, only neutral feelings accompany the subtle mind of death. Although the Vaibhāṣikas assert that the subtle mind of death may be any of the three kinds—virtuous, nonvirtuous, or neutral—the higher Buddhist schools, the Cittamātra and the Madhyamaka, assert this to be only neutral. Also the *Treasury of Knowledge* says:

> During death, transference, and rebirth, they are neutral.[324]

This quote explains that from among the three—pleasant, unpleasant, and neutral feelings—the feeling that accompanies the subtle mind of death is only neutral. Further, the *Treasury of Knowledge Autocommentary* says:

> "When death is gradual, the four." When death takes place gradually, four faculties cease—body faculty, life-force faculty, mental faculty, and the faculty of equanimity; they do not cease

> separately. One should know that it is in such a manner that afflictive and neutral minds [cease] at the point of death. When in a virtuous mind state at the time of death, "In a virtuous [state], all [those] and the five," which means that when in a virtuous mind state at the point of death, all the above-mentioned faculties and the remaining five such as faith cease, for they most certainly exist in virtuous minds.[325] [223]

The *Autocommentary* explains that in the Vaibhāṣika tradition, a virtuous, an ethically neutral, and a nonvirtuous afflictive mind can serve as the mind of dying. *Levels of Yogic Deeds*, however, says:

> *Question:* How about having virtuous minds during death and transference?
>
> *Answer:* It is like this. When dying, those who are faithful remember earlier virtuous dharmas, or another person may cause them to remember. At that time, the dying person mentally engages in any virtuous dharma, such as faith, for as long as there is gross discernment. When discernment becomes subtle, virtuous minds are blocked, and the state of mind is only neutral. For at that time, that dying person cannot make any effort in the virtue previously cultivated, and even another person cannot cause them to remember it.
>
> *Question:* How about having nonvirtuous minds during death and transference?
>
> *Answer:* It is like this. When dying, those who are accustomed to nonvirtuous dharmas remember them, or another person may cause them to remember. At that time, the dying person mentally engages in any nonvirtuous dharma, such as attachment, for as long as there is the activity of gross discernment. This is in every respect the same as the above case of virtue.[326]

Here Asaṅga states that no matter how accustomed one is to virtue and so on before the time of death, at the time of the subtle mind of death, the mind is ethically neutral. [224]

Also, in terms of the realms, the mind of the desire realm is considered to be gross because it has both inquiry and analysis, and it has feelings that are pleasant, unpleasant, and neutral. Compared with that, the mind of the first absorption is subtler because it is free of painful feeling. The mind of the second absorption is even subtler because it is free of both inquiry and analysis. The mind of the third absorption is subtler than that because it is free of joy. The mind of the fourth absorption is free even of happiness and is endowed only with neutral feeling, so it is even subtler than the third. As for the four formless states, they are totally free of any discernment of form and are posited in terms of the order of subtlety of their objects: the first observes limitless space, the second observes limitless consciousness, the third observes nothing at all, and the final one—the mind of the peak of cyclic existence—is said to be the most subtle of all since its object and mental image are unclear. As it says in the *Treasury of Knowledge Autocommentary*: "Regarding the statement 'The subtle that follows upon the subtle,' the discernment that occurs during meditative absorption at the peak of cyclic existence is subtle."[327] Therefore, with regard to the desire-realm mind, the four absorptions, and the four formless-realm minds, the earlier ones are grosser and the later ones are subtler.

Furthermore, the scriptures teach that the outwardly distracted mind is drawn inward in reliance upon calm abiding. As the result of being freed from the faults of subtle and gross laxity and excitation, the mind becomes progressively subtler. [225]

16

Gross and Subtle Minds in Highest Yoga Tantra

ONE OF THE MOST important presentations on psychology found in Indian Buddhist sources emerges in the treatises of highest yoga tantra in general, and in particular in the cycle of teachings pertaining to *Guhyasamāja Tantra*—as outlined, for example, in Nāgārjuna's *Five Stages*, Āryadeva's *Lamp on the Compendium of Practices*, Nāgabodhi's *Presentation of the Guhyasamāja Sādhana*, and Nāropa's *Clear Compilation of the Five Stages*. These texts present many points that do not come up in the explanations of psychology found in the scriptures common to the Buddhist Hīnayāna and Mahāyāna traditions. For example, these texts explain that the consciousnesses have wind as their "mount" or medium and that, since at the subtle level wind and mind constitute the same entity, matter and consciousness constitute the same entity at the subtlest level. Similarly, many of the activities of consciousness are explained in terms of its medium, the wind. Also, if we take the example of something like the relationship between the external objects and the internal mind, there are explanations of the relationship between the internal and external elements from the perspective of the functioning of the winds. [226] Since these and other explanations are taught in such detail, here we offer a brief explanation about the presentation of mind as found in the texts of highest yoga tantra.

In general, the texts of highest yoga tantra distinguish consciousness into two types: adventitious consciousness and primordial consciousness. *Adventitious consciousness* refers to several types of consciousness: the sense consciousnesses such as the visual consciousness, which depend on a physical sense faculty; the gross mental consciousnesses such as recollections of something from the past, present, or future, which depend on the mental sense faculty; the root and secondary mental afflictions; and the minds of the *three luminosities*. Such adventitious consciousnesses exist when there is a gross bodily basis, so when the gross bodily basis ceases, the adventitious consciousnesses also gradually cease. [The second type of consciousness, *primordial consciousness*, is not adventitious.] It is explained that, just as water and its wetness are inseparable, so too both the very subtle primordial innate mind of clear light and the very subtle wind that is its mount abide inseparably at all times within sentient beings.

Moreover, the gross adventitious consciousnesses arise from that primordial innate consciousness, which has arisen continuously without beginning. They occur on the basis of the gross life-sustaining wind until the subtle mind of death manifests. During the death process, the gross adventitious consciousnesses that are based on the gross life-sustaining wind gradually become subtler [227] and dissolve into the very subtle primordial innate consciousness itself. The state during the time of abiding within the subtle primordial mind is called the *clear light of death*. After that, the gross winds once again manifest out of the subtle life-sustaining wind, which is the basis of the primordial consciousness. Simultaneously, the consciousness having exited from the physical basis of this life establishes the intermediate state (known in Tibetan as the *bardo*), and the gross adventitious consciousnesses arise from the primordial consciousness. All this is stated in the tantric texts.

Now if we take a human being as an example, when the external breathing stops at the time of death, the blood flow to the brain is halted, and the brain stops functioning. According to general medical accounts, the person is acknowledged as having died. However, the texts of highest yoga tantra, or secret mantra, say that what happens next is a sequence of gradual dissolution: first into the mind of *luminous appearance*, then the mind of *luminous radiance*, then the mind of *luminous imminence*, and finally the dawning of the very subtle consciousness—the mind of the *clear light of death*. At this point, all the sense consciousnesses have ceased, and given that all those varied appearances normally experienced by ordinary gross mental cognitions have also ceased at this point, it must be accepted that such objective appearances no longer remain.

This way of identifying the stages of these gross and subtle consciousnesses is mentioned in such texts as Nāropa's *Clear Compilation of the Five Stages*:

> Entities have a twofold mode of being:
> that of the mind and that of the body. [228]
> In terms of their states,
> there is the coarse, the subtle, and the very subtle.
> Their common mode of being is their indivisibility.[328]

In terms of consciousness in general, the sense consciousnesses are gross, the eighty conceptions and the root and secondary mental afflictions are subtle, and the minds of the four empty states are very subtle. Furthermore, it is said that, among those states, the fourth state, the mind of the clear light of death, is very subtle. In terms of the mental consciousness, it is said that thoughts that are part of the eighty conceptions are gross; the first three empty states at the time of the basis—namely, the luminous appearance, the luminous radiance, and the luminous imminence—are subtle; and the fourth empty state—the mind of the clear light of death—is a very subtle mind.

How these consciousnesses become gross or subtle is also determined by the winds that are the basis of each consciousness, inseparably mixed with them as one entity, being gross or subtle or moving to a greater or lesser degree. That consciousness has wind as its mount is stated as follows in Nāgārjuna's *Five Stages*:

> It is the mount of consciousness.[329]

This means that it is the wind that makes consciousness move. [229]

In what sense is consciousness moved by the wind? The texts explain that without relying on the wind, consciousness has no ability to engage with an object; it can engage with an object only if it is activated along with the wind. For example, Āryadeva's *Lamp on the Compendium of Practices* says, "Although that is so, [the three types of consciousness] have appearances; thus they function when they are conjoined with the wind element."[330] This means that although the three—luminous appearance, luminous radiance, and luminous imminence—are not physical, because they have objects, they fluctuate with the wind. Therefore "engaging with an object by way of the wind" means that the mind engages with an object by mounting the wind. Within that there are two possibilities according to whether or not the wind diverts the mind to the object.

How wind and mind, which are inseparably mixed, function with regard to an object is expressed in Āryadeva's *Lamp on the Compendium of Practices*: "All supramundane activities are to be fulfilled by means of the subtle element and the minds of luminous appearance and so on, since they are nonphysical, having become mixed, like butter poured into butter."[331] Here it states that since they are mixed like butter poured into butter, wind and mind perform their activities regarding the object by way of constituting the same entity. Also, when the gross winds have gradually dissolved and are functioning with increasing subtlety, then the object and the mental image within the consciousnesses dependent on them likewise become gradually more subtle too. [230] The *Vajra Garland Tantra* says:

> The winds gradually disintegrate,
> the former into the latter ones respectively,
> dissolving into their very own nature
> and then again into the *mucilinda*.[332]

At the time of death, the preceding winds gradually dissolve into the subsequent ones and then continue to dissolve until reaching the subtle life-sustaining wind, here called *mucilinda*. Āryadeva's *Lamp on the Compendium of Practices* says: "Having done all this in the body, they finally enter into the indestructible and . . ."[333] At the final moment of death, the winds having dissolved into the very subtle indestructible wind at the heart center, one enters into clear light that is an absence.

According to the highest yoga tantra tradition in general, when each of the four elements—earth, water, fire, and wind—within one's own continuum dissolves into the next, the movement of wind lessens. Finally, when the winds that move the indicative conceptions together with consciousnesses dissolve into the first luminosity, then the indicative conceptions cease, owing to which the three empty states (the minds of luminous appearance, luminous radiance, and luminous imminence) and the fourth empty state, the mind of the clear light of death, arise in that order. They appear because the movement of each wind has gradually become weaker than the preceding one. Then, at the time of birth, when a tiny movement of wind emerges from the wind of clear light, the luminous imminence, the luminous radiance, and then the luminous appearance arise, because each wind has progressively become stronger than the preceding one. [231]

When, out of the wind of the luminous appearance, the movement of wind becomes stronger, then the indicative conceptions arise.

The Eighty Conceptions Indicative of the Minds of the Three Luminosities

According to the Guhyasamāja cycle of teachings, the three luminosities are, as we have seen: (1) the luminous appearance, (2) the luminous radiance, (3) the luminous imminence. In the context of explaining these three very subtle minds, a presentation is made of their corresponding eighty conceptions, the consciousnesses indicating the three luminosities, which are: the 33 in the nature of the luminous appearance, the 40 in the nature of the luminous radiance, and the 7 in the nature of the luminous imminence. These 80 conceptions, as well as the 108 conceptions similarly mentioned in *Vajra Garland Tantra*, are very subtle consciousnesses because they depend on the slightest movement of their own mount, the principal subtle life-sustaining wind. How these conceptions are moved by the wind that is their mount is explained in Nāgabodhi's *Clarifying the Meaning of the Five Stages*, which says: "To present the response to the opponent's objection, 'Wind divorced of consciousness cannot know an object,' [Nāgārjuna's *Five Stages*] states, 'It is the mount of consciousness,' where the word *it* refers directly to the wind. [232] So it is through being mounted on a wind that the respective consciousness knows an object."[334]

As for the identification of each of the eighty conceptions, first the thirty-three conceptions in the nature of the first *luminous appearance* are listed in *Five Stages* as follows:

> Their intrinsic natures elaborated, I will explain:
> nondesire, moderate [nondesire],
> and great [nondesire],
> mental going and coming,
> the three sorrows [lesser, moderate, and great],
> likewise, peace, conceptualization, fear,
> moderate fear, great fear,
> craving, moderate craving,
> great craving, grasping,
> nonvirtue, hunger, thirst,

feeling, moderate feeling,
instances of great feeling,
cognizer, cognizing, and basis [cognized],
fine investigation, a sense of shame,
compassion, the three degrees of affectionate love,
hesitation, accumulation,
jealousy; such are
the thirty-three intrinsic natures,
the self-knowing minds of embodied beings.[335] [233]

These are enumerated as follows: (1–3) are called *nondesire*, which is a state of not wanting an object and which has three degrees—lesser, moderate, and great; then (not counted separately) there is *mental going*, which is the mind going out toward an external object, and *mental coming*, which is the mind engaging an internal object; (4–6) are *sorrow*, which is the mental torment when separated from what one loves and which has three degrees—lesser, moderate, and great; (7) is *peace*, which is the mind in a peaceful state or the mind resting naturally; (8) is *conceptualization*, which is a state of excitation or the mind thinking coarsely; (9–11) are *fear*, which is the mind afraid of encountering unpleasant circumstances and which has three degrees—lesser, moderate, and great; (12–14) are *craving*, which is mental adherence to an object and which has three degrees—lesser, moderate, and great; here "craving" does not have a joyous aspect—unlike *desire*, listed among the forty conceptions in the nature of the luminous radiance—but it is called *craving* since it yearns for its object; (15) is *grasping*, which is a tight mental holding on to objects of sense desire; (16) is *nonvirtue*, which is one's own mental aversion, or mental ambivalence, toward virtuous action; (17) is *hunger*, which is a desire for food; (18) is *thirst*, which is a desire for drink; (19–21) are *feelings*, which are pleasant, unpleasant, or neutral, and of which there are three degrees—lesser, moderate, and great; (22) is the *cognizer*, (23) is the *cognizing*, and (24) is the *cognized*—these three being three conceptualizations; (25) is *fine investigation*, which is the analysis of what is correct and not correct; (26) is *a sense of shame*, which shuns wrongdoing for personal or religious reasons; [234] (27) is *compassion*, which is wanting beings to be free from suffering; (28–30) is *affectionate love*, which is wanting to protect the object of focus, holding it as beautiful, and wanting to be with it, and which has three degrees—lesser, moderate, and great; (31) is *hesitation*, which is a mind of uncertainty; (32) is *accumulation*, which is a mind wanting to collect things; (33) is *jealousy*, which is a mind disturbed by other people's success.

According to this method of counting the thirty-three conceptions in the nature of the first *luminous appearance*, *mental going* and *mental coming* are not counted because they are common to all.

As for the forty in the nature of the second, *luminous radiance*, it says in *Five Stages*:

> Desire, adherence, joy,
> moderate joy, great joy,
> rapture, ecstasy,
> amazement, excitement,
> satisfaction, embracing,
> kissing, sucking,
> stability, diligence, pride,
> activity, stealing, power,
> enthusiasm, foolhardiness,
> moderate foolhardiness,
> great foolhardiness, vehemence,
> flirtation, animosity,
> virtue, verbal clarity, truth,
> untruth, certainty,
> nongrasping, giver,
> exhortation, bravery,
> shamelessness, pretense, malice,
> brutality, trickery; [235]
> any of the forty in the nature [of *luminous radiance*]
> is a moment of the very empty.[336]

They are enumerated thus: (1) is *desire*, which is a mental attachment toward an unattained object; (2) is *adherence*, which is a mental attachment toward an attained object; (3–5) are minds of *joy* having seen an attractive object, of which there are three degrees—lesser, moderate, and great; (6) is *rapture*, which is a mind of bliss having obtained an object of desire; (7) is *ecstasy*, which is a rapturous mind's experiencing again and again; (8) is *amazement*, which is thinking of something that has never occurred before; (9) is *excitement*, which is a mind distracted upon seeing something attractive; (10) is *satisfaction*, which is a thought contented with the object; (11) is *embracing*, (12) is *kissing*, and (13) is *sucking*, which are minds desiring to do those three activities; (14) is *stability*, which is an unfluctuating continuum of mind; (15) is *diligence*, which is applying oneself to virtue; (16) is *pride*, which is thinking oneself superior; (17) is *activity*, which is completing the usual activities; (18) is *stealing*, which is wanting to steal wealth; (19) is *power*, which is wanting to destroy the forces of others; (20) is *enthusiasm*, which is a mind cultivating the path of virtue; (21–23) is *foolhardiness*, which is a mind wanting to engage in nonvirtue out of conceit and which has three degrees—lesser, moderate, and great; (24) is *vehemence*, which is wanting to dispute with holy beings without cause; (25) is *flirtation*, which is wanting to be playful or put on airs when seeing someone attractive; (26) is *animosity*, which is a mind of resentment; (27) is *virtue*, [236] which is wanting to make effort in virtuous actions; (28) is *verbal clarity*, which is wanting to speak so that others understand; (29) is *truth*, which is wanting to speak in a way that does not pervert the meaning and allows others to discern it; (30) is *untruth*, which is wanting to speak in a way that perverts the meaning and prevents others discerning it; (31) is *certainty*, which is fully trusting in the meditational deity; (32) is *nongrasping*, which is not wanting to hold on to the object; (33) is being a *giver*, which is wanting to give away possessions; (34) is *exhortation*, which is wanting to motivate others who are lazy; (35) is *bravery*, which is wanting to be victorious over the enemy that is the mental afflictions and so on; (36) is *shamelessness*, which is engaging in nonvirtue rather than shunning it for personal or religious reasons; (37) is *pretense*, which is deceiving others with fabrications; (38) is *malice*, which is training in evil views; (39) is *brutality*, which is injuring others; and (40) is *trickery*, which is being dishonest.

As for the seven in the nature of the third *luminous imminence*, it says in *Five Stages*:

> Instances of indifference,
> forgetfulness, error,
> nonspeaking, disenchantment,
> laziness, and doubt.[337]

First (1) is *indifference*, which is in between—neither wanting nor not wanting an object; (2) is *forgetfulness*, which is deterioration of mindfulness; (3) is *error*, which is a thought such as apprehending a mirage to be water and so on; (4) is *nonspeaking*, which is not wanting to speak; (5) is *disenchantment*, which is mental weariness; (6) is *laziness*, which is to have no delight in virtue; (7) is *doubt*, which is to be of two minds. [237]

This division of conceptions into three groups can be further subdivided into three levels of subtlety based on whether, through the force of winds, the strength of the conceptions' movement toward an object is lesser, middling, or great. Regarding that, in the reverse order, from the clear light there arises the luminous imminence, from that there arises the luminous radiance, and from that there arises the luminous appearance. Following upon which there is an increasing movement of the winds, resulting in the arising of those conceptions in the nature of the three luminosities. With regard to those conceptions, when any of the thirty-three arise, it is the trace of the luminous appearance; when any of the forty arise, it is the trace of the luminous radiance; and when any of the seven arise, it is the trace of the luminous imminence. Therefore it is taught that the conceptions are the results of the three luminosities.

Calling them "the natures of the three luminosities" means they characterize the three luminosities. As for how they characterize the three luminosities, they do not characterize them by virtue of being the same thing as them, in the way that being "bulbous" is taken to be a defining characteristic of a pot. Rather, as in "a house with a raven" where the presence of the raven is characterizing the house, an ontologically distinct characterizer is posited as the defining characteristic. Moreover, the fact that the conceptions in the nature of the three luminosities have three levels of movements of wind (lesser, moderate, or great) indicates that their causes, the three luminosities, also have three levels of movements of wind (lesser, moderate, or great). The differentiations of the conceptions into three classes is not due to the differences in their objects or their way of apprehending their objects; rather, it is due to the difference in the force of the winds having three levels of strength (lesser, moderate, or great) to move the conceptions toward their objects. And these differences in the force of the winds arise owing to the minds of the three luminosities. The indicative conceptions [238] include all three ethical valences—virtuous, nonvirtuous, and neutral. And since they include both conceptual and nonconceptual mental states, they should not be understood to refer only to conceptual thoughts that apprehend their objects as associable with words.

Presentation of Subtle and Very Subtle Mind and Wind

As for identifying subtle and very subtle minds and winds, the luminous appearance, the luminous radiance, and the luminous imminence, along with the winds that are their mounts, are considered to be subtle; and the primordial mind with the wind that is its mount is considered to be very subtle. However, in general, the three types of consciousness—the luminous appearance, luminous radiance, and luminous imminence—are very subtle. The reason they are considered to be very subtle is that (a) their dominant condition, the mental sense faculty, is subtle, and, (b) they are mental conceptions wherein gross appearances have subsided in dependence on the very slight movement of their mount, the subtle life-sustaining wind.

Both the subtle mind and the subtle wind operate inseparably as one entity. Consciousness is the part to which the object appears, and wind is the part that moves toward the object. Having given the example of how a cripple with clear eyesight and a blind man with strong legs can travel as a pair from one place to another, the Kashmiri master Lakṣmī mentions in his *Commentary on the Five Stages* that in the case of the consciousness, moving and traveling is done by the wind that is its mount and seeing the object is done by the consciousness. There is no way for these two to be separated, as we explained in the first section.

Five Stages says:

> Having explained these three types of mind—
> the luminous appearance, the luminous radiance, [239]
> and the luminous imminence . . .[338]

Thus the defining characteristics of the three—the luminous appearance, the luminous radiance, and the luminous imminence—presented in sources such as the above excerpt are as follows. The definition of the *mind of luminous appearance* is: a mental consciousness that occurs between the movement and dissolution of conceptions that has an inner sign of there not appearing any gross dualistic appearances apart from the dawning of a purely empty white appearance, like a clear autumn sky filled with moonlight. As for the etymology, it is called *luminous appearance* due to its appearing like moonlight when the winds that move the conceptions have dissolved. And it is called *empty* as well because it is empty of the eighty conceptions together with their winds. Āryadeva's *Lamp on the Compendium of Practices* says:

> What are the defining characteristics of the *luminous appearance*? Without body or speech, it has a nature without a shape. Having the nature of luminosity, it perceives all things without exception, just as the illumination from moonbeams pervades the stainless autumn sky. Thus it is the luminous appearance. It is the first emptiness, the luminous appearance of the wisdom that is the ultimate mind of enlightenment.[339] [240]

The definition of the *mind of luminous radiance* is: a mental consciousness that occurs between the movement and dissolution of conceptions that has an inner sign of there not appearing any gross dualistic appearances apart from a purely empty red appearance, like a clear autumn sky filled with sunlight. As for the etymology, it is called *luminous radiance* due to its appearing very radiant, like sunlight; and it is called *very empty* as well because it is empty of the luminous appearance together with its winds. *Lamp on the Compendium of Practices* says:

> What are the defining characteristics of the *luminous radiance*? Without body or speech, it has no shape and is free of subject and object. It is very clear and perceives all things without exception as stainless in nature, just as the illumination from sunbeams pervades an autumn sky. It is the second completely excellent mind of enlightenment.[340]

The definition of the *mind of luminous imminence* is: a mental consciousness that occurs between the movement and dissolution of conceptions that has an inner sign of there not appearing any gross dualistic appearances apart from the dawning of a purely empty black appearance, like a clear autumn sky filled with the dense darkness of space. As for the etymology, it is called *luminous imminence* because it appears very unclear, like the darkness of space at midnight, and because it is drawing near to the clear light. And it is called *great empty* as well because it is empty of the luminous radiance together with its winds. [241] Because it is unclear, it is also dubbed as *ignorance*—a name of one facet of the luminous imminence. *Lamp on the Compendium of Practices* says:

> What is the *luminous imminence*? Likewise, without body or speech, it has the form of an absence, as is characteristic of space, naturally pervading like the midnight darkness; subtle and selfless, with life-force stilled, it is motionless; without awareness, unmoving, an entranceway dependent on the seed syllables of speech, it is called *thoroughly established*. This luminous imminence, which has the characteristic of ignorance, is the great empty. These are explained as the three types of consciousness.[341]

When the black luminous imminence dawns, conditioned by the winds having abruptly ceased, the object that appears arises in the image of blackness, like a very pristine autumn sky filled with the blackness of midnight. In the earlier part of this stage, the mind has mindfulness; however, as mindfulness of subjectivity gradually diminishes, mindfulness ceases to exist.

To explain the very subtle primordial mind, Āryadeva's *Discernment of the Self-Consecration Stage* says:

> Earth dissolves into water,
> water dissolves into fire,
> fire enters the subtle element,
> and wind dissolves into mind;
> mind dissolves into the mental factors,
> and mental factors dissolve into ignorance, [242]
> and that too dissolves into clear light,
> whereupon all three saṃsāric existences completely cease.[342]

As this passage states, when its cause, the wisdom of luminous imminence, along with its mount, the winds, have dissolved, the clear light dawns. One awakens out of the later part of the luminous imminence that is devoid of recollection, and without even the slightest element of gross dualistic appearance, an utterly pure and completely empty appearance arises. This resembles the natural color of the sky at dawn without any of the three faults—sunlight, moonlight, and darkness—tainting the pristine autumn sky. This mental consciousness of the basic state that is supported by its mount, the very subtle primordial wind, is posited as the extremely subtle primordial mind of the basic state. It has the potentiality to give rise to all good qualities and faults. It is present in every sentient being's continuum, flowing continuously since beginningless time without a break for even a single moment. Its nature remains untainted by pollutants. Since the pollutants arise from inappropriate attention as their condition, they are adventitious, in that they are separable from the clear-light mind. This primordial mind is the subtlest of any mind of the basic state.

The very subtle primordial mind and its mount—the very subtle wind that is inseparable from that primordial mind—are a very subtle and inseparable wind and mind, which are together referred to as the *primordial body-mind*. The wind that is the mount of the primordial mind is called the *life-sustaining wind*. In general, [243] of the two types of life-sustaining wind—gross and subtle—the wind that is the mount of the primordial mind is the subtle life-sustaining wind. The life-sustaining wind that is the basis of the extremely subtle mind is explained as dwelling at the heart center. The heart center referred to here, in the case of a human being, is situated at the central point between the two breasts and closer to the back. At this point, there are knots around the central channel, both above and below, which are knots formed by the *lalanā* and *rasanā* (the right and left channels). Between these knots and within the central channel dwells the indestructible drop. This is what is being referred to as the "heart" in this context.

The gross life-sustaining wind during the basic state flows in and out of the nostrils and dwells in the area of the channel at the heart center, though it does not dwell inside the *dhūtī*, the central channel, at the knot at the heart center explained above. The subtle life-sustaining wind is inseparable from the very subtle mind of clear light; being the mount of that mind, it is said to be a very subtle wind that is indestructible and that emanates light rays of five colors and dwells inside the *dhūtī*, the central channel, at the knot at the heart center. A condensed explanatory tantra, *Enquiry of the Four Goddesses*, says:

Supremely subtle, merely half the size [of a white mustard seed],
is the form of the drop that is the basis of the mind,
always dwelling in the center of the heart
and radiating magnificent rays of light.[343]

The innate wind that dwells in its own place after the dissolution of the other winds is precisely this very subtle life-sustaining wind. [244]

The Stages of Dissolution of Mind and Wind and Those of Death and Dying

How does the gross mind dissolve into the subtle, and the subtle mind give rise to the grosser levels of mind? First, according to the tradition of the *Kālacakra Tantra*, in the case of a human being with the six elements, when the elements dissolve in stages at the time of death, first the water element puts out the fire element, then the earth element is subsumed into the water element and vanishes, then the water element is dried up by the wind element, then the wind and space elements gradually dissolve and disappear. At that time, when the knot in the central channel has unraveled on its own and the movement of winds in both the right and left channels has ceased, then the winds enter the central channel and dissolve, at which point all object-based conceptualizations will have ceased. The mind of natural clear light, with an aspect of being empty like the sky and free from all limitations of proliferations, appears by its own nature. This is called "the dawning of the clear light of death." For example, it says in the "Method of Accomplishment" chapter of the *Stainless Light: Commentary on Kālacakra*:[344]

> "By water" and so on indicates that fire is destroyed by water at the time of death of those born from a womb in this world of human beings. Therefore, by means of just this meditative concentration, the contemplative should first eliminate fire in the body by means of water. Next, since there is no longer any fire, [245] the earth that has entered the water loses its solidity and dissolves like salt into water. Then the wind, once it has dried up all the water, disappears into space. In this way, the collection of elements quickly disperses. Then the mind—the foundation

> consciousness—should be placed in fire and the final darkness. That is, it should be placed in the central ground, the space element, the form of all aspects that is devoid of objects.[345]

Here "foundation consciousness" refers to the clear-light mind. And in the phrase "fire and the final darkness," "darkness" refers to the luminous imminence at the time of death, and "fire" refers to the clear light of death that arises from the luminous imminence. This clear light of death is experienced by ordinary people, but they are unable to recognize it. Thus it is taught that when the sense consciousnesses are manifesting during the basic state, the sense consciousnesses are gross. Then, from when the sense consciousnesses have ceased until the dissolution of the winds that move the conceptions, the mind is subtle. Next the winds that move the conceptions dissolve, and from this point until the winds that are the mounts of the three luminosities dissolve, the mind is very subtle. And when the clear-light mind arises upon dissolution of the winds that are the mounts of the three luminosities, the mind is extremely subtle. [246]

According to the tradition of highest yoga tantra in general, when a womb-born human being endowed with the six elements dies, the gradual dissolution of the gross consciousnesses occurs as a result of the gradual dissolution of the elements and so on, which function as the basis of the gross consciousness. How the inner and outer signs and so forth appear is explained in the *Guhyasamāja Subsequent Tantra*, which says:

> The elements of the signs are five in number.
> This Bodhivajra has explained.
> The mirage-like is the first,
> the smoke-like, the second,
> that like fireflies is third,
> the fourth is like an oil lamp,
> the fifth is constant clarity
> like a cloudless sky.
> By way of the immutable vajra path,
> they emanate throughout the realms of space.[346]

The stages of dissolution are like this. When the form aggregate begins to dissolve, the earth element becomes less able to function as a basis of consciousness; together with the weakening of the ability of the earth element, the ability of the water element to function as a basis for consciousness becomes clearer and clearer. On account of this ordering, it is said that the earth element dissolves into the water element; however, the earth element does not become the water element. A similar process continues until fire dissolves into wind. The outer signs are said to be that one's limbs become thinner, sunken, and lusterless, [247] one's body seems to be dissolving underground, one's eyes become unclear, and so on. The inner signs, according to the *Guhyasamāja Tantra*, are said to be the arising of mirage-like appearances and so forth. Nāgabodhi's *Presentation of the Guhyasamāja Sādhana* says:

> From that, an adept should know the signs of entering into the clear light in the case of the form aggregate: all the limbs shrink, and the body becomes feeble and loose. When the exalted mirror-like wisdom dissolves, hair-like hallucinations occur. When the earth element stops functioning, the entire body becomes dry. When the eye sense faculty stops functioning, the eyes change and shrink. When physical objects cease, one's body loses its luster and becomes completely emaciated.[347]

The aggregate of feeling dissolves after the form aggregate. At this point, the water element becomes less able to function as a basis of consciousness, whereby the ability of the fire element becomes more obvious. Among the outer signs is the fact that the moisture of the body dries up. For example, the mouth becomes dry, and even the liquid inside the eyes dries slightly. It is also said that the power of movement of the eyes lessens. Inner signs of this, according to *Guhyasamāja Tantra*, are said to be the arising of smoke-like appearances. Nāgabodhi's *Presentation of the Guhyasamāja Sādhana* says: [248]

> When the aggregate of feeling dissolves, physical feeling arising from wind, bile, phlegm, and their assembly is no longer experienced. When the exalted wisdom of equality dissolves, the three types of mental feeling are no longer recalled. When the water element dries up, the saliva, blood, seminal fluid, and so on within the body becomes dry. When the ear sense fac-

> ulty dissolves, external and internal sounds are no longer heard. When sound objects dissolve, the sounds of one's own body are no longer heard.[348]

The aggregate of discernment dissolves after the aggregate of feeling. At this point, the fire element becomes less able to function as a basis of consciousness, and the ability of the wind element becomes more obvious. Outer signs are that, through the gathering of the body's heat, a feeling of heat is no longer experienced. And memory becomes lost in that one is unable to recognize even one's close relatives. Inner signs of this, according to *Guhyasamāja Tantra*, are said to be the arising of appearances like fireflies or scattering sparks. *Presentation of the Guhyasamāja Sādhana* says:

> When the aggregate of discernment dissolves, there is no recognition of sentient beings such as humans and so on. When the exalted wisdom of fine investigation dissolves, one does not remember the names of any beings at all, including one's father, mother, brother, sister, and so forth. [249] When the fire element dissolves, one cannot ingest any food. When the nose sense faculty dissolves, the upper winds, stacked one on top of another, go into the ground. When the objects of smell dissolve, the objects of smell within one's own body are no longer smelled.[349]

The aggregate of conditioning factors dissolves after the aggregate of discernment. At this point, the wind element becomes less able to function as a basis of consciousness. The outer sign is the cessation of the breath, and the inner sign is said to be the arising of an appearance like a flickering oil lamp. The earlier appearance like scattering sparks has become subtler, and only a vivid reddish appearance remains. At this point, since the heartbeat has stopped and the breathing has ceased, the event is commonly identified as death. *Presentation of the Guhyasamāja Sādhana* says:

> When the aggregate of conditioning factors ceases, engagement in all physical activity stops. When the exalted wisdom of accomplishing activities dissolves, recollection of worldly activities and purposes ceases. When the wind element dissolves, the

> ten winds, such as the life-sustaining wind, move from their place of abiding. When the tongue sense faculty dissolves, the tongue thickens and shortens, and its root turns blue. When the physical tastes dissolve, experience of the six types of taste ceases.[350] [250]

After the aggregate of conditioning factors, the grosser levels of consciousness dissolve into the subtler levels of consciousness; this process is explained as follows. When the winds that are the mounts of the eighty conceptions in the nature of the three luminosities dissolve, the first luminosity, called *empty*, dawns, and an appearance arises like moonlight shining in a cloudless sky. Then, after that luminosity has dissolved, the luminous radiance, called *very empty*, dawns, and a reddish or orange appearance arises like sunlight shining in a clear sky. Then, after that luminosity too has dissolved, the luminous imminence, called *great empty*, dawns, and a black appearance arises like a pristine sky pervaded by midnight darkness, and mindfulness also declines and ceases. Then, when the lack of mindfulness due to the black oblivion has cleared, the clear light, called *all empty*, dawns, and an appearance arises like a pristine sky at daybreak that is free of any aspect of darkness, moonlight, or sunlight tainting the sky. This is the actual clear light.

Where the text says "Wind dissolves into consciousness," it means the following. The wind that directly moves the conceptions in the nature of the three luminosities loses the ability to function as a mount of consciousness. Then, because its ability transfers to the wind of the luminous appearance, it is taught that wind dissolves into the luminous appearance but not that all the winds dissolve. At that time, only the subtle innate wind functions as a basis of consciousness; there are no gross winds functioning as a basis. When the elements withdraw, this is when the wind dissolves into the luminous appearance, then dissolves into the luminous radiance, then dissolves into the luminous imminence, and finally dissolves into the all-empty clear light. [251]

The way in which the subtle mind gives rise to the gross minds is explained in *Guhyasamāja Compendium of Vajra Wisdom* as follows:

> Any consciousness that arises from the clear light is itself called *mind*, or *mental cognition*, or *consciousness*. It is the root of all

> phenomena, for out of the nature of the pure and the afflictive arise the dual conceptions of self and others. Wind is the mount of that consciousness; then from wind comes fire; from fire comes water; from water comes earth; from those come the five aggregates, the six bases, and the five objects. All of these are a mixture of consciousness and wind. Then there follows the experience of the clearly active three luminosities that have the nature of the three consciousnesses; and with the luminosities as their cause, the indicative conceptions fully arise.[351]

When a slight movement of wind swells up from the clear light, first the luminous imminence, then the luminous radiance, and then the luminous appearance arise. Then the reverse process of the four occurs—from the flickering like an oil lamp through to the mirage-like appearance. All four occur immediately prior to conception in the womb. However, in terms of the gross elements serving as a basis for consciousness, first wind, fire, water, and earth arise in serial order, then the aggregates of form and so on and the bases, such as the eye sense faculties, arise. [252]

According to the Kālacakra texts, the way in which a human being's body, which is endowed with the six elements and born from a womb, develops should be understood as follows. When the extremely subtle mind of clear light—the mind referred to by the term *foundation consciousness*—emerges after the cessation of the intermediate state, it enters into the center of the egg and sperm generated from the sexual union of the mother and father, whereby the three things—consciousness, egg, and sperm—come together in the mother's womb, and the sentient being's body comes to be formed. In the case of this globule—this mixture of consciousness, egg, and sperm—the earth element of the mother's womb supports it, the water element makes it cohere, the fire element matures it, the wind element increases it, and the space element, the empty aspect of the womb, allows it to grow. This globule is a mixture of consciousness, egg, and sperm in the womb, and its earth element makes it heavy and solid, its water element makes it moist, its fire element makes it mature, its wind element (which is the seed of the ten winds that will come about in the future) causes its body to expand, and its element of space allows the body to grow.

The above tantric explanation, referencing primarily the cycle of teachings of *Guhyasamāja Tantra*, about how the stages of death occur in the basic state, how the grosser stages dissolve into the subtler stages, and how the subtler levels give rise to the grosser levels, applies not only to how birth and death occur in cyclic existence. Even within the cycle of a single day and night, during the three states of waking, dreaming, and deep sleep, there is the process of consciousness moving from the grosser to the subtler levels and back. Likewise, even with respect to the arising of a single moment of consciousness, [253] the texts speak of how one can differentiate between (1) an empty state, as in deep sleep, (2) a state in which consciousness is oriented toward becoming active, as in dreaming, and (3) a state of full manifest consciousness, as in the waking state. Also, the primordial mind of clear light, or the innate state, which is utterly beyond language and conceptualization, beyond subject and object duality, can arise during the basic state. This is pointed out in Buddhaśrījñāna's *Two Stages of Meditating on Reality*:

> The limitless perfect joy of the dharmakāya
> is experienced just for a moment
> at death, fainting, sleep, yawning, or coitus.
> What if one were to meditatively cultivate it?[352]

Even in the case of ordinary beings, when one experiences death, fainting, deep sleep, yawning, and coitus (specifically when experiencing orgasm), there arises "just for a moment"—that is, for a short time—something similar to the naturally arisen primordial mind. [255]

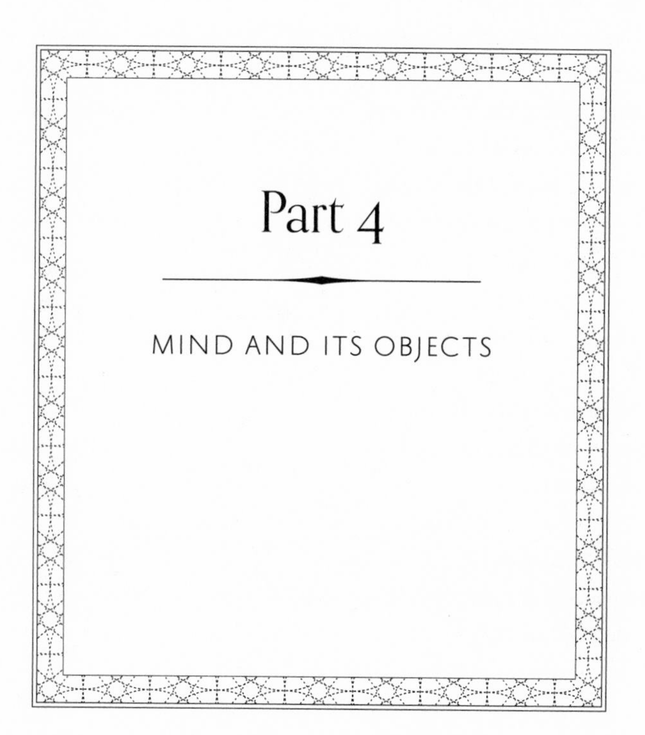

Part 4

MIND AND ITS OBJECTS

BUDDHIST EPISTEMOLOGY

OVER THE CENTURIES, Tibetan scholarship on the Indian Buddhist classics has been greatly facilitated by the creation of literary genres that extract and organize materials on particular topics in the Indian Buddhist texts. The previous discussion of mental factors, for example, reflected the "mind and mental factors" literature (Tib., *sems dang sems byung*), which extracts and organizes materials from an array of Abhidharma texts. In their examination of "mind and its object," our authors' efforts are similarly informed by another such genre known as "mind and cognition" (Tib., *blo rig*).[353] Drawing on the main themes of that genre, our authors now have the opportunity to explore in greater detail some issues we have already encountered. To do so, our authors focus on Indian Buddhist epistemological texts, especially those of Dharmakīrti and his followers. Yet the way that material is structured and interpreted reflects nuances and innovations articulated by Tibetan scholars themselves. Here, we encounter especially the influence of the renowned scholar Chapa Chökyi Sengé (1109–69), whose work on numerous issues figures prominently in this discussion. Our authors note that Chapa's opinions were at points controversial, and his views received especially trenchant criticism from Sakya Paṇḍita Kunga Gyaltsen (1182–1251) and his followers. Our authors remark on these differences at crucial points, and as with other topics treated in this volume, they aim to present a range of theories. Nevertheless, Chapa's approach has advantages for unpacking central aspects of the Buddhist epistemological tradition, and our authors thus organize this part around key concerns for Chapa. In the first half, they examine the varieties of objects, along with related issues such as the role of images in cognition and the nature of conceptuality. In the second half of this part, our authors then turn to a sevenfold typology of cognition. To set the stage for the appreciation of these topics, I will now examine some of their key features.

IMAGES

Although earlier precedents can be cited, the Indian Buddhist epistemological tradition (Skt., *pramāṇa*) is generally traced back to Dignāga and Dharmakīrti in the sixth and seventh centuries. Through the works of various commentators, critics, and interpreters, Dharmakīrti becomes particularly influential for later Indian Buddhism and its transmission to Tibet. To understand the Dharmakīrtian account, it is helpful to begin with the relatively straightforward example of an ordinary person's visual sense perception. Dharmakīrti and his followers hold that sense perception is a causal process, and several conditions are required for visual perception to occur: some material, visible stuff must be present; other external conditions such as adequate light must be involved; various mental factors including a basic level of attention (*manasikāra*) must be active; contact (*sparśa*) between the object and the sense faculty must occur; and so on. When all the requisite preconditions are in place, an *image* (*ākāra*) or phenomenological form of the object is generated in visual consciousness. Dharmakīrti indicates that this image is not just a mirror image of the object, since it varies across individuals, owing to such factors as the acuity of their sensory faculties and their currently active interests and affective states. Simultaneous with this *object image* (*grāhyākāra*), a *subject image* (*grāhakākāra*) must also arise. This subject image accounts for the phenomenal sense of consciousness or knowing that accompanies the experience, and it also is part of the subject-object structure—the sense of "in here / out there." These images of the subject and of the object arise simultaneously in the moment of visual perception, and for a tiny fraction of a second, those images are presented without any categorization or conceptualization. Here, we should add that this involves a particular meaning of "conceptualization" that we will discuss further below. Importantly, our authors note that earlier models in the Abhidharma do not accept the notion that perception is mediated by images or phenomenal forms in this way, but for the epistemological tradition established by Dignāga and

Dharmakīrti, an object image and a subject image are necessarily present in any moment of ordinary consciousness.

Some crucial features of this model may already be evident. First, in the most straightforward account of Dharmakīrti's model, what is directly presented in a moment of sense consciousness is not the visible thing itself. Instead, it is an image or phenomenal form presented in consciousness. When I see an object that I identify as "blue," for example, the blue color that I am directly experiencing is not a thing outside my mind; it is an image within consciousness. Second, my identification of that object as "blue" does not occur in the moment of perception itself; that conceptualization occurs as a perceptual judgment or *subsequent cognition* (Tib., *bcad shes*) that conceptualizes the image *after* the initial perception. The perceptual image and the consciousness in which it occurs are thus completely nonconceptual. A third key feature here is that perceptions do not occur in a vacuum; they are filtered and defined by the interests, goals, and dispositions active in the mind. For Dharmakīrti and his followers, a perception can count as a "valid cognition" only if it can produce a subsequent cognition that provides epistemically reliable information about the object in a way relevant to my goals. Thus, even though my perception is an image in consciousness, my perception must enable me to act on the cause of that image—a causally efficacious object—that is relevant to my goals. In a sense, what I directly see is just an image in my mind, but no organism is just interested in mental images; we wish to encounter opportunities and avoid dangers. If the image caused by the object cannot lead me to act on causally relevant stuff (and not just images in my mind) in ways that enable me to achieve those goals, then perception would be pointless, at least on the Dharmakīrtian account.

One important feature of the Dharmakīrtian system is that it posits a distinction between the image presented in sense consciousness and the object that caused the image. In other words, this model involves a kind of gap between the phenomenal image and its cause (an approach that, in Western philosophy, would be akin to a sense-data theory). The image might be something such as the presentation of a red object in awareness with a particular shape that we conceptualize as an "apple," and the cause would be the actual material stuff that we wish to eat. Some interpreters of Dharmakīrti such as Chapa attempt to close this gap, even to the point that, by reinterpreting Dharmakīrti's notion of the object image in particular ways, the image's role as a mediator—a bridge between the material stuff and immaterial consciousness—is reduced or eliminated. As a result, for these interpreters, sense consciousness ends up engaging more-or-less directly with objects (akin to versions of "direct realism" in Western philosophy). Other interpreters, such as Sakya Paṇḍita, seek to preserve this gap in their interpretations of Dharmakīrti's model of perception. There is thus a spectrum of interpretations, some that seek to close the gap between the object and the object image and some that seek to preserve it. An exploration of this spectrum's full range would be complex indeed, and in the interest of simplicity, our authors decide to favor Chapa's end of the spectrum, where the gap between the object and the object image in consciousness is reduced or eliminated. However, as Georges Dreyfus has noted, Dharmakīrti's earliest Indian interpreters tend to assume a stronger version of this gap, and thus in historical terms, the typology of objects discussed by our authors arose in response to interpreters who assumed a clear gap between object and object image. Let us now examine the typology of objects our authors present, with an eye to clarifying how it seeks to reduce that gap.

Types of Objects

Based on Dharmakīrti's account and fully elaborated by later Indian and Tibetan interpreters, the typology of objects presented by our authors is fourfold: (1) appearing object, (2) observed object, (3) conceived object, and (4) engaged object. The interpretation of these categories varies, and not all Buddhist epistemologists used this typology or accepted it. Dharmakīrti and his earliest interpreters do not explicitly use this typology, and they lack the technical term *appearing object*; however, they do refer to the object image as an "appearance" (Skt., *pratibhāsa*) in awareness. This stands in contrast to the *observed object* (*grāhya*)—the thing that causes an appearance or object image to arise in a moment of perception. Thus, if Dharmakīrti's earliest Indian interpreters were to use this fourfold model, they would say that the first two types of objects—the appearing object and the observed object—should be distinct. But following Chapa, our authors collapse the gap between object and object image, and the appearing object and the observed object are thus synonymous. In lieu of being interpreted as what causes an object image in a perception, the observed object is precisely what is directly apprehended by a sense consciousness, and that object is also the appearing object in that it is cognized by that consciousness by way of appearing to it.

So far, we have been discussing the typology of objects in relation to *perception*, which is necessarily nonconceptual for the Buddhist epistemologists. To understand the third type of object—the conceived object—we must examine *conceptual* cognition. In brief, conceptual cognitions involve mental objects such as the conceptual image of a pot presented in the thought "This is a pot." This conceptualized "pot" is both the appearing object and observed object of that conceptual cognition, since it is what that cognition directly apprehends. That conceptual pot, however, is presented as referring to—or simply being the same as—some real pot in the world. As such, that real pot is the "conceived object" of that conceptual cognition. In other words, it is what that conceptual cognition is guiding us to. In part this means that the thought of a unicorn does not truly have a conceived object, since there is no real thing to which it can refer in this way.

The fourth type of object, the engaged object, is what a cognition is prompting us to act upon, whether through physical actions or additional mental activity. A conceptual cognition prompts us to engage in this way precisely by presenting its mental objects as the conceived object—as some real thing in the world. A perceptual cognition, being nonconceptual, has no conceived object; instead, it directly presents its content—the observed object—as something for us to act upon, and in comparison to a conceptual cognition, a perceptual cognition has a kind of vividness that is especially relevant to prompting actions on its engaged object. From a strictly epistemological standpoint, this category of the engaged object may seem redundant, since it is not directly connected to questions of truth or justification, but Buddhist thinkers are interested in more than just the conditions that make a cognition epistemically reliable. Their theories are also informed by a keen interest in the mechanisms involved in our behaviors and the role that cognitions play in prompting—or changing—those behaviors.

On Concept Formation through "Exclusion"

Already in part 1, our authors pointed out that conceptual cognition is deceptive. On the one hand, we ordinary beings must rely on concepts—as expressed in language and thought—to make our way in the world, yet our thoughts and statements about the world, while useful for engaged action, invariably mislead us. In the introduction to part 1, I noted the two ways in which conceptual cognitions are mistaken. First, they present their content—such as the concept "fire"—as identical to some real fire in the world, but unlike an actual fire, the thought of fire cannot truly burn anything. Second, the concept or thought of fire involves a universal (Skt., *sāmānyalakṣaṇa*), which amounts to an essence or "fire-ness" that characterizes every fire. Yet Buddhist epistemologists maintain that this notion of "fire-ness" is simply a mental construct; in fact, there is no such universal, no entity that is exactly the same in any two things that we call "fire." Every thing that we call "fire" is actually unique in every way, not at all the same, even if by using that concept or expression we somehow are able to successfully achieve our goal of becoming warm and do not confuse those entities with stuff that will make us cold.

If conceptual cognitions are always mistaken, how can they reliably guide action? The key to this question lies especially in the problem of accounting for some sameness that enables us to use a single concept or expression such as "fire" for multiple unique things. If, as Buddhist epistemologists maintain, there is in reality no such sameness, how are concepts or expressions such as "fire" still successful in guiding our actions? The answer to this question is the theory of exclusion (*apoha*) developed by Dignāga and Dharmakīrti with further elaboration by many generations of commentators.

The full details of exclusion theory are complex, and to set the stage for the in-depth summary provided by our authors, it will be helpful to examine the term *exclusion* itself. As we have noted, Buddhist epistemologists reject the notion of any real sameness—what in Western philosophical terms would be called a *universal*—that characterizes every instance of, for example, a fire. Yet even though there is no *real* sameness, our cognitive systems can construct an *unreal* sameness. The key here is that our cognitive systems are not simply constructing concepts out of some bad habit, as Dharmakīrti puts it. Instead, these concepts help us avoid what we think will inhibit our flourishing while also helping us obtain what we believe will promote it. In short, we are concerned with desirable outcomes—identifying something as "fire" is tied to my desire to get warm and to engage with an object capable of making me warm. And with that context in place, my cognitive systems construct something that is, *in practical terms*, the same for all fires, even though there is no sameness in the world.

Although every individual instance of fire is completely unique, our cognitive systems can ignore the variations among those instances and instead focus on the way that each thing we call a "fire" is "excluded" (Skt., *apoḍha*, *vyāvṛtta*) from all things that do not have the expected or desired effects. And this "exclusion of that which does not have the expected goal-oriented, causal capacity" (*atadarthakriyāvyāvṛtti*) constitutes a sameness. Thus, even though there is no real "fire-ness" that characterizes all fires, in my experience each thing I call "fire" is different from other things that don't do what I expect a fire to do. A somewhat tongue-in-cheek way of putting this is to say that, in terms of the things that we call a fire, we can say that they are all the same in that they are "not a non-fire." This statement, however, is not as a simple as it looks, and this process certainly cannot be reduced to a mere logical double-negation, which would be both trivial and pointless. Instead, this notion of exclusion only makes sense when we understand that it falls within a behavioral context, where the formulation and use of concepts (whether expressed through words or not) is ultimately tied to our goal-oriented behaviors that end in a concrete experience of, for example, feeling warm.

The example of fire that I have chosen is a traditional one, but it also has the advantage of easily invoking an experience. Nearly everyone has had the experience of being cold and then seeking some means to become warm, whether with a "coat," a "heater," a "fire," or the like. And while the vast array of things that we can fit under these categories are varied indeed, it is easy to understand how the point of that categorization is to achieve a goal: the question of the exact sameness of every coat or heater is not a practical concern. Of course, we can—and do—develop higher-order concepts that are less concrete in this way, but the claim of the exclusion theory is that even those higher-order concepts emerge out of a more basic system that serves to guide our embodied actions toward the concrete goals that we seek. For again, the thought of a fire (or some other abstraction) cannot make us warm. The claim here is that what counts is the actual, perceptual experience.

Seven Types of Cognitions: A Model of Transformation

Conceptual cognitions involve distortions that do not come into play for perceptual cognitions. This is not to say that perceptions are simply pure, unconstructed encounters with the world. At all levels of analysis, Buddhist epistemologists acknowledge that perception is a highly conditioned process influenced by numerous factors, such as the particularities of our embodiment, the context formed by our expectations, and the acuity (Skt., *pāṭava*) of our attentional and volitional capacities. Still, perception holds a special place in Buddhist epistemology. As we have seen, at a basic level our conceptual system operates in a way that facilitates our goals, which themselves involve concrete perceptual experiences, such as feeling warm. The mere thought of feeling warm is not enough: we want to actually experience warmth. In this way, perception is said to be vivid (*spaṣṭa*) in ways that thought or conceptuality is not. Importantly, this also means that perception can be harnessed to the project of personal transformation, precisely because its vividness includes a visceral, embodied encounter with whatever is being perceived. To take a key Buddhist example, it is certainly laudable to have a conceptual understanding of the notion of personal selflessness (*pudgalanairātmya*), such that I intellectually understand that, even if it feels like I have some unchanging, absolute, and autonomous identity that constitutes my "self," I do not in fact have that kind of self at all. This intellectual understanding of selflessness can be helpful in reforming some of my dysfunctionality, but an actual perceptual experience of selflessness, where I viscerally feel that lack of any such essentialized and unchanging self, will have a much stronger impact on my beliefs and behavior. This insight into the primacy of perception underlies the sevenfold typology of cognition that figures prominently in our authors' discussion.

The sevenfold typology again harkens back to Chapa, who is renowned for presenting it. Here again, there are disagreements among Tibetan philosophers about some of these categories, but our authors choose to set aside those disagreements in favor of a presentation that echoes an analysis made by the present Dalai Lama. In this account, the sevenfold typology traces a kind of developmental arc from being caught in the delusion of, for example, a fixed and essentialized self, and gradually moving toward a transformative perceptual experience of selflessness. The first stage involves (1) a *distorted cognition*, namely, the belief that my identity is immutable, unchanging and autonomous. But then, perhaps under the influence of something I have read or someone I have met, I begin to question this idea. Eventually, a state of (2) *doubt* arises, and since the issue of my own identity is of considerable importance, I begin to study this question. Through my rational analysis, I reach a point of (3) *correct assumption*. My analyses have not uprooted my doubt, but I am starting to recognize that my previous belief in a fixed, essentialized self is untenable. Finally, as I engage with this problem more, I am able to have a truly valid moment of (4) *inference*, where I am fully confident that I have come to the correct conclusion—namely, that selflessness most accurately describes a key feature of my experience.

These first four types of cognitions are all very much in the conceptual realm, and His Holiness's way of interpreting them connects to a typical sequence within Buddhist contemplative practice that involves studying (Skt., *śruti*, literally, "hearing"), contemplating (*cintā*), and meditating (*bhāvanā*). In a sense, the presence of persistent doubt is what gets us started in the serious study of an issue, and as we mull it over in contemplation, we start to get an idea of where our analysis is headed. Finally, we reach a point of conviction in that analysis, which here would correspond to a well-formed inference. But at this point, we are still in the realm of concepts, and we have not had a visceral, perceptual experience that will truly impact our cognitive schema about the self. As a result, our behavior will not change significantly. Instead, we must begin to meditate or focus on the conclusion to our analysis, and this enables us to have (5) a sustained *subsequent cognition*, in which the first moment of our inferential insight into selflessness is sustained in thought. We then apply additional meditative techniques to focus one-pointedly on that sustained insight, and this eventually leads to (6) a *direct perception* of selflessness.

That moment of the *direct perception* of a transformative insight is visceral in ways that the intellectual understanding cannot be, and it recruits our entire embodied experience—not just our thoughts—to the task of change. Dharmakīrti attempts to explain this point by citing cases that many of us would understand. For example, if I wake up in the middle of the night and hear a thief in my house, the visceral reaction that I feel at the moment is dramatically different from merely thinking about a thief coming that night. The point of Dharmakīrti's example is not that a thief is actually in my house; instead, Dharmakīrti means that we wake up, perhaps from a dream of thieves, and then we have this visceral experience. In a similar way, the long and intense process of contemplating selflessness, when combined with meditative techniques, can result in a visceral, perceptual experience that truly reorients our beliefs and behaviors. This stands in contrast to (7) an *indeterminate perception*, which in this context would be a meditative experience that, while vivid and even dramatic, has no visceral impact on our beliefs or behaviors, precisely because it is not emerging from—or contextualized by—a careful inquiry into a crucial issue such as the question of personal identity.

His Holiness the Dalai Lama's insightful way of interpreting the sevenfold typology of cognition sets aside some of its thornier details, including especially the very status of indeterminate perception itself. In a direct and instructive way, it evokes precisely the underlying concern of this model: in what way can our beliefs and experiences be cultivated to reduce suffering and enhance human flourishing?

John Dunne

Further Reading in English

For an account of the Indian Buddhist epistemological tradition's approach to perception, see chapter 2 of John D. Dunne, *Foundations of Dharmakīrti's Philosophy* (Boston: Wisdom Publications, 2004). See also book 2, part II, of Georges B. Dreyfus, *Recognizing Reality: Dharmakīrti's Philosophy and Its Tibetan Interpretations* (New York: State University of New York Press, 1997).

Chapa's "new epistemology" is discussed at length by Pascale Hugon and Jonathan Stoltz in *The Roar of a Tibetan Lion: Phya pa Chos kyi seng ge's Theory of Mind in Philosophical and Historical Perspective* (Vienna: Austrian Academy of Sciences Press, 2019).

For Dharmakīrti's theory of concept formation through *apoha*, or exclusion, see the essays in *Apoha: Buddhist Nominalism and Human Cognition*, edited by Mark Siderits, Tom Tillemans, and Arindam Chakrabarti (New York: Columbia University Press, 2011). The first essay, "Key Features of Dharmakīrti's *Apoha* Theory," offers an accessible overview.

For the background to the Tibetan formulation of the seven types of cognition, see Pascale Hugon, "Tibetan Epistemology and Philosophy of Language," in *The Stanford Encyclopedia of Philosophy*, edited by Edward N. Zalta (Metaphysics Research Lab, Stanford University, 2015), https://plato.stanford.edu/entries/epistemology-language-tibetan/. A translation and study of a well-known Tibetan work in this genre is found in Lati Rinbochay and Elizabeth Napper, *Mind in Tibetan Buddhism: Oral Commentary on Ge-shay Jam-bel-sam-pel's Presentation of Awareness and Knowledge, Composite of All the Important Points, Opener of the Eye of New Intelligence* (Valois, NY: Snow Lion Publications, 1980).

17

How the Mind Engages Its Objects

OBJECTS OF MIND

HAVING COMPLETED THE presentation of types of mind in the preceding chapters, we will now discuss how a mind engages its object. The definition of an object is: that which is known by a mind. When categorized, there are four types of objects of mind: (1) appearing object, (2) observed object, (3) conceived object, and (4) engaged object.[354]

The appearing object and the observed object are synonymous, and every consciousness has both. As for the appearing object and the observed object of a *nonconceptual consciousness*, they are the object of nonconceptual consciousness by virtue of their image clearly appearing to that consciousness. As for the appearing object and the observed object of a *conceptual consciousness*, they are the appearance that is the direct object of a conceptual cognition when its object occurs to it; they are thus the appearance that was explained previously to be the linguistic referent. Even though the direct object is not in itself a real thing (i.e., something involved in a causal process), according to the explanations of some earlier Buddhist epistemologists it is not contradictory for a conditioned thing to be a direct object of conceptual cognition. Thus a distinction needs to be made between the appearing object and the direct object of conceptual cognition. [256]

The *conceived object* refers to the object in terms of how a person or a conceptual cognition conceives it when apprehending it, and it exists in the case of both a conceptual cognition and a person. For example, a pot is the conceived object of a conceptual cognition or a person thinking "This is a pot." In cases where the way of conceiving accords with reality—such as a conceptual cognition apprehending sound to be impermanent—the conceived object exists. But in cases where the way of conceiving does not accord with reality—such as a conceptual cognition apprehending sound to be permanent—the conceived object does not exist. The conceived object and the engaged object are synonymous in the case of a conceptual consciousness. Since nonconceptual consciousnesses do not conceive in such terms as "This is such-and-such," a conceived object is not posited for them.

The engaged object exists in the case of a person and of a cognition that accords with reality. A pot is posited both as the engaged object of an eye consciousness apprehending a pot and of a conceptual cognition apprehending a pot. According to the theory that verbal statements too possess conceived objects and engaged objects, the expressed referents of a verbal statement should be considered to be its conceived object as well as its engaged object.

We may wonder, "Do all instances of awareness have four objects?" No. There are many variations. Some types of consciousness have four objects, whereas others have two or three. A conceptual cognition apprehending a pot has four objects: the *appearance as a pot* to a conceptual cognition apprehending a pot is both the appearing object and the observed object of a conceptual cognition apprehending a pot, and the *pot* is both the conceived object and the engaged object of a conceptual cognition apprehending a pot. An eye consciousness apprehending yellow has only three objects: [257] *yellow* is all three—the appearing object, the observed object, and the engaged object—of an eye consciousness apprehending yellow. A conceptual cognition apprehending a rabbit's horn has just two objects: the *appearance as a rabbit's horn* to a conceptual cognition apprehending a rabbit's horn is both the appearing object and the observed object of that conceptual cognition. Even though a *rabbit's horn* is the conceived object and the engaged object of that conceptual cognition, it doesn't exist even conventionally.

Moreover, in the case of objects, there are also objects as cognized, explicit objects, implicit objects, focal objects, and so forth. A pot is the *object as cognized* for an eye consciousness apprehending a pot and for a conceptual cognition apprehending a pot. Impermanent sound is the object as cognized for a direct perception and for an inferential cognition realizing sound to be impermanent. Likewise, permanent sound is the object as cognized for a conceptual cognition apprehending sound to be permanent, and a rabbit's horn is the object as cognized for a conceptual cognition apprehending a rabbit's horn. Thus, in the case of a cognition that accords with reality, since things are the way it apprehends them, its object as cognized exists. And in the case of a mind that does not accord with reality, since things are not the way it apprehends them, its object as cognized does not exist.

As for the *explicit object* and *implicit object*:[355] The explicit object of a cognition is an object that appears or whose image arises in that cognition. The implicit object of a cognition, [258] although an object of that cognition, does not appear or arise as an image in that cognition. For example, since a conceptual cognition apprehending a pot explicitly realizes the pot and implicitly realizes the exclusion of non-pot, it is said that the *pot* is the explicit object of the conceptual cognition apprehending a pot and the *exclusion of non-pot* is the implicit object of the conceptual cognition apprehending a pot. Likewise, an inferential cognition realizing that sound is impermanent explicitly cognizes sound as impermanent and implicitly cognizes sound as not permanent, so it is said that *sound as impermanent* is the explicit object of that inferential cognition and *sound as not permanent* is the implicit object of that inferential cognition.

As for the *focal object* of that cognition, there are different ways of positing this depending on context. It may be the basis on which the mind acts to eliminate false superimpositions, or it may be the basis or ground on which various false superimpositions are imputed by the mind. For example, sound is the basis on which an inferential cognition realizing sound to be impermanent acts to eliminate false superimpositions, and it is the basis or ground of false superimpositions imputed by a conceptual cognition apprehending sound as permanent. Therefore it is said to be the focal object of both cognitions. Also, both the inferential cognition realizing sound to be impermanent and the conceptual cognition apprehending sound to be permanent must be posited as cognitions whose ways of apprehending are directly contradictory (i.e., as impermanent and as permanent), having observed the same focal object (i.e., sound).

In general, whether an instance of awareness has realized a given object is determined on the basis of whether it has obtained that object. It is those states of mind that remain steady, like a tent stake planted into dry ground, that realize their object. But minds that are unsteady, like a stake plunged into a bog, cannot realize their object.

There are also different modes in which cognition of an object takes place, such as the following: directly cognizing through perception, such as an eye consciousness cognizing the pot in front of it; [259] and cognizing by means of inference or reasoning, such as a mind cognizing from afar the presence of fire on a mountain pass through directly seeing smoke issuing from that place.

Does an Image Appear to Sensory Cognition?

The Vaibhāṣikas say that when an eye consciousness apprehends its object—blue, yellow, and so on—an image does not appear. Instead, the eye consciousness apprehends its object nakedly without an image. The other Buddhist schools, beginning from Sautrāntikas up to the higher systems, say that consciousness sees by means of an image appearing to the mind. Also, the Vaibhāṣikas accept that the object, a visible form and so on, appears to the sense faculty, though since the consciousness lacks form, it cannot have an image of the form appearing in the manner of a reflection. For this reason, they do not accept that an image of the object, such as blue or yellow, appears to the consciousness. Subschools of those who accept that the consciousness bears images explain this in different ways. Some say that while the image appearing to a consciousness is an image in that consciousness, it is not consciousness itself. Others say that because it is an object-image in that consciousness, it is consciousness itself.

Those who accept sense consciousnesses to have an image call the image of the object that appears to the sense consciousness the *object-image*. Consider the example of an eye consciousness apprehending a form, which has two parts—a part that is the appearing image of the external form and a part that is clear cognition experiencing itself. The first is called the *object-image* of the eye consciousness. Since there is no difference between this, the eye consciousness apprehending a form, and an image of form appearing to the eye consciousness apprehending that form, the object-image is the eye consciousness itself. [260] What is meant by the phrase "an image appearing to the eye consciousness" is that the eye consciousness arises from its own cause in the image of the object. There is no separate thing that arises as an image other than the eye consciousness itself. The second—the part that is clear cognition experiencing the eye consciousness—is called the *subject-image* of the eye consciousness. According to those who accept reflexive awareness, there is no difference between this and a reflexive perception experiencing the eye consciousness.

Both the Sautrāntikas and the Cittamātras agree in maintaining that in the phrase "the object-image of an eye consciousness apprehending blue" the *object* refers to *blue* and *image* refers to *the observed object of an eye consciousness apprehending blue that arises in the image of blue.* Nevertheless, they also disagree in the following way. The Sautrāntikas assert that in the case of an eye consciousness having a visible form's image, it arises from the external material form as its cause. The Cittamātras assert that it arises in the nature of an image through the force of latent potencies within the mindstream as its cause. Also, those who accept external objects say that when blue is seen by an eye consciousness apprehending blue, that blue-like image itself that appears to it is projected inwardly by the blue—just as when, for example, the design of a seal is stamped on a piece of paper, that design itself is fixed onto the paper from the side of the seal. For if that similar image itself were not an image projected by the external object, then this would entail the fault that an eye consciousness apprehending blue would arise even in places where there is no blue. Those who do not accept external objects say that the blue-like image itself that appears to the eye consciousness is not projected inwardly by the blue. [261] Instead, like the appearance of blue in a dream, it appears through the force of the ripening of latent potencies within the continuum of the mind.

The Vaibhāṣikas say that forms and so on are seen nakedly by the physical sense faculty that functions as a support of visual cognition and that there is no need for an image to act as a link between the object and subject. So it is not the case that the sense consciousness sees the object. Rather, for example, when a pot is seen, the particular individual pot right in front of one is seen nakedly by the active sense faculty. And when a visible form is seen by an active eye sense faculty, it is said that the eye consciousness apprehends or knows the form, thus a distinction is drawn between "seeing" and "knowing." Sautrāntikas, on the other hand, draw no such distinction between seeing and knowing, for they say such things as "that which knows a visible form is that which sees that form." Thus they consider there to be no difference between seeing and knowing. The Vaibhāṣikas then respond that according to the explanation of the Sautrāntikas and others, if the eye consciousness alone sees visible forms, then since consciousness is nonobstructive there would be the fault that it must see even visible forms that are obstructed by walls and so on. Presenting the views of the Vaibhāṣikas, the *Treasury of Knowledge* says:

> They say: "The eye, acting as a basis, sees forms,
> not the eye consciousness based on it,
> because forms that are behind barriers [262]
> are not seen."[356]

Also, *Clarifying the Meaning of the Treasury of Knowledge* says: "According to those who say it is the consciousness, since the consciousness has no obstructivity, consciousness would arise even regarding something concealed; but it does not arise. So this proves that the eye sees, not the consciousness."[357]

The Vaibhāṣikas say that the image of either the subtle atoms or their aggregation does not appear to the eye consciousness, and that although the eye consciousness does not know subtle atoms individually, it explicitly knows a gross mass of aggregated subtle atoms. For example, we do not see an individual hair or grain of sand when it is placed separately all by itself far away, but we explicitly see a mass of many gathered together. Jitāri's *Distinguishing the Scriptures of the Sugata* says:

> Consciousness that has arisen from a sense faculty
> directly knows aggregated subtle atoms;
> the masters maintain that this is the view
> of treatises of the Kashmiri Vaibhāṣikas.[358] [263]

Also, the same master's *Distinguishing the Scriptures of the Sugata Autocommentary* says:

> Is the position of these [Vaibhāṣikas] the same as that of the Sautrāntikas, for whom consciousness is endowed with images that have an external object as its content? This is not so. Instead, "Consciousness arisen from a sense faculty without images directly knows aggregated subtle atoms."[359] What are called "directly perceiving minds that arise from empowering sense faculties" are minds without an image of the object, so according to them [the Vaibhāṣikas], it is not that [the connection with the object] is mediated by images, as is theorized [by the Sautrāntikas]. Rather, masses that are aggregated subtle atoms or tiny particles are explicitly cognized or known, not the

> wholes as others maintain. One might object: if subtle atoms were beyond the sense faculties, then how could they possibly appear? The answer is that individually they are beyond the sense faculties, but when they come together with others of similar type, they become objects of the sense faculties. It is not the case that what cannot appear separately also cannot appear in a collection, because while individual hairs and grains of sand are not seen from a distance, they are perceived collectively.[360]

Thus the Vaibhāṣikas say that when a sense consciousness realizes its object, it realizes it nakedly without any mediation by an image. In response, the Sautrāntikas and other schools focus on establishing sense consciousness as endowed with an image on the understanding that if the sensory cognitions are proven to possess object-images, then it is easy to understand that other cognitions induced by them [264] must bear an image of the object. Furthermore, they assert that when the sense consciousnesses determine their objects, they do so with images. For if they were to cognize their objects in a clear manner without images, this would imply that, like consciousness itself, those very objects would have the nature of clarity.

Those who accept consciousness to be with images consider that the way in which a sense consciousness cognizes its object is as follows. For example, when one looks at a pure crystal that reflects a color (such as when it is placed on a colored surface), although both the color and the crystal are alike in being seen, the crystal is cognized as appearing from its own side and the color by way of a reflection. The Sautrāntikas and others who accept that sense consciousnesses have images agree that, in this way, an image similar to the object appears without an intermediary to the sense direct perception, whereas visible forms, sounds, and so on appear in the manner of a reflection. When they say, "The image of the object appears to an eye consciousness," they understand this to mean that a specific eye consciousness arises in the image of its corresponding object, and they maintain that such an image of the object that appears to the eye consciousness is nothing separate from the eye consciousness itself. Bodhibhadra's *Explanation of the Compendium of the Essence of Wisdom* says:

> In that case, according to these Sautrāntikas, when a person sees a pure crystal change color according to the hue of something nearby, both the crystal and the color are perceived by the eye—the crystal itself is perceived by direct perception, and the color is perceived by way of a reflection. Thus the person perceives two observed objects. [265] As illustrated by this example, the appearance to perception is an image of consciousness only, while the bases that appear as shape and color to the consciousness are other than the consciousness. Since these bases are asserted to be a collection of atoms with nothing between them and not touching, they maintain that there are two observed objects.[361]

One might ask, "If the Sautrāntikas accept that consciousness is endowed with images, then what is the difference between them and the Cittamātras?" The Sautrāntikas and the Cittamātras do not differ in denying that consciousness lacks images and in declaring that it has images. But it is taught that they disagree about whether there exist, or do not exist, material forms that are bundles of aggregated atoms that have natures different from that of consciousness. This same text of Bodhibhadra says: "If consciousness arises bearing images, how does this [Sautrāntika view] differ from that of the Yogācāra system? It differs in maintaining that it is not the case that forms that are aggregations do not exist. For [this Sautrāntika system] maintains it is not the case that outside of consciousness there are no forms that are bundles of aggregated atoms."[362]

According to the traditions that assert external objects, even though blue and an eye consciousness apprehending blue are different substances, nevertheless blue and the reflection cast by blue [which is in the nature of consciousness], or the image of blue, are similar in that they are cause and effect. Therefore, referring to the arising of an image similar to blue as "blue appearing" is legitimate. [266] They say that just this suffices for explaining that a blue object is also sensed or experienced. Śāntarakṣita's *Ornament for the Middle Way* says:

> According to those who accept consciousness to have images,
> although the two are actually distinct,
> since that thing and its reflection are similar,
> simply through imputation "[the object] is sensed."[363]

This text further says:

> According to those who do not accept consciousness
> to be affected by an image of the object,
> even the externally cognized
> features [of the object] would not exist.[364]

As long as one accepts the existence of external objects, one must definitely accept sense consciousness to arise in the image of the external object. For those who do not accept this, there would not exist any *likeness* that, when it appears clearly, constitutes the clear appearance of the external object. So although they might imagine themselves to be accepting cognition of the external object, they will not be able to posit anything that represents external reality. This is what Śāntarakṣita explains in the above lines. Śāntarakṣita, being a Yogācāra-Mādhyamika himself, accords with the Cittamātra system when it comes to presentations of the conventional level of reality. It is important to understand that the above statement is being made from such a standpoint. [267]

How Do Images Appear to Sensory Cognition?

With respect to the view that when a sense consciousness realizes its object it does so by means of an image, some raise the following objection. In this view, when one is, say, viewing a painting with various shades of color, there must exist images matching those various shades of color. For without an equal number of images, this would mean that cognition does not perceive images commensurate with the object, which in turn suggests that consciousness lacks the ability to realize its object just as it is. Furthermore, since the consciousness cannot be a different entity from the images, if images matching the variety in the object were to exist, then there should also exist an equally manifold consciousness matching the plurality of images.

Historically the following three standpoints emerged in response to this fundamental objection: (1) *Proponents of an Equal Number of Subjects and Objects* say that when manifold images of an object appear to a consciousness, that consciousness is also manifold. (2) *Half-Eggists* say that just one unitary image appears to one consciousness. (3) *Nonpluralists* say that manifold images appear to one consciousness. These three viewpoints will be explained in more detail later in volume 3. Most Cittamātra and Madhyamaka thinkers accept the Nonpluralist view—that manifold images appear to one consciousness—to be the correct standpoint. [268] This view maintains that there is no need for two distinct eye consciousnesses and so on to be present in order to apprehend manifold colors based on some such multicolored object. The single eye consciousness apprehends the many hues all at once.

Engagement via Affirmation versus Engagement via Exclusion

To understand how the mind engages its object, it is crucial to recognize that there are two different ways in which it can do so: it can engage the object via exclusion, and it can engage the object via affirmation—or through the capacity of the real object itself. Given the great importance of understanding the difference between these two ways, we offer below a brief explanation of the two modes of engagement.

Conceptual cognitions, which engage their objects through exclusion, have the following features: (1) *causal feature*—they are not produced by objective conditions, such as the presence of an external object nearby, but through the inner force of latent potencies or habits of mind; (2) *objective feature*—they do not depend on unique particulars to act as their appearing objects but take on a universal or something constructed by conceptual cognition as their appearing objects; (3) *functional feature*—they engage their objects in such a way that their location, time, and nature appear conflated. Such cognitions are known as cognitions that engage via exclusion. [269]

How does engaging via exclusion take place? Say a pot appears to a conceptual cognition apprehending a pot. Not all of the aspects, such as being produced, being impermanent, and so on that are identified with the pot in terms of location, time, and nature, appear to that conceptual cognition [as they do to a nonconceptual cognition]. Instead, only one of these features appears. Even a single object such as a pot has innumerable features that are its own attributes, such as the pot being impermanent, or being produced, or arising from effort, and so on. When a conceptual cognition apprehending a pot engages its object, the pot, it engages it through excluding that which is not a pot, what we might label "non-pot." In this case, it does not engage the pot as impermanent or the pot as produced and so on through having excluded the pot as permanent or the pot as unproduced and so on. Such cognitions are therefore known as "engaging via exclusion" or "engaging through differentiating the attributes." This topic is explained clearly by the Buddhist master logician Dharmakīrti. Here is a somewhat lengthy excerpt from his *Exposition of Valid Cognition Autocommentary*, which says:

> If inferential cognition apprehended real things,
> then when having ascertained one attribute,
> it would apprehend all attributes.
> In the case of exclusion, this fault does not arise.

> The problem of the nonoccurrence of other instances of valid cognition does not apply only to something that has been directly perceived. An inference that causes a cognition of a real thing in an affirmative manner does not perform an exclusion; when it determines one quality, it will determine all qualities because they are not really different [than the determined quality]. Hence, in this case of inference as well, no other instances of valid cognition would occur [to determine anything else because everything about the thing has been determined]. For when one quality is determined, it doesn't make sense for something of the same nature [namely, the other qualities of the object] also to be not determined. [270] When an inferential cognition excludes superimpositions, another inference can still occur [in order to remove other superimpositions] because the superimpositions excluded by one cognition are not the superimpositions excluded by another.
>
> *Objection*: Ascertainment of something not yet realized is not necessarily preceded by error. For instance, when we suddenly realize that there is fire because smoke is present, we cannot superimpose non-fire there. Therefore it is not a matter of excluding superimpositions in every case.
>
> *Response:* Regarding this, the following has already been stated. In that case, when the subject is realized, all [its

attributes] would be realized, since they are not different [from the subject itself]. Were [the subject and its attributes] different, then there would be no cognition [of the attributes] as present in the subject, since they would not be related to the subject. Therefore, in the present example, the one who is seeing that place would not have ascertained that the place has that nature [of possessing fire]. Why? Because of the error [caused by that place's similarity to places without fire]. That person thus determines that the place is devoid of fire due to having a cognition that is not oriented toward the possibility that fire exists; in doing so, how could that person not be wrong? One who has not superimposed [the conceptual semblance of non-fire onto fire] or who has no doubt [about the presence of fire] will not pay attention to the presence of evidence in that place in order to know that [the fire is present]. Nor will that person pay heed to the positive and negative concomitance [that this evidence has with the presence of fire].

Therefore it is clearly established that inferential reasons
have exclusions as their objects.
Otherwise, when the subject [of a proposition] is
established,
what [predicate] other than that subject would not also be
thereby established?[365] [271]

This same text also says: "On what grounds is it known that an exclusion is made known through language and inferential evidence and that a real thing is not made known directly by them? This is known because additional valid cognitions and statements are employed [to know the object further]."[366]

Consider a nonconceptual consciousness such as a direct sense perception. For example, when an eye consciousness apprehending a pot sees the pot, it necessarily sees simultaneously all the factors that are identified with the pot in terms of location, time, and nature, such as its impermanence and its having been produced. Although it sees such attributes, that cognition contacts the object, the pot, which is its objective condition, without depending on either the label *pot* or the thought "that's a pot." The appearance of the pot via the objective condition presenting itself to cognition comes about as if it were transferred into that sense consciousness. Therefore minds that engage affirmatively, without conflating location, time, and nature, necessarily take the unique particular object itself as an object through its appearing just as it is. *Exposition of Valid Cognition* says the following primarily from the perspective of perception:

> Given that the nature of the object is unitary and perceptible,
> what other portion [of that object]
> that was not already perceived
> could there be that is cognized by [other] valid cognitions?[367] [272]

Moreover, as already explained in the context of describing the difference between conceptual and nonconceptual consciousness, when any cognition engages an object, if it engages by way of qualifying it in terms of a type or universal, qualities or attributes, difference or nondifference, having a location, time, and so on, then it must be posited as engaging via exclusion. Thus Candrakīrti's *Clear Words* says: "For this is to demonstrate merely that the five sense consciousnesses are dumb."[368] A distinction is made between dumb sense consciousnesses and the conceptual consciousnesses that are comparatively smarter.

Furthermore, "engaging an object through differentiating its attributes" and "engaging via exclusion" have much the same meaning. For example, when a conceptual consciousness apprehends a tree, it engages its object, a tree, by excluding from its object the tree's attributes—such as the tree's impermanence, its production, and other factors—and engages only the "tree" itself. Also, when one brings to mind the conceptual cognition of a pot, for example, it may feel as if a single cognition perceives and apprehends simultaneously the pot's lip, base, and belly, but this is due to the speed with which cognitions operate. Were we to deeply investigate how our conceptual cognitions engage the object, then when we have the conceptual cognition "the pot's lip," we would see that we are not having the conceptual cognition "its base" or "its belly." [273] When we have the conceptual cognition "the pot's base," we are not apprehending anything other than its base. When the mind focuses on simply a pot in general, which is a collection of parts, and thinks "This is a pot," it evidently does not consider its individual parts and think "This is its base and this is its belly."

Thus when conceptual cognitions engage an object, they do not engage it through a real object presenting an image of itself to the mind. Rather, they engage it by a process of subjective delineation. They engage something that is isolated from other features or just the attribute to which the mind is directed. They do not engage anything else. Thus *Exposition of Valid Cognition* says:

> The ascertainments and verbal expressions
> that serve to remove [false superimpositions]
> are as numerous as those superimpositions of [false] attributes.
> Hence they all have distinct referents.[369]

In contrast, nonconceptual consciousnesses [such as sense consciousnesses] are said to engage their objects without differentiating their attributes. Generally speaking, when nonconceptual consciousnesses engage their objects, they engage through the object presenting an image of itself to consciousness, just as a reflection appears in a mirror. They do not engage in subjective delineation or analysis. Whatever object is engaged, all the visible parts or qualities of the object are engaged. This is known as "engaging without differentiating the attributes." *Exposition of Valid Cognition* says: [274]

> Therefore, all the qualities of
> a seen thing are seen.[370]

Moreover, engaging by way of a linguistic convention is also characterized as an engagement via exclusion. In general this term *linguistic convention* is mentioned frequently in texts on valid cognition, which speak about "linguistic convention, within the subdivision of name and linguistic convention," and "applying a linguistic convention to an object," and "engaging an object through a linguistic convention." In the context of explaining the theory of *exclusion of other*, "a conceptual cognition that engages through a linguistic convention" refers to a mind that extends out to its object through a process involving a universal that is dependent on linguistic conventions (i.e., semantic content), and this is an engagement via exclusion. Conversely, a mind that engages by way of the object presenting an image of itself is an engagement via affirmation.

Accordingly, engagement via exclusion and engagement via affirmation can be defined in the following manner. The definition of engagement via exclusion is: to engage an object through differentiating its attributes. The definition of engagement via affirmation is: to engage an object without differentiating its attributes. In general there are two things that engage via exclusion: language and cognition. Language that engages via exclusion encompasses all intelligible sound. Cognition that engages via exclusion encompasses all conceptual cognition. As to whether or not persons can be characterized as engaging via exclusion, there are two opinions: one that posits persons to engage via exclusion and one that does not. All nonconceptual consciousnesses are cognitions that engage via affirmation. [275]

In brief, language and conceptual cognitions that engage through exclusion, through other-exclusion, through the force of desire, through categories, through universals, through differentiating the attributes, and so on, have much the same meaning. Likewise, minds that engage through establishing the object, through an affirming image, through the causal capacity of a real thing, without depending on a linguistic convention, with a clearly appearing object, and without differentiating the attributes and so on have much the same meaning.

In the case of a direct perception of a pot, all the attributes based on the pot, such as produced, impermanent, and so on, appear and are engaged. But a conceptual cognition apprehending a pot, having made hairsplitting distinctions among the pot's attributes, engages only one feature of the pot, such as the pot itself isolated from what is not the pot. In this way one should understand the natures and attributes of engagement via exclusion and engagement via affirmation. *Exposition of Valid Cognition* says:

> Even when something has been directly perceived,
> the cognition [immediately following that perception] has a
> universal as its object; it is conceptual.
> As a cognition [of an object] that has not been superimposed
> with other attributes,
> it has as its object just that [exclusion].[371]

Also, speaking of the contrary view the same text says:

> For whom words denote through the capacity of reality
> and do not depend on a desire to express . . .[372] [276]

The explanation of many points related to linguistic reference by Buddhist masters can be understood on the basis of this view that language and conceptual cognitions engage their objects through exclusion. For example, having applied the linguistic convention *cow* to a white cow's assembled conglomeration of a hump, dewlap, and so on, when one later sees a black cow, a cognition of the universal thinking "This is a cow" arises naturally. This is in contrast to the explanation of the non-Buddhist Vaiśeṣika school, which maintains this is possible because those two cows are pervaded by an objectively distinct universal "cowhood" (a permanent entity that is established altogether separately from the two particular cows). For the Buddhists, the linguistic convention is applied on the basis of the capacity to give rise naturally to a cognition of resemblance having seen a white cow and a black cow and, likewise, on the basis of the two cows similarly performing the function of a cow, in that they create a similar effect. Applying the linguistic convention in this way accounts for the ability to think "This is a cow" when later another cow is seen.

Thus, according to Buddhist epistemologists, when this "cognition of the universal" is deeply investigated, it is posited to be a "cognition of resemblance." Furthermore, this cognition of the resemblance of phenomena of similar type arises from our innate, latent predispositions. In the case of the linguistic convention *pot* and the referent pot, in no way do they posit an objective relation between language and its referent. Such a relation comes about solely through the capacity of customary classificatory conventions. Thus they maintain that language indicating its referent is highly contingent on what the speaker wishes to say. [277]

Regarding this theory of exclusion and linguistic reference, the masters of Buddhist epistemology, Dignāga and Dharmakīrti, made an incomparable contribution to philosophy in general, and particularly to classical Indian philosophy. We will discuss this in greater detail in volume 4.

18

The Sevenfold Typology of Cognition

We will now discuss, using the example of a person engaging in a line of inquiry, how the mind engages an object in progressive stages. In general, it is from this perspective that the sevenfold typology is primarily presented. This typology is understood to have been introduced by Chapa Chökyi Sengé (1109–69), a great Tibetan logician, who established it as comprehensively sevenfold in the following manner.

In general, a mind engages an object in two ways:

(a) Engaging one standpoint without differentiating it from an opposing standpoint

(b) Engaging one standpoint while differentiating it from an opposing standpoint

Option (a) is (1) *doubt*: when doubting that something is impermanent, we can also doubt that it might be permanent, so this is engaging without differentiation from that opposing standpoint. Option (b) can be of two types:

(i) Engaging in a way that does not accord with the actual state of affairs [278]

(ii) Engaging in a way that accords with it

Of these two types, option (i) is (2) *distorted cognition*. Option (ii) has two types:

(a) Engaging an object without counteracting contradictory superimpositions

(b) Engaging an object by way of counteracting them

Of these two types, option (a) is (3) *indeterminate perception*. Option (b), in turn, can be of two types:

(i) Engaging an object already realized
(ii) Engaging an object not realized before

Of these two types, option (i) is (4) *subsequent cognition*. Option (ii) consists of two types:

(a) Counteracting superimpositions by clearly seeing a particular
(b) Counteracting them without clearly seeing the particular

Of these two types, option (a) is (5) *direct perception*. Option (b) can be of two types:

(i) Following a correct reason
(ii) Not following one

Of these two types, option (i) is (6) *inference*. Option (ii) is (7) *correct assumption*.

Among these seven, five are not valid and two (direct perception and inference) are valid cognitions. In this way, the typology of cognition in terms of seven categories is presented on the basis of how the mind engages its object.[373]

Now let us explain these seven categories of cognition by relating them to how these types of cognition engage their objects in progressive stages in the case of a single person. First, say there is a cognition within our own mental continuum that either reifies or denies an object's features owing to a mis-knowing confused about the way that object exists; this is *distorted cognition*. Then that strong distorted conviction is undermined through the use of consequential reasoning, and a mind of *doubt* arises. Next, thinking about the topic rationally, using a correct reason characterized by the three modes and so on,[374] that mind of doubt is removed, and *correct assumption* arises. Not satisfied with that alone, [279] we contemplate again and again with the wisdom of fine investigation, and a valid *inference*—a stable ascertainment of what is to be inferentially established—arises. But this mere ascertainment is not enough, for we need to continuously cultivate familiarity with what has been inferentially ascertained so that this knowledge becomes almost like second nature within our mind. Thus there arises the *subsequent cognition* that is the continuity of the mind that has familiarized itself in that way. In the end, the knowledge derived from the three modes of a correct reason too arrives at perception. That is, the mind that familiarizes itself with the certainty gained by the inferential valid cognition, through repeated acquaintance, must in the end transform into *perception*—a cognition whose object appears clearly without being mixed with a universal. But a mere *indeterminate perception*, where that perception does not ascertain its object even though it clearly appears, is not the endpoint of this process. All this constitutes a presentation of the sevenfold typology in accordance with the approximate order in which they might arise in an individual.[375]

1. Distorted Cognition

As in the case of the sevenfold typology of cognition, distorted cognition can be explained in relation to a single person's stages of knowing an object that exists in a hidden manner. Owing to a cognition that does not understand the way in which its object exists, there arises inappropriate attention that views its object in a distorted manner. That inappropriate attention, having exaggeratedly imputed certain qualities such as being attractive or repulsive, strengthens one's attachment or aversion. Through the force of attachment or aversion, one engages in nonvirtuous conduct of body, speech, and mind, [280] causing oneself and others to be bound up in suffering. These "distorted cognitions"—cognitions that engage their objects in a distorted manner—induce mental afflictions such as attachment and become causes and conditions of unwanted suffering. Therefore, to stop the cognition that initiates the causal process of suffering and so on, one must first recognize this distorted or mis-knowing mind and then one must counteract it.

Thus one should ask: What is a correct cognition? And what is a distorted or mis-knowing cognition? These are differentiated based on whether the mind has the support of valid cognition—whether its object exists in the way that the mind holds it. For example, if upon seeing a bright light in the far distance, a cognition arises thinking that this light is light reflected from a crystal when in fact it is light reflected from ice, then this cognition is a distorted cognition. If one approaches the place and looks, and a cognition arises understanding that this light is light reflected from ice, immediately canceling out the earlier thinking, then this cognition is a correct cognition. In this way, when the later cognition functions to counteract the object as cognized by that previous cognition, valid and nonvalid minds are demarcated. *Exposition of Valid Cognition* says:

> Where there is valid cognition about something,
> that will counteract other [nonvalid cognitions].[376]

And Jñānagarbha's *Distinguishing the Two Truths* says: [281]

> If a valid cognition could be refuted by another valid cognition,
> then the very criteria of valid cognition would lose credibility.[377]

The definition of a distorted cognition is: a mind that engages its main object in a distorted manner. The *Descent into Laṅkā Sūtra* says:

> When a cognition mis-knows,
> it is a distorted cognition.[378]

Thus distorted cognitions are those holding what is existent to be nonexistent and holding what is nonexistent to be existent, holding what is not the case to be the case and holding what is the case to be not so. When categorized, distorted cognitions are of two kinds: distorted conceptual consciousness and distorted nonconceptual consciousness. Examples of distorted conceptual consciousness are a conceptual cognition apprehending it to be dawn when there is a full moon, a conceptual cognition apprehending a rabbit's horns, and a conceptual cognition apprehending sound as permanent. Examples of distorted nonconceptual consciousness are a visual consciousness to which a snow mountain appears blue, a sense consciousness that perceives a white conch shell as yellow, a sense consciousness to which one moon appears as two, a visual consciousness that perceives a mirage as water, and a sense consciousness to which trees appear to be moving [when instead it is oneself that is moving]. [282]

2. Doubt

Doubt may be explained by way of the following illustration. Suppose within one's own mindstream there is a strong distorted conviction that sound is permanent. When one investigates whether there is any valid cognition that refutes the object as cognized by a wrong view such as this, and when someone else presents a correct consequence of one's own position that shows that sound is impermanent, and when one reads and critically reflects on treatises that establish sound to be impermanent, the conviction that sound is permanent loosens. At this point there arises a doubt not tending toward fact—the thought that sound is only probably permanent. When one again analyzes in detail whether there is a refutation of the object as cognized by a cognition holding sound to be permanent, there arises a balanced doubt—a thought that sound may or may not be permanent. Then when one analyzes the meaning of that more deeply than before, and when someone else posits a correct proof statement showing that sound is impermanent, there arises a doubt tending toward the fact—the thought that probably sound is impermanent.

The definition of doubt is: a cognition that on its own vacillates between two standpoints with uncertainty about the object. In the example "sound is impermanent," in the phrase "vacillates between two standpoints with uncertainty" *is impermanent* is one standpoint and *is not impermanent* is the other standpoint. The *Compendium of Ascertainments* says: "What is doubt? It is any uncertainty in the sense of having two mental attitudes or [283] two objects."[379] When categorized, there are three kinds: (1) doubt not tending toward the fact, (2) balanced doubt, and (3) doubt tending toward the fact. The first is a mental factor thinking, for example, that sound is either permanent or impermanent, though probably permanent. The second is a mental factor wondering, for example, whether sound is permanent or impermanent. The third is a mental factor thinking, for example, that sound is either permanent or impermanent, though probably impermanent.

3. Correct Assumption

Correct assumption may be explained by way of the following illustration. There are methods for transforming a doubt in one's mind that tends toward the fact, thinking sound is probably impermanent, into the nature of a mind convinced that sound is impermanent. One can depend on a correct reason put forward by someone else, or one can think it through for oneself. From this arises a correct assumption thinking that sound definitely is impermanent. A mind observing that sound is impermanent that is *convinced* that sound is impermanent, until it ascertains this with valid cognition, is said to be a correct assumption. This is primarily from the perspective of how a correct assumption arises on the basis of a reason with regard to a hidden object. [284] [Other types of correct assumption are described below.]

The definition of a correct assumption is: a construing awareness that conceives with conviction the main thing that is its engaged object but does not obtain a realization of its object of scrutiny. The phrase "the main thing that is its engaged object" excludes distorted cognition from overlapping with correct assumption. "With conviction" excludes doubt from overlapping with correct assumption. "Conceives" excludes direct perception from overlapping with correct assumption. "Does not obtain a realization of its object of scrutiny" excludes subsequent cognition as well as inferential cognition from overlapping with correct assumption.

The term *correct assumption* does not appear explicitly in the texts of Dignāga and Dharmakīrti, but the following appears in Śāntarakṣita's *Compendium of Reality*:

> Others declare that a cognition that realizes an object
> by way of cognizing a linguistic statement is language-based
> cognition.
> Such cognitions derive either from unproduced [eternal
> words]
> or from the words of a trustworthy person.
>
> Since it is an apprehension of what is hidden,
> it is not perception,
> and it is not inferential cognition
> because it lacks its characteristics.[380]

These lines appear to implicitly present the notion of "correct assumption." This part of the text presents the Mīmāṃsā assertion of a type of valid cognition arising from speech, a type that is other than valid perception and valid inference.[381] Even our own tradition [285] accepts that there are cognitions derived from speech (such as from the testimony of a reliable witness) that apprehend objects hidden to our senses. It would be difficult to posit these as either direct perceptions or inference, so these cognitions cannot be other than correct assumptions. Therefore, just as this citation from the *Compendium of Reality* has been cited in some of the writings of Tibetan masters as a source for the notion of correct assumption, we have cited it here.[382]

When categorized, there are five types of correct assumption: (1) correct assumption without any reason, (2) correct assumption with a contradictory reason, (3) correct assumption with an inconclusive reason, (4) correct assumption with an unestablished reason, and (5) correct assumption with a reason that is not properly understood. Examples of these are as follows. First is thinking without any reason that a pot is produced. Second is thinking that a pot is produced because it is not produced. Third is thinking that a pot is produced because it is an object of comprehension. Fourth is thinking that sound is produced because it is an object of eye consciousness. Fifth is thinking that sound is impermanent because it is produced without having first ascertained (a) that sound is produced and (b) the pervasion that if something is produced it must be impermanent.

The meaning of this last one is as follows. Although "being produced" is a correct reason proving that sound is impermanent, the person to whom the reason is being presented has no idea that sound is produced nor that if something is produced, it must be impermanent; nonetheless he has a correct assumption apprehending that sound is impermanent because it is produced.

There is also an alternate way of categorizing correct assumption into three types: correct assumption without any reason, correct assumption not understanding the reason, and correct assumption with a fallacious reason. [286]

In brief, take the case of a person in whom understanding derived from learning has arisen with respect to the aggregates and so on being impermanent though the understanding derived from critical reflection has not yet arisen. His understanding derived from learning that the aggregates and so on are impermanent is a true conviction about its object, but since a valid ascertainment of this has not been obtained, this cognition and many others of its kind can only be considered as correct assumption and not as any of the other six types of cognition.

4. Inferential Cognition

Inferential cognition may be explained by way of the following illustration. Take the case of someone who has generated within her mindstream a correct assumption observing that sound is impermanent, as described above. When that person, out of a desire to know that sound is impermanent, engages in powerful analytical meditation through relying on mindfulness and meta-awareness, the result is an inferential cognition. Furthermore, such an inference is grounded in perception: in its origins it is grounded in perceptual auditory consciousness apprehending sound, and in the end it is grounded in a yogic direct perception realizing sound to be impermanent.

The definition of inferential cognition is: a construing awareness that through depending on its basis, a correct reason, is nondeceptive regarding its object of comprehension. The *Compendium of Valid Cognition* says: [287]

> An inference for one's own sake is this:
> perceiving the object via a tri-modal reason.[383]

When one infers the thesis from seeing the reason and remembering the relation, that is called *inferential cognition.* When categorized in terms of the object of comprehension that is validly known, there are three types of inferential cognition: inferential cognition based on empirical fact, inferential cognition based on popular convention, and inferential cognition based on trustworthy testimony.

The definition of inferential cognition based on empirical fact is: a construing awareness that through depending on correct evidence based on empirical fact is nondeceptive regarding its object of comprehension, a slightly hidden object. An example is an inferential cognition realizing that sound is impermanent because of being produced. *Exposition of Valid Cognition* says:

> It is stated that, with respect to its own object,
> an inferential cognition does not rely on scripture.[384]

This states that an inferential cognition based on empirical fact does not rely on scriptural testimony.

The definition of inferential cognition based on popular convention is: a construing awareness that through depending on correct evidence based on popular convention is nondeceptive regarding its object of comprehension, a language-based popular convention; for example, an inferential cognition realizing that the word *moon* can be used to refer to the "rabbit bearer" (a popular epithet of the moon in ancient India). Also, *Exposition of Valid Cognition* says: [288]

> Therefore inferential cognition [based on empirical fact]
> has a different object from that based on popular
> convention.[385]

The definition of inferential cognition based on trustworthy testimony is: a construing awareness that through depending on correct evidence based on trustworthy testimony is nondeceptive regarding its object of comprehension, a very hidden object. An example is an inferential cognition that realizes its object of comprehension, a very hidden object, in dependence on its reason, a scripture that has passed the threefold analysis. *Exposition of Valid Cognition* says:

> Because trustworthy statements are generally nondeceptive,
> [cognition from those statements is considered to be]
> inferential.[386]

5. Subsequent Cognition

Subsequent cognition can be explained by way of the following illustration. Consider the case of someone who has a valid inferential cognition realizing that sound is impermanent. In that person's mindstream, later moments of the continuum of valid cognition observing that sound is impermanent are subsequent cognition. The definition of subsequent cognition is: a cognition that realizes again, by the force of a previous valid cognition, an object already realized by the earlier valid cognition that induced it. *Exposition of Valid Cognition* says: [289]

> Since it apprehends the apprehended, that conventional
> [cognition]
> is not accepted [to be valid cognition].[387]

This means that subsequent cognition, which is a conventional cognition remembering something, is a consciousness that, through recollection, apprehends the object that the earlier valid cognition that induced it had already apprehended and fully understood. Thus it is subsequent cognition, but it is not accepted to be valid cognition.

When categorized, subsequent cognition has two types: perceptual subsequent cognition and conceptual subsequent cognition. Examples of the first are a subsequent perceptual sense consciousness, such as the second moment of an eye consciousness apprehending a pot, and a subsequent perceptual mental consciousness, such as the second moment of a clairvoyant perception knowing someone else's mind. Conceptual subsequent cognition is of two types: a conceptual subsequent cognition induced by perception and a conceptual subsequent cognition induced by an inference. An example of the first is an ascertaining consciousness knowing blue that is induced by a sense perception apprehending blue. An example of the second is an ascertaining consciousness knowing a pot to be impermanent that is induced by a valid inference realizing that a pot is impermanent.

The above explanation of later moments of perception and inference as subsequent cognition is based on Dharmottara's *Explanatory Commentary on the Ascertainment of Valid Cognition*. This text says: "Because the first moment of perception and inference is able to engage the object through ascertaining the continuum of a thing able to perform a function, [290] later ones that are not separate from the establishment and abiding of its continuum are not valid cognition."[388]

6. Direct Perception

Direct perception can be explained by way of the following illustration. Consider the case of someone who has a subsequent cognition realizing that sound is impermanent. Having thoroughly accustomed herself to sound being impermanent by way of analyzing it again and again, finally at the end of the gradual process of moving through the nine stages of mental stabilization and so on, there arises a valid perception realizing clearly that sound is impermanent, unmixed with a universal. That is a yogic direct perception.

The definition of a direct perception is: a cognition that is unmistaken and free of conceptuality. For example, an eye consciousness apprehending a pot. The *Compendium of Valid Cognition Autocommentary* explains the meaning of a citation from the *Abhidharma Sūtra* and states: "Someone with eye consciousness [291] cognizes blue but does not think 'This is blue.'"[389] The phrase "Someone with eye consciousness cognizes blue" indicates that it is an unmistaken cognition. The phrase "but does not think 'This is blue'" indicates that it is free of conceptuality. Thus this quotation completely presents the definition of perception. Also, Dharmakīrti's *Drop of Reasoning* says: "Direct perception is unmistaken and free of conceptuality."[390]

"Free of conceptuality" does not mean being free of the substance of thought or free of thought in itself.[391] It means being free of a construing awareness that apprehends word and referent as suitable to be associated. Furthermore, since this cognition itself is not a construing awareness that apprehends word and referent as suitable to be associated, the manner in which it is free of conceptuality was explained earlier. "Unmistaken" means being unmistaken with regard to the appearing object. The nature of the conceptuality that perception is free of, and the way in which conceptual cognition operates with linguistic referents and so on, was discussed when we explained how mind is categorized as either conceptual or nonconceptual (in chapter 3).

When categorized, there are four kinds of direct perception: (1) sense perception, (2) mental perception, (3) reflexive awareness, and (4) yogic perception. The definition of *sense direct perception* is: a cognition that is unmistaken and free of conceptuality and that arises from its uncommon dominant condition, a physical sense faculty. This has five types, which range from a sense direct perception apprehending a form [292] to a sense direct perception apprehending a tangible object. The definition of a sense direct perception apprehending a form is: a cognition that is unmistaken and free of conceptuality and that arises in dependence on its objective condition, a form, and its uncommon dominant condition, the eye sense faculty. There are three kinds: a sense direct perception apprehending a form that is a valid cognition, one that is a subsequent cognition, and one that is an indeterminate perception. Examples of these are, respectively:

the first moment of a sense direct perception apprehending a form, the second moment of a sense direct perception apprehending a form, and a sense direct perception apprehending a form that directly induces a doubt thinking "Did I or did I not see a form?" The same pattern should be extended to the remaining four sense direct perceptions.

The definition of *mental direct perception* is: a cognition that is unmistaken and free of conceptuality and that arises in dependence on its own uncommon dominant condition, a mental sense faculty. The *Compendium of Valid Cognition Autocommentary* says: "The mental cognition that perceives form and so on, being nonconceptual and engaging with the image of an experience, is also [a type of perception]."[392] In general, there are diverse opinions on the question of what exactly is a mental perception of an ordinary person and also whether such a mental perception of an ordinary person is capable of giving rise to ascertainment of its object. These will be discussed in volume 4.

A *perceptual reflexive awareness* is the subject-image that is of the nature of reflexive awareness. The definition of a perceptual reflexive awareness is: a subject-image awareness that is unmistaken and free of conceptuality. Reflexive awareness [293] and perceptual reflexive awareness are synonymous. Reflexive awareness is a consciousness directed exclusively inward. Devendrabuddhi's *Commentary on Difficult Points in the Exposition of Valid Cognition* says: "However, the cognition that illuminates the previous cognition, inasmuch as it has the nature of illuminating itself, is established to be aware of itself reflexively and not by virtue of being experienced by some other cognition."[393] The presentation of reflexive awareness and the differences between whether reflexive awareness is accepted or not accepted according to subsects within the Buddhists' own tenet systems will be discussed in volume 4.

The definition of *yogic direct perception* is: a mental perception bearing an image of the truth and free of conceptuality that arises in dependence on its own uncommon dominant condition, meditative concentration that is a union of calm abiding and special insight. *Exposition of Valid Cognition* says:

> The yogic cognition mentioned above
> arises from meditation;
> the nets of conceptions having been destroyed,
> it appears only clearly.[394]

As for that, it is mentioned earlier, in the second chapter of *Exposition of Valid Cognition*, "Establishing Validity" (*pramāṇasiddhi*), that yogic perception is a consciousness in a yogin's mindstream that directly realizes the truth about reality. [294] This is because (a) it arises through the power of meditative concentration that is a union of calm abiding and special insight perceiving ultimate reality and (b) it is a consciousness free of conceptuality and unmistaken about ultimate reality.

This explanation of the definition and divisions of direct perception is from the perspective of the Sautrāntika school. According to the Cittamātra and Yogācāra-Madhyamaka traditions, the definition of perception is: a cognition that is free of conceptuality and is not mistaken on account of any temporary causes of error. The definition of sense perception is: a cognition that arises from stable latent potencies and is free of conceptuality, and that arises from its own uncommon dominant condition, a physical sense faculty. The other types of direct perception are like those posited above.

Also, Dharmakīrti's *Ascertainment of Valid Cognition* says: "Other [cognitions, i.e., those different from unreliable ones] have a consistent continuity (*aviśliṣṭānubandha*) for as long as saṃsāra lasts because the latent potencies for them are stable; from the perspective of their reliability for conventional action, they are valid cognitions."[395] They are said to "arise from stable latent potencies" in that they arise through the activation of the latent potencies of a similar type and so on, which are their causes; those latent potencies are stable in the sense that they persist continuously.

When direct perception is categorized, there are the four of sense direct perception and so on. When sense direct perception is categorized, there are the five, from sense direct perceptions apprehending forms to sense direct perceptions apprehending tangible objects. This much is similar to the Sautrāntika tradition. [295]

7. Indeterminate Perception

According to the Sautrāntika tradition, the definition of a cognition that is an indeterminate perception is: a cognition having a clear appearance of a unique particular that is its engaged object but is unable to induce an ascertainment or determinate cognition of it. Since it does not ascertain the unique particular that is its engaged object, even though it appears clearly, it is called *indeterminate perception*. *Exposition of Valid Cognition* says:

> A consciousness that is attached to some other object
> lacks the ability [to produce a discordant subsequent cognition] because it does not apprehend a different object.[396]

When categorized according to the Sautrāntika tradition, a cognition that is an indeterminate perception can be of three kinds: a sense direct perception that is an indeterminate perception, a mental direct perception that is an indeterminate perception, and a reflexive awareness that is an indeterminate perception. An example of the first is a sense direct perception apprehending a form on an occasion when the ear consciousness is listening to a sound with great attachment. And the same pattern of positing indeterminate perception is applied to the nose consciousness and so on. An example of the second is a mental perception apprehending a form that arises following a sense perception apprehending a form within the mindstream of an ordinary person. An example of the third is a perceptual reflexive awareness experiencing a mental perception apprehending a form within the mindstream of an ordinary person. [296]

We can identify these seven types of cognition by applying them in a particular scenario as follows. A cognition thinking that there is no fire on a particular smoky mountain pass is a *distorted cognition*. A cognition thinking that there may or may not be a fire on that pass, though probably there is, is *doubt*. A cognition with the conviction that there is a fire on that pass but without ascertaining it to be there is *correct assumption*. A cognition ascertaining that there is a fire on that pass by reason of there being smoke is *inferential cognition*. The second moment of such an inferential cognition is *subsequent cognition*. A cognition clearly realizing the presence of fire on that pass is *direct perception*. A cognition with a clear appearance of a fire on that pass that does not ascertain it to be a fire is *indeterminate perception*.

Alternatively, suppose a cowherd goes out to search for a lost cow. In fact the cow is in the area to the west, but someone tells him incorrectly, "It is in the east." This causes the cowherd to generate a contrived conviction that the cow is in the east. That type of mind is a *distorted cognition*. Then, however much he searches for the cow in the east, he cannot find it at all. So he generates a *doubt* thinking "My cow is probably not in this area." When someone else tells him correctly, "Your cow is in the area to the west," he generates a *correct assumption* unequivocally holding the cow to be in the west. When he goes to explore, he eventually sees a cow's fresh footprints, hears it lowing, and so on. Upon finding reliable valid ascertainment of the cow being there, he generates a *valid inferential cognition*. The second moment of that is a *subsequent cognition*. [297] Gradually after that, from a far distance, his visual consciousness sees the cow without ascertaining it. This is an *indeterminate perception*. When he gets closer, he generates a *valid direct perception* seeing the cow. In this way we can also understand the function of the seven types of cognition by examining how a cowherd's mind goes through gradual stages with respect to a single cow. [299]

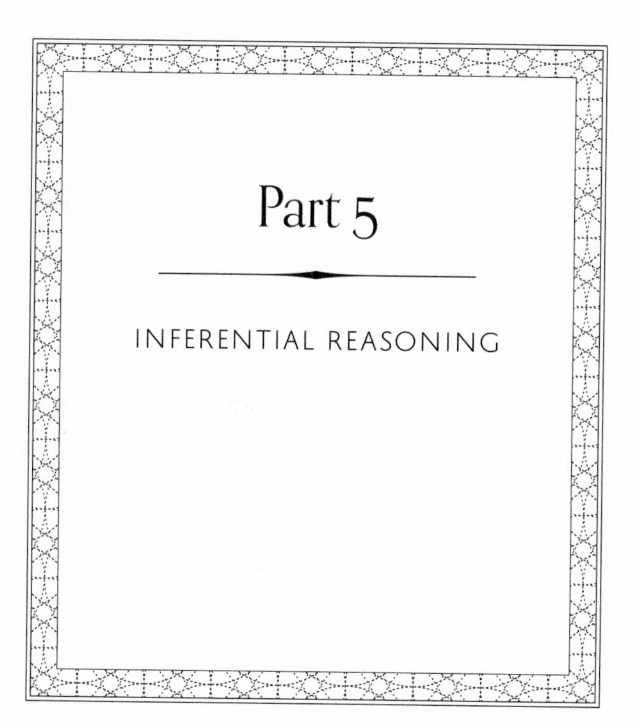

Part 5

INFERENTIAL REASONING

The Crucial Role of Reasoning

One of the keen insights that runs throughout the Buddhist tradition is that the ultimate authority for knowledge claims must be one's own experience. While an enormous corpus of scripture is readily available to any Buddhist philosopher, no scriptural passage—even those attributed to the historical Buddha himself—can in itself be used to establish empirical claims, at least according to the Buddhist epistemologists. Even reasoning, which inevitably requires the use of concepts, must eventually give way to direct experience, in part because conceptual cognitions are always distorted, in ways that we have previously discussed. And as we saw in the previous part, for the purposes of personal transformation and behavior change, only direct experience—not inferential or conceptual cognition—has the kind of visceral impact necessary for effective change.

Yet this emphasis on direct experience stands alongside another key theme: ordinary experience prompts intuitions that do not conform to the way things really are. At the very least—as noted previously—our ordinary experience suggests that we have an enduring, autonomous self that is the perceiver of perceptions, the feeler of feelings, and the owner or controller of the mind-body system. Thus, even though in the end the Buddhist paradigm would have us arrive at a direct experience of, for example, our mind-body system as devoid of any such absolute self, we are starting from a place where a direct experience of selflessness is inaccessible. To get to that point, we must engage in intensive observation and analysis guided by reason. And as a result, the Buddhist traditions examined in this volume have placed a tremendous emphasis on the crucial importance of rigorous, irrefutable reasoning.

Three Domains

One way to understand the role of reason in these Buddhist traditions is to examine the notion of three domains of inquiry articulated by Dignāga, Dharmakīrti, and subsequent Buddhist epistemologists. The first domain includes things that are directly available to our perceptual experience, or more precisely, knowledge claims we can justify directly based upon our experience. To take again the typical Indian example of the pot or water jug, my visual experience of an item on the table in front of me is enough for me to justify the claim "There is a pot on the table." There are other claims, however, that I may not be able to make directly from my perception alone. For example, when I see black smoke billowing out of a house, I cannot directly see that there is a fire present; the fire is epistemically hidden (Skt., *viprakṛṣṭa*) for me, in that it is not directly available to my senses. Nevertheless, I can rationally infer—based on what I can perceive—that there is indeed some incendiary source in that location. This is the second domain: knowledge claims about epistemically remote objects that I can justify by using an inference—or a chain of inferences—based in what is directly perceptible.

These first two domains—the directly perceptible and the remote—constitute what we might call an "empirical" sphere of knowledge, inasmuch as the evidence of the senses is what we fall back on when we are making knowledge claims in these contexts. There is, however, a third domain that, transcending the empirical, concerns objects that are extremely hidden (*atyantaparokṣa*). These are things that, at least given one's current capacities, cannot be examined in any empirical way. For the Buddhist epistemologists such as Dharmakīrti and his followers, the only way to justify claims about such trans-empirical things is through the testimony of a reliable person (*āptavacana*), which paradigmatically consists in the words of the Buddha himself as transmitted in the Buddhist scriptures. Importantly, however, Dharmakīrti maintains that *scripture cannot be used to justify empirical claims*. In other words, if some matter of concern can be adjudicated by ordinary perception and empirical inference, then one cannot use scripture to support one's claims. And if there is a conflict between some claim made in scripture and claims that can be empirically justified, then *the scriptural claim must be rejected*. Moreover, according to at least some interpretations, Dharmakīrti holds that claims justified by reference to scripture are not truly justified at all. In other words, according to this interpretation, scripture cannot be used to definitively prove or disprove trans-empirical claims. They only provide a provisional justification that enables one to address some important issue relevant to contemplative practice that, as Dharmakīrti puts it, require some answer but cannot be resolved any other way.

For our purposes, the upshot of the schema of these three domains is that, if we are to detect and correct that false intuitions that emerge from our ordinary experiences, then we cannot simply rely on scriptural claims. Of course, the teachings of the Buddha or some other reliable person can help direct us to an area of inquiry, such as a false sense of an absolute self, but from the Buddhist perspective articulated in this volume, that inquiry itself must eventually become a purely empirical one. And again, since our ordinary perceptions prompt intuitions that are not the way things truly are, we become especially reliant on empirical inferences that, using the reliable aspects of our perceptions, can help us see more clearly what we are actually experiencing.

Inference and Its Structure

Given the importance of empirical inference (Skt., *vastubalapravṛttānumāna*), it is no surprise that an entire section of this volume is devoted to inferential reasoning. This topic appears in early Buddhist sources, but Dignāga and Dharmakīrti are the key sources for the material presented here. To lay the groundwork for our authors' articulation of the facets of inferential reasoning, it may be helpful to summarize the core theories concerning the structure and content of inferential reasoning.

An *inference*, or strictly speaking an *inference for oneself* (*svārthānumāna*), is a type of cognition that enables one to know something by virtue of knowing something else. For Buddhist epistemologists, the basic structure of an inferential cognition looks like this:

S is P because E

Here "S" stands for a Subject or, more literally, a *property possessor* (*dharmin*) to which some Predicate or *property* (*dharma*) is being attributed. The statement or conceptual cognition "S is P" is thus the *thesis* (*pratijñā*) or proposition that is being established by the inference. The justification for formulating that thesis—the warrant for attributing that predicate to that subject—is the presence of another property that acts as the Evidence (*hetu* or *liṅga*), namely, the property that indicates the presence of the property to be proven. An example of all this, presuming that I am looking at some house with smoke billowing out of it, would be:

> This house (S) is a locus of fire (P) because of being a locus of smoke (E).

In order for an inference of this kind to be valid, three relations must be in place: S must have E as its predicate or property (the house must have smoke billowing out of it); the presence of the evidence E must invariably indicate the presence of the predicate or property P (wherever smoke occurs, fire must also occur); and the absence of predicate P is invariably concomitant with the absence of the evidence E (where there is no fire, there can be no smoke). These three relations constitute the three "modes" (*naya*) of a reliable inference.

The Three Modes

Introduced by Dignāga and refined by Dharmakīrti, the theory that a reliable inference must exhibit the three modes of relation among its terms is a distinctive feature of their account of inferential reasoning. The theory emerges in part from Dignāga's interest in categorizing the cases where inferences succeed, and our authors acknowledge the importance of Dignāga's efforts by concluding this part with an entire section on his schema of successful and unsuccessful cases. In the interest of simplicity, however, we understand this issue by examining just the three modes that underlie Dignāga's cases.

As noted above, the first mode amounts to a requirement that the property or thing being used as evidence (E) can be construed as a predicate or quality of the subject (S) of the thesis "S is P." In the example above, it must be true that the house is indeed a locus of smoke. One way this can go awry is if there actually is no smoke present at all; in that case, the evidence (namely, the smoke) is unestablished (*hetvasiddha*). Another way that this mode can fail is that, while the evidence may indeed be present, we are mistaken about the subject that bears that evidence as a property. In the case of our house-fire example, perhaps I have mistaken some other structure for a house, and my inference actually should be about a shed. This problem becomes more acute when we consider the cases where we need to construct inferences about things—such as the kind of self that Buddhist philosophers refute—that have never existed. Here, we run the danger of having no subject at all (*āśrayāsiddha*), and these kinds of cases are part of what motivates the extensive analysis of consequential reasoning provided by Buddhist theorists, as we will see below.

Moving on to the other two modes, they concern the relationship between the evidence (E) and the predicate to be proven (P)—the smoke and the fire, in our example. Both of these modes fall under the general rubric of the *pervasion* (*vyāpti*) between the evidence and the predicate. In short, every case where the evidence is present necessarily is "pervaded" by the presence of the predicate. The metaphor of one thing pervading or spreading though another suggests a Venn diagram:

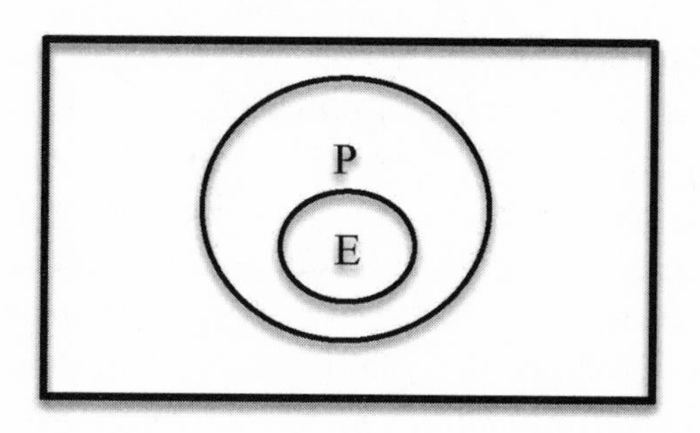

Applying this diagram of overlapping sets to our example, all cases where fire (P) is present pervade all the cases where smoke (E) is present. Thus, even though there are cases where there is fire without smoke, any time there is smoke, there must be fire. And this means that smoke can act as evidence or as an indicator (*gamaka*) of fire even when we cannot perceive the fire directly.

This relationship of pervasion encompasses two of the three modes, and they are articulated as the positive pervasion (*anvayavyāpti*) and negative pervasion (*vyatirekavyāpti*). The simplest way to articulate them is simply to say, for the positive pervasion, "Wherever the evidence is present, the predicate is present." And for the negative concomitance, one can say, "Where the predicate is absent, the evidence is absent." Students of logic will recognize that these statements are logically equivalent, and while this is true, one must be cautious here because the relationship between the evidence and predicate is not simply a logical one. Instead, it concerns the very natures of these things themselves, and thus it is a nature relation (*svabhāvapratibandha*).

Without going into great detail, we can note that the notion of a nature relation emerges from a fundamental insight articulated especially by Dharmakīrti: namely, that the relationship between the evidence and the predicate is not a matter of mere co-occurrence (*sahabhāva*). Here, Dharmakīrti is concerned with cases where we mistakenly believe that something can stand as evidence for something else. For example, it might happen to be the case, on a particular mango tree, that every mango we picked from a particular branch was ripe. And we might just infer that the next fruit (S) on that branch is ripe (P) because it is on that branch (E). But this inference is not, for Dharmakīrti, a reliable one because it fails to eliminate the possibility that it is mere happenstance that the ripe fruit were on that branch. Instead, the nature of the evidence must relate to the predicate's nature in such a way that it is *impossible* for the evidence to be present while the predicate is absent. In short, evidence and predicate stand in a relationship of what Dharmakīrti calls "the necessary relation of unaccompanied nonoccurrence" (*avinābhāvaniyama*). If the evidence is not "accompanied" by the predicate, then the evidence simply cannot occur.

For Dharmakīrti and all later Buddhist epistemologists, this necessary relation between the evidence (such as smoke) and the predicate (such as fire) occurs in just two ways: either the evidence is the effect of the predicate or the evidence is a property whose essential nature requires the presence of the predicate. These are known respectively as *effect evidence* (*kāryahetu*) and *nature evidence* (*svabhāvahetu*). Our authors discuss these at some length, and effect evidence is perhaps fairly straightforward. Of course, determining exactly when a causal relation is actually in place poses many problems, but we need not concern ourselves with that issue here. Instead, let us turn to the case of nature evidence. The meaning of this term is perhaps less obvious than effect evidence, yet this form of evidence plays an especially important role in Buddhist practice.

A typical example for nature evidence would be something like this: This thing (S) is a tree (P) because of being an oak (E). To unpack this in straightforward terms, we can say that the relationship between evidence and predicate amounts to this: something that has the property of being an oak invariably has the property of being a tree because the features of the object that are required for us to correctly refer to it as an oak already include all of the features that are needed for us to correctly call it a tree. Although much more can be said about nature evidence, the main point is that, as Dharmakīrti puts it, this form of evidence is especially helpful for those of us who are "confused" (*muḍha*) about the way we are using certain concepts. For example, on one level we may be quite correct to know that a pot is something that is causally produced, but we may not recognize that when we use the term *causally produced*, we already have grounds for saying that a pot is also necessarily *impermanent*. Of course, unless we have a problematic attachment to pots, it may not be so crucial to understand the full implications of their causal nature. But when we turn to, for example, the constituents of our mind-body system, we may readily acknowledge that they are causally produced, but we still somehow believe that they are not impermanent. Or to use a parallel example, we may not fully recognize that if we can speak of ourselves as "born," which we readily acknowledge, then the very use of this term means that we must die. And given this example, it is perhaps obvious why nature evidence plays a critical role in practice. Recall that, while direct experience is the bedrock of our knowledge, our ordinary experience prompts numerous delusions. Nature evidence plays an especially prominent role in enabling us to see how what we readily acknowledge in our own experience has implications that we have failed to uncover. In another key example, while I may readily acknowledge that I engage in causal actions in the world, I may not recognize that, merely by seeing myself as *causally engaged*, I am thereby necessarily *interdependent* and not an independent, autonomous self.

Much more can be said about the various aspects of inference unpacked by our authors. Numerous issues of great importance come to mind: the type of entailment required by the notion of pervasion; the requirement to ground inference in concrete, observable cases; the problem of determining when a necessary relation of unaccompanied nonoccurrence is in place; and so on. To this list should also be added the presence of a third type of evidence, *nonperception* (*anupalabdhihetu*), which uses the principles of effect evidence and nature evidence to negate, rather than establish, some thesis. But these issues can be explored through our authors' efforts themselves, and with that in mind, let us turn briefly to examining two other key aspects of inference: proof statements and consequences.

Proof Statements

So far, we have been discussing inference as a type of cognition. In contrast, a proof statement is a way of *inducing* an inferential cognition in another person, and as such, it is also called *inference for another* (*parārthānumāna*). In the classical formulation found in Indian Buddhist texts, a proof statement has an intriguing feature: the thesis to be proven is not explicitly stated. Instead, one lays out the ingredients, so to speak, that should lead one's listener to themselves to have a kind of "aha" moment in which they come to realize the truth of the unstated—but clearly implied—thesis.

The proof statement begins with a statement of the pervasion, and generally, just the positive pervasion is stated. Consider the case where one wishes to induce an inference in another person such that their mortality is recognized. One would begin by stating the positive pervasion: "Whoever is human (E) is mortal (P)." Importantly, the pervasion itself must be accompanied by a concrete case that is acknowledged by the discussants as exemplifying the pervasion. And so one might add "as in the case of Socrates." One of the purposes of offering an exemplifying case is that it compels a consensus about the pervasion, but it also prevents one from appealing to types of reasoning that cannot be grounded in experience. Having stated the pervasion and its exemplification, one then states the relationship between the subject of the evidence: "You (S) are human (E)." If one's interlocutor has fully accepted the pervasion ("Whoever is human is mortal"), then simply by saying "You are human," one should be able to directly cause them to experience an inferential cognition whose content is "I am mortal." Traditionally, one then ends the proof statement by noting the type of evidence used; in this particular case, one would say, "The evidence used here is nature evidence."

A proof statement, or inference for another, is often deployed in the context of philosophical debate, and historically, it seems that kings and other benefactors would actually sponsor formal debates between philosophers from different traditions. Since institutions large and small—temples and monasteries, for example—required considerable support, the motivation to win such debates was considerable. Indian philosophers in multiple traditions, including Buddhists, discussed at length the ways one can win debates by any means possible, even by confusion and misdirection. Yet the Buddhist approach to proof statements suggests that the main goal is not simply to win. Instead, the point of the proof statement is to provide another person with the information that they can use to come to a particular understanding on their own. In other words, to return to the example above, one does not directly say, "You are mortal!" Instead, one provides the conditions for the other person to conclude, "Ah, I see—I must be mortal." The goal is thus not to simply win the debate by proving the interlocutor to be wrong. Rather, one seeks to induce an understanding in the other person in a fashion that might even be characterized as therapeutic.

One problem, however, is that this approach to proof statements does not enable one to engage with persons who hold beliefs about completely unreal entities. Recall that the evidence (E) must be observed as a quality or feature of the subject (S) or thing under discussion. But what if that subject is a unicorn? Or a square circle? Or what if the pervasion—the relation between the evidence and the predicate to be proven—that a person holds to be true actually has no exemplifying case, precisely because it is impossible? I might absurdly believe that if something is a square circle, then it is necessarily blue, but the theory of proof statements presented here does not enable one to dissuade someone from holding such beliefs. In such cases, one must turn to another mode of argumentation: the consequence.

CONSEQUENCES

As we have noted, the Indian Buddhist traditions that form the core of the volumes in this series are often structured in terms of an ascending scale of analysis, and the highest or subtlest level of analysis is a version of the antirealist Madhyamaka known as the Prāsaṅgika—literally, the Consequentialists. The problem of trying to bring others to a particular understanding of one's own position is especially acute for the Consequentialists because, to put it in simple terms, Consequentialists hold that all ordinary persons are caught up in the delusion that things truly exist in the way that they appear. Thus, as an ordinary person, when I see a pot on the table in front of me, it seems to be objectively real—from its own side, so to speak—without in any way being dependent on the way I see it, or even on other causes and conditions. It simply exists on its own. For the Consequentialists, however, that kind of objectively real, autonomous pot does not exist at all, yet if they are to lead me to that understanding, how could they use a proof statement? For me, there is a truly real thing called a pot in front of me, but from their perspective, I might as well be saying that there is a square circle or a married bachelor in front of me. How then can we have a discussion about it? The fault of having no subject at all (*āśrayāsiddha*) would continually apply. This is where the style of reasoning that employs consequences (*prasaṅga*) plays a crucial role.

The case of the Consequentialists is an especially obvious one where consequential reasoning must be employed, but even issues at a lower level of analysis may best be conveyed to a conversation partner by using consequences. Our authors thus choose a simpler example—the failure to see that a causal entity, which comes from causes and produces effects, must necessarily be impermanent or in flux. Using this example, our authors offer a detailed discussion of reasoning through consequences, including a schematic analysis of eight different ways the pervasion (the relationship between the evidence and the predicate) can be configured. Without reproducing their rich analysis, we can simply note some key features of this style of reasoning.

Overall, consequential reasoning begins with the context of a false belief, for example, that a pot or water jug is permanent or unchanging. This harkens back to our ordinary intuition: when we see a carafe of water on the table, we do not generally believe that it is constantly changing in each moment. The Consequentialists would focus especially on the problem of the pot itself—why do we think that there is an objectively existent entity called a pot in the first place? But at a lower level of analysis, where the existence of something we call a pot is accepted without much contention, consequential reasoning addresses especially the relationships among the various properties that we attribute to it. For example, in a quite straightforward way, we might believe that a pot is permanent, and yet it is capable of changing state, such that in one moment it is empty, and a few moments later, it is filled with water. In this context, consequential reasoning brings us to an "aha" moment in which we recognize the way that this belief is incompatible with what we experience: obviously, a jug filled with water and one that is empty are not quite the same, even in terms of being a jug. The point of consequential reasoning, however, is not to force this reasoning upon one's interlocutor directly. Instead, and again in an arguably therapeutic fashion, one simply enables one's interlocutor to arrive, on their own, to the realization that their belief was false. Here again we see how the Buddhist approach to many issues, including styles of reasoning, are inextricably linked to the practical task of personal transformation.

John Dunne

Further Reading in English

For an overview of the features of inference according to the Indian Buddhist epistemological tradition, see chapter 4 of Jan Westerhoff, *The Golden Age of Indian Buddhist Philosophy* (Oxford: Oxford University Press, 2018).

For details of the relations and the "three modes" involved in inference, see chapters 1 and 3 of John D. Dunne, *Foundations of Dharmakīrti's Philosophy* (Boston: Wisdom Publications, 2004).

A prominent text in the Tibetan genre that underlies this presentation of inferential reasoning is presented by Katherine Rogers in *Tibetan Logic* (Ithaca, NY: Snow Lion Publications, 2009).

19

Reasoning and Rationality

THUS FAR WE HAVE discussed how, when conceptual and nonconceptual cognitions engage their objects, they do so via an image, as well as how, respectively, they engage via exclusion and via affirmation. Now one might ask, "When unimpaired subjects engage their objects, what means are there for comprehending an object's mode of existence just as it is?" In response, it is said that the means by which the mind engages with reality are: (1) the four forms of rationality, (2) consequential reasoning, which is used as a way to eliminate distorted conviction, (3) proof statements, which are used as a way to eliminate doubt oscillating between two points of view, and (4) inferential evidence, which is used as a way to explicitly ascertain the object once those distorted cognitions have been dispelled. These four means are encompassed within the science of inferential evidence. What is called the "science of inferential evidence" means the same as the science of inferential reasoning. [300]

THE FOUR FORMS OF RATIONALITY

First, to explain the four forms of rationality, as cited earlier Buddha himself stated in the sūtras:

> Monks and scholars, just as you test gold
> by burning, cutting, and polishing it,
> so too well examine my speech.
> Do not accept it merely out of respect.[397]

As these words suggest, when the Buddha's followers engaged with the meaning of the scriptures, they didn't emphasize faith and belief but rather began with finding certainty based on correct rational analysis. This required establishing a realistic account of the object.

An important method propounded by the masters of Nālandā Monastery for analyzing such an object is known in the Buddhist scriptures as the four forms of rationality. First, based on the way in which things exist in reality, one investigates a thing's own character or nature. Second, based on the inquiry into its nature, one investigates how each thing functions naturally according to its own character. Third, based on that, one investigates how each depends on another. That is, one establishes the way they exist based on relations such as the relation between cause and effect, or the relation between part and whole, or the triad of agent, action, and object, and so on. Fourth, when analyzing on the basis of those three forms of rationality, one sets forth inferential reasons such that: if this is the case, then that must be the case; if this exists, then that must exist; [301] if this is not the case, then that cannot be the case; if this does not exist, then that cannot exist. This type of inferential reasoning is able to prove whatever point is to be proven. Thus the four forms of rationality concern: (1) Based on how things actually exist in reality, what is each thing's own essential nature (*rationality of nature*)? (2) Based on that nature, how does it function (*rationality of functionality*)? (3) Based on that, how does one thing exist in dependence on another (*rationality of dependence*)? (4) Based on those kinds of relations, how should inferential evidence be applied (*rationality of inferential proof*)?

Each of these four forms of rationality is identified in the *Unraveling the Intention Sūtra* as follows: "Rationality should be understood to have four aspects: rationality of dependence, rationality of functionality, rationality of inferential proof, and rationality of nature."[398] Also, the *Compendium of Knowledge* says: "What reasoning is used to analyze phenomena when analyzing phenomena diligently? There are four forms of rationality: rationality of dependence, rationality of functionality, rationality of inferential proof, and rationality of nature."[399] When commenting in detail on each of the four forms of rationality based on the *Unraveling the Intention Sūtra*, Asaṅga's [302] *Śrāvaka Levels* first speaks about the rationality of dependence as follows:

> What is the rationality of dependence? Dependence has two aspects: dependence in terms of production and dependence in

> terms of designation. With regard to dependence in terms of production, the production of the aggregates is dependent on those causes and conditions through which the aggregates are produced. The designation of the aggregates is dependent on the collections of names, collections of phrases, and collections of letters through which the aggregates are designated. These are respectively called *dependence in production* and *dependence in designation* in the case of the aggregates. Dependence in production and dependence in designation are a form of rationality—a process or method—for examining the production and designation of the aggregates. Therefore they are called the *rationality of dependence.*[400]

Second, *Śrāvaka Levels* speaks about the rationality of functionality as follows:

> What is the rationality of functionality? The aggregates, which are produced by their own respective causes and conditions, engage in their own respective functions. For example, the eye functions to see forms, the ear functions to hear sounds, and so on, up to the mind [303] functions to know phenomena. Visible form is construed as the eye's object of experience, sound is construed as the ear's object of experience, and so on, up to phenomena being construed as the mind's objects of experience. In ways similar to these cases, things thus function causally in relation to each other, and the rationality—the process or method—for examining this is called the *rationality of functionality.*[401]

Third, concerning the rationality of inferential proof, it is as stated in the *Śrāvaka Levels* in the passage that begins: "What is the rationality of inferential proof? The aggregates being impermanent, or dependently arisen, or in the nature of suffering." The passage ends with "This is called the *rationality of inferential proof.*"[402]

Fourth, concerning the rationality of nature, *Śrāvaka Levels* says:

> What is the rationality of nature? It is as follows. Why are the aggregates like this and why does the world exist in this way?

> Why is solidity the defining characteristic of earth, wetness the defining characteristic of water, heat the defining characteristic of fire, and mobility the defining characteristic of wind? Why are the aggregates impermanent? [304] Why is nirvāṇa peace? Similarly, why is the defining characteristic of form "that which is able to obstruct," the defining characteristic of feeling "that which is experienced," the defining characteristic of discernment "that which distinguishes," the defining characteristic of conditioning factors "that which directly conditions," and the defining characteristic of consciousness "that which knows"? Such is the nature (Skt., *svabhāva*) of those phenomena, their character (*prakṛti*); it is the way they are (*dharmatā*). That nature is itself here a form of rationality—a process or method. Similarly, in order to comprehend and bear in mind "Either it is like this, or otherwise, or not at all," in all cases one relies just on the nature of things. This rationality of just the nature of things is called the *rationality of nature*.[403]

Śrāvaka Levels speaks about three kinds of rationality of nature: the rationality of nature commonly known in the world, such as fire being hot and water being wet; the rationality of inconceivable nature, which lies beyond the comprehension of most people; and the rationality of ultimate nature, such as the ultimate reality of things. This text says: "Through the rationality of nature, in regard to just how things are, or their commonly known nature, their inconceivable nature, [305] or their abiding (ultimate) nature, one attends to it; one does not think about it nor conceptualize it. In this way one seeks rationality."[404] In these ways *Śrāvaka Levels* explains in detail each of the four forms of rationality.

Thus the *rationality of dependence*, in the case of causally conditioned things such as a sprout, is that for it to be produced depends on its cause—a seed and so on. Dependence in production means that something that is a result has the nature or character of depending on its own causes and conditions. The same principle applies to dependence in designation, which concerns parts and wholes, or collections and members that are parts of those collections. Consider, for example, a pot and the components that are parts of the pot, or consider letters, words, phrases, and statements, which must be understood as things having a nature where one depends upon another.

The *rationality of functionality* needs to be understood as follows: conditioned things that arise from causes and conditions have the intrinsic quality or natural disposition to perform their own specific functions. For example, the eye functions to see forms, fire functions to burn, wind functions to move, and so on. [306]

The *rationality of inferential proof* is an investigation into cause and effect, nature, and so forth—something that needs to be proven that is not perceptually evident. It is inferential evidence that establishes this without contradicting perception and so forth (the three types of valid cognition).[405] Examples include proving something's impermanence because it is produced and proving the presence of fire from that of smoke.

The *rationality of nature* concerns, for example, that fire is hot, that water is wet, and that the aggregates and such like are conditioned by causes and conditions. These are the essence of those things. Since their essence is also naturally there, that kind of character or nature is called the *rationality of nature*. Also, when one appeals to the rationality of nature, it is like the final limit of explanation or understanding. If someone asks, "Why is it the nature of fire to be hot?" the answer is simply, "It is fire's nature to be hot—that is how things are." Also such questions as "Why is a physical object obstructive?" or "Why is it the nature of clear and cognizing consciousness to experience?" can only be answered by saying that it is the way it is. There is no other reason. [307]

Consequential Reasoning

Consequential reasoning is the means by which a distorted conviction is eliminated. In the explanations of the eight or sixteen categories of logic discussed by classical Indian epistemologists, consequences are referred to as "refutations." Dignāga's *Introduction to Entering into Valid Reasoning*, for example, says:

> Proof statements and refutations,
> along with their fallacious forms, serve to make others
> understand.
> Perception and inferential cognition,

> along with their fallacious forms, serve to make oneself understand.[406] [308]

This identifies the eight categories of reasoning: (1) correct and (2) fallacious proof statements, (3) correct and (4) fallacious refutations, (5) correct and (6) fallacious perceptions, (7) correct and (8) fallacious inferences. Among these, the latter four categories are methods that enable oneself to ascertain the object of knowledge. Moreover, since a reliable cognition ascertaining a hidden object of knowledge is an inference and a reliable cognition ascertaining a perceptually evident object of knowledge is a valid perception, there are two valid cognitions: perception and inference. Correspondingly, given that when one follows other types of cognition one is lead astray, their counterparts—fallacious perception and fallacious inference—are taught in order to eliminate such sources of error. The first four categories are methods to induce ascertainment in others after having understood it for oneself. Refutations are taught in order to demolish a distorted conviction. Proof statements are taught in order to clear away a doubt that oscillates between two points. The two fallacious kinds are taught for the purpose of eliminating sources of error.

Nāgārjuna's *Finely Woven* says: "Valid cognition and the object known are connected."[407] This passage identifies sixteen categories of logic: (1) valid cognition, (2) object of knowledge, (3) doubt, (4) purpose, (5) example, (6) tenet, (7) components of inference, (8) reasoning, (9) ascertainment, (10) debate, (11) presentation, (12) refutation, (13) fallacious inference, (14) quibble, (15) futile rejoinder, (16) defeat. [309]

In general, there are two types of cognition: one that realizes its object and the other that does not realize it. The latter has three types: simply not realizing the object, distortedly conceiving it, and doubting. Within distorted conceiving there is *reification*, which involves viewing something nonexistent to be existent, and *denial*, which involves viewing what is existent to be nonexistent. As for reification, there are the following two types: *intellectually acquired*, which grasps objects owing to certain reasoning or philosophical beliefs, and *innate naturally arising*, which do not grasp in the manner just described. Among these, the kinds of distorted conceptions that are dispelled by logical consequences are primarily the intellectually formed reifications and denials.

The Meaning of "Consequential Reasoning" and Its Types

Suppose someone posits as evidence something that his opponent[408] accepts, saying, "If what you accept were so, then that would follow," conveying a contradiction in the opponent's fallacious thesis by drawing out unwanted consequences. This is called a *consequence*. For example, say someone is convinced that sound is permanent. You can put forward the consequence "It follows that the subject, sound, is unproduced because it is permanent." Since the opponent perceives that sound is produced by causes and conditions, he or she accepts that it is produced. So in this case, the thing that the opponent is asserting—permanence—is put as the evidence "If sound were permanent just as you say, then it would be unproduced." [310] An unwanted meaning, that is, sound is unproduced, is an absurd consequence (given the opponent's other assumptions). Therefore, in terms of the etymology, it is called a *consequence* because an unacceptable conclusion follows from the opponent's assertions, or because a certain consequence is expressed.

To identify the subject of debate, the predicate, and the evidence in a consequence, let us use the following as an example: "It follows that the subject, sound, is unproduced because it is permanent." In this consequence "sound" is the *subject of debate*, "unproduced" is the *predicate* or the unwanted consequence, and "permanent" is the *evidence*. The *thesis* of this consequence is "sound is unproduced"; the actual *pervasion*[409] is "if it is permanent then it must be unproduced"; the *opposite of the predicate* is "produced," and the *opposite of the evidence* is "impermanent."

As explained in volume 1 of this series, a disputant may respond to a consequence with one of the following replies: (1) "I accept." (2) "Why?" (3) "The evidence is not established." (4) "There is no pervasion."[410]

Consequences are of two types: correct consequences and fallacious consequences. The definition of a *correct consequence* is: a consequence that, (a) through using evidence and a pervasion established from the opponent's perspective, entails a consequence unacceptable to the opponent, and (b) the opponent cannot answer coherently. Mokṣākaragupta's *Language of Logic* says: "'What is it that is being referred to here as a *consequence*?' It is a form of reasoning that yields a consequence unacceptable to the opponent by means of a logical pervasion that has been established by valid cognition."[411] [311]

Correct consequences are of two types: a correct consequence implying a proof and a correct consequence not implying a proof. The definition of a *correct consequence implying a proof* is: a correct consequence in which the inverse is characterized by the three modes. For example, consider an opponent who accepts that sound is permanent and has established by valid cognition that whatever is permanent must be unproduced; the opponent also knows that sound is produced. For this type of opponent, we can posit a consequence such as, "It follows that the subject, sound, is unproduced because it is permanent." This consequence brings to mind the three modes of a correct inference: "The subject, sound, is impermanent because it is produced." Here "produced," which is the inverse of the consequence's predicate, is now put as the evidence; and "impermanent," which is the inverse of the consequence's evidence, is now put as the predicate of the thesis. This form of consequence is thus called *implying a proof*. In a correct consequence implying a proof, the opponent must accept the evidence ("because it is permanent," in the example above). But he cannot have established this by valid cognition. For if he had established the evidence by valid cognition, then the thesis to be proven ("sound is impermanent," in the above example) by the inference implied by the consequence would be contradicted by valid cognition. Also, the opponent must have established the pervasion by valid cognition. For if he had not established it by valid cognition, then the negative pervasion of the inference that it implies would not be established. Also, the opponent must have established the negation of the consequence's thesis by valid cognition (e.g., that it is not the case that sound is unproduced). [312] For if not, then recognition of the evidence being an attribute of the subject (e.g., that sound is produced) in the inference implied by the consequence would not be established.

The definition of a *correct consequence not implying a proof* is: a correct consequence in which the inverse is *not* characterized by the three modes. Consider, for example, someone who has established by means of valid cognition that sound is produced and accepted that it is permanent; he also has established by means of valid cognition that whatever is produced must be impermanent. From these features of that person's position, we can draw out a consequence such as, "It follows that the subject, sound, is impermanent because it is produced." This consequence does not imply a proof. This is because there is no correct inferential evidence in which the opposite of the predicate in this consequence, "permanent," is put as the evidence, and the opposite of evidence, "unproduced," is put as the predicate of the thesis.

In brief, the function of a correct consequence implying a proof is as follows. It is a correct consequence that cannot be rebutted by the opponent, given that the opponent has accepted the evidence, established the pervasion by means of valid cognition, and the thesis is refuted by valid cognition. Thus it directly dispels the opponent's distorted conviction, which is the object to be eliminated. And it indirectly induces in the opponent's mind a valid cognition realizing the thesis of an inference on the basis of an implied proof statement that is characterized by the three modes of a correct proof. In ways such as these, this type of consequence is a method to help others. The function of a correct consequence not implying a proof is as follows. It directly dispels what is to be eliminated—the opponent's distorted conviction—by way of its being a correct consequence that cannot be rebutted by the opponent. [313]

The definition of a *fallacious consequence* is: a statement of a consequence that is unable to overturn a manifest distorted conviction that the consequence is supposed to eliminate. There are three types: (1) a consequence that states only the predicate, owing to the lack of expertise of the proponent who is positing it, (2) a consequence that states only the evidence, and (3) a consequence where the predicate and the evidence are the same. Examples of these are, respectively: "It follows that it is a conditioned thing," "Because it is a conditioned thing," and "It follows that the subject, a pot, is a functioning thing because it is a functioning thing."

A consequence is presented in order to clearly bring three points of direct contradiction to the opponent's mind. Since an appropriate opponent (for the correct consequence) has accepted the evidence of that consequence to be established, he cannot reply, "The evidence is not established." And since he has established the pervasion by means of valid cognition, he cannot reply, "There is no pervasion." And since the thesis is the opposite of what he believes the subject to be, he also cannot reply, "I accept." Thus he cannot counter the consequence with a reply. When the opponent ends up with "three points of direct contradiction," none of the three replies can be given. At that point it is a correct consequence. Here when the opponent, having seen the three points of direct contradiction, drops his thesis, the posited consequence is referred to as one that "has become a correct consequence." It should also be understood that at this point, since the distorted conviction has been eliminated, the previous consequence is no longer a correct consequence for that particular person. [314]

The Eight Modes of Pervasion of a Consequence

Moreover, there is an explanation of eight modes of pervasion in terms of how the pervasion of a consequence is ascertained.[412] The eight modes of pervasion of a consequence are: (1) the actual positive pervasion, (2) the inverse positive pervasion, (3) the actual converse pervasion, (4) the inverse converse pervasion, (5) the actual negative pervasion, (6) the inverse negative pervasion, (7) the actual contradictory pervasion, and (8) the inverse contradictory pervasion. Thus, among these eight, four are actual and four are inverse.

These eight are described as follows. (1) The actual positive pervasion: if something has that consequence's evidence as its property, then it must also have the consequence's predicate as its property. (2) The inverse positive pervasion: if something has that consequence's evidence as its property, then it cannot have the consequence's predicate as its property. (3) The actual converse pervasion: if something has that consequence's predicate as its property, then it must have the consequence's evidence as its property. (4) The inverse converse pervasion: if something has that consequence's predicate as its property, then it cannot have the consequence's evidence as its property. (5) The actual negative pervasion: if something does not have that consequence's predicate as its property, then it cannot have the consequence's evidence as its property. (6) The inverse negative pervasion: if something does not have that consequence's predicate as its property, then it cannot *not* have the consequence's evidence as its property. (7) The actual contradictory pervasion: if something has that consequence's evidence as its property, then it cannot have the consequence's predicate as its property.[413] (8) The inverse contradictory pervasion: if something has that consequence's evidence as its property, then it cannot *not* have the consequence's predicate as its property. [315]

[316] Furthermore, fallacious consequences can be categorized in many ways. These include (1) where both the pervasion and the evidence are not established, (2) where only the pervasion is not established, and (3) where

THE EIGHT MODES OF PERVASION OF A CONSEQUENCE	
1. The actual positive pervasion $E(x) \rightarrow P(x)$ If some x has the consequence's evidence as its property, then it must have the consequence's predicate as its property. *Example*: "It follows that a pot, the subject of predication, is impermanent because it is produced."	*2. The inverse positive pervasion* $E(x) \rightarrow \neg P(x)$ If some x has the consequence's evidence as its property, then it has the negation of the consequence's predicate as its property. *Example*: "It follows that a pot, the subject of predication, is impermanent because it is not produced."
3. The actual converse pervasion $P(x) \rightarrow E(x)$ If some x has the consequence's predicate as its property, then it has the evidence as its property. Example: "It follows that a pot, the subject of predication, is impermanent because it is produced."	*4. The inverse converse pervasion* $P(x) \rightarrow \neg E(x)$ If some x has the consequence's predicate as its property, then it has the negation of the evidence as its property. Example: "It follows that a pot, the subject of predication, is impermanent because it is not produced."
5. The actual negative pervasion $\neg P(x) \rightarrow \neg E(x)$ If some x has the negation of the consequence's predicate as its property, then it has the negation of the evidence as its property. *Example*: "It follows that a pot, the subject of predication, is impermanent because it is produced."	*6. The inverse negative pervasion* $\neg P(x) \rightarrow \neg(\neg E(x))$ If some x has the negation of the consequence's predicate as its property, then it does *not* have the negation of the evidence as its property. *Example*: "It follows that a pot, the subject of predication, is impermanent because it is not produced."
7. The actual contradictory pervasion $E(x) \rightarrow \neg P(x)$ If some x has the consequence's evidence as its property, then it has the negation of the consequence's predicate as its property. *Example*: "It follows that the subject, a pot, is permanent because it is produced."	*8. The inverse contradictory pervasion* $E(x) \rightarrow \neg(\neg P(x))$ If some x has the consequence's evidence as its property, then it does *not* have the negation of the consequence's predicate as its property. *Example*: "It follows that the subject, an impermanent thing, is permanent because it is an unproduced phenomenon."

only the evidence is not established. Within the category where both the pervasion and the evidence are established, the following types are possible: (1) both the pervasion and the evidence are established on the basis of a commitment to a theory, (2) both are established by means of valid cognition, (3) the pervasion is established by a commitment to a theory while the evidence is established by means of valid cognition, and (4) the pervasion is established by means of valid cognition while the evidence is established by a commitment to a theory. Given that each of these can be subdivided in some manner, there are many ways of categorizing consequences. Those who wish to know these in greater detail should learn them from the Collected Topics texts.[414] [317]

Proof Statements

When someone has a doubt oscillating between two standpoints, which occurs once his distorted conviction has been removed by means of a consequence, that doubt must be eliminated through the use of a proof statement. A *correct proof statement*, a *flawless statement of evidence*, and an *inference for the purpose of others* are synonymous. It is a "correct proof statement" because it is a statement expressing a correct proof. It is a "flawless statement of evidence" because it is a statement of correct evidence characterized by the three modes. It is an "inference for the purpose of others" because its purpose is to make someone else realize the thesis and it is the cause of the resulting inference. Mokṣākaragupta's *Language of Logic* says: "'For the purpose of others' indicates that which is for the sake of others. Inference for others is a statement. It is a statement that, demonstrating the three modes [of correct evidence], causes others to realize—that is, understand—it. So even though it is a verbal statement, it can metaphorically be referred to as an 'inference.'"[415] [318]

The definition of an inference for the purpose of others is: a flawless two-part statement demonstrating the three modes of the evidence seen by the proponent's own valid cognition to the opponent without any fault in wording, factual reference, and cognition. The two parts are the part expressing the pervasion and the part expressing the evidence as an attribute of the subject.

When it is taught that a correct proof statement must have "four distinctive features" those four are: (1) *Distinctive causal feature*: there must be a prior valid cognition that has already realized what is to be demonstrated. (2) *Distinctive feature of the subject matter*: it must express its subject matter—the three modes of correct evidence—without any fault of excess or omission. (3) *Distinctive feature of its character*: it must be free from any faults of the three—wording, factual reference, and cognition. (4) *Distinctive functional feature*: it must produce its effect, an inference.

The faults in wording, factual reference, and cognition refer to the following. When a proof statement demonstrates the three modes, if the three modes are established with respect to the facts but not established by the valid cognition of the proponent and the opponent, then this is a *fault of cognition*. If the three modes are not established with respect to the facts, then this is a *fault of factual reference*. If the three modes are expressed excessively or insufficiently, then this is a *fault of the wording*. The *Compendium of Valid Cognition* says:

> Inference for the purpose of others
> makes clear the facts seen by oneself.[416]

The phrase "seen by oneself" indicates that it is not enough for the three modes demonstrated by a proof statement to be merely accepted by the opponent; rather, they must be ascertained by the valid cognition of both the proponent and the opponent. The word "facts" indicates that it is not enough for the three modes demonstrated by a proof statement [319] to be posited by scriptural citation alone; rather, the empirical facts must be established as an object of valid cognition. The phrase "makes clear" indicates that when demonstrating the three modes to an opponent by means of a proof statement so as to produce a recollecting awareness that recalls the three modes, it must be free from the faults of exceeding the three modes and of not satisfying the three modes. *Exposition of Valid Cognition* also says:

> Some say that since it is others who are to be instructed
> [through the proof statement],
> the [evidence] that is understood by the other,
> even if it is not seen by oneself, is a means of proof.
> The words "seen by oneself" serve to counter this [false
> opinion].[417]

Passages such as these show extensively that a correct proof statement must be free from the faults in wording, factual reference, and cognition.

An example of a proof statement with faulty wording is: "Sound is impermanent because it is produced; an example is a pot. And just as a pot is produced, so sound is produced. Therefore sound is impermanent." A proof statement such as this has the fault of being both an excessive and an insufficient statement. By explicitly expressing its thesis, "sound is impermanent," this proof statement has the fault of excessive wording. And by not expressing the pervasion, it has the fault of insufficient wording. Also it has the fault of repetition because the application [of the example to the case at hand], "and just as a pot is produced, so sound is produced," repeats the expression of the [evidence as an] attribute of the subject "because it is produced." And the conclusion, "Therefore sound is impermanent," repeats the expression of the thesis, "Sound is impermanent." It is said, "This kind of [faulty] proof statement has five parts." [320] The five parts express: (1) the thesis, (2) the [evidence as an] attribute of the subject, (3) the concordant example, (4) the application [of the example to the case at hand], and (5) the conclusion. According to both masters Dignāga and Dharmakīrti, a proof statement having five parts is faulty. They explain that a proof statement has just two parts, expressing (1) the pervasion and (2) [the evidence as an] attribute of the subject.

There are two types of correct proof statements: a correct proof statement expressed in terms of a homologous case and a correct proof statement expressed in terms of a heterogeneous case. The definition of the former is: a flawless two-part statement explicitly demonstrating—without any fault in the wording, factual reference, and cognition—that the evidence is present only within the set of homologous cases of the thesis it establishes. An example is the proof statement "If something is produced, it is pervaded by being impermanent; an example is a pot; sound is also produced." Through adducing the example, this statement applies the example and the evidence as homologous instances of the predicate to be proven; hence, it is called "a correct proof statement expressed in terms of a homologous case." The definition of the latter (that is, in terms of a heterogeneous case) is: a flawless two-part statement explicitly demonstrating—without any fault in the wording, factual reference, and cognition—that the evidence is necessarily absent in the set of heterogeneous cases of the thesis it establishes. An example is a proof statement such as: "If something is permanent, it is pervaded by being unproduced; an example is space, which is not causally conditioned; sound is produced." Through adducing the example, this statement applies the example and the evidence as heterogeneous instances of the predicate to be negated; [321] hence, it is called "a correct proof statement expressed in terms of a heterogeneous case." Dignāga's *Entering into Valid Reasoning* says:

> A homologous example is a case that, similar to the inferential evidence, exists [only] where the thing to be proven is present. An example is "Whatever is produced is seen to be impermanent, such as a pot and so on." A heterogeneous example is a case that shows that the inferential evidence does not exist where the thing to be proven is absent. An example is "Whatever is permanent is seen to be unproduced, such as space and so on."[418]

The purpose of using a correct proof statement expressed in terms of a homologous case, such as proving that sound is impermanent as evidenced by being produced, is to remove the doubt that "being produced" is contradictory evidence in the proof that sound is impermanent. The purpose of using a correct proof statement expressed in terms of a heterogeneous case, such as proving that sound is impermanent as evidenced by being produced, is to remove the doubt that "being produced" is inconclusive evidence in the proof that sound is impermanent. *Ascertainment of Valid Cognition* [referring to Dignāga] says: "This is the procedure: both should be stated as antidotes to exclude contradictory and inconclusive evidence."[419] [322]

With regard to there being many different recipients of, and purposes for, positing an object in a proof statement, *Exposition of Valid Cognition* says:

> Regarding the example, the fact that [the thesis] is of the
>   nature [of the evidence]
> or that it is the cause [of the evidence]
> is demonstrated to someone who does not know [that
>   relationship];
> for those who know, the evidence alone is stated.

> Therefore, if one knows the relation [between evidence and predicate],
> the statement of either one of the two [homologous or heterogeneous examples]
> will, by implication, recall to mind the other one.[420]

Thus Dharmakīrti explains how a proof statement needs to be made to those people who have no knowledge of the relation between the evidence and predicate with regard to the example (either the intrinsic relation or the causal relation between the evidence and the predicate) so that they may gain an understanding of the two evidence-predicate relations. To those opponents who are versed in both forms of the pervading relation yet do not know that the evidence is an attribute of the subject, one must present a proof statement so that they understand just the evidence or just the attribute of the subject. However, to those who already know the attribute of the subject and both forms of the pervading relation, one need not posit a proof statement for the purpose of their understanding something they do not yet know. Rather, Dharmakīrti states that one must posit a proof statement expressed either in terms of homologous cases or in terms of heterogeneous cases in order to induce a simultaneous recollection of all three modes. Also, since a correct proof statement expressed in terms of a homologous case shows that the evidence is present only in the set of homologous instances of what is proven, one can implicitly know that the evidence is not present in the set of heterogeneous cases of that proof. Conversely, since a correct proof statement expressed in terms of a heterogeneous case explicitly shows that the evidence is only absent in the set of heterogeneous instances of what is proven, [323] one can implicitly know that the evidence is present in just homologous cases. Therefore, for a single opponent, a proof statement expressed in terms of either homologous cases or heterogeneous cases individually is sufficient; so it is said that there is no need to directly posit both.

In brief, for someone who has ascertained and not forgotten that the evidence is an attribute of the subject, the pervasion is expressed in order to establish the positive and negative pervasions. For someone who has ascertained and not forgotten the positive and negative pervasions, one states that the evidence is an attribute of the subject in order to establish that the evidence is an attribute of the subject. For someone who has ascertained and not forgotten that the evidence is an attribute of the subject, along with the positive and negative pervasions, both are expressed.

Also, there are cases in which the proof needs to be expressed first and the three modes are established only afterward. When someone who has understood the three modes individually is able to recollect them simultaneously, the proof statement posited earlier becomes a correct proof statement for that person.

The function of a proof statement is to give rise to an understanding that simultaneously recollects all three modes. *Exposition of Valid Cognition* says:

> The intrinsic ability [to prove the thesis]
> lies really in the three modes;
> and the ability to give rise to the recollection of these
> lies in the correct statement [of the tri-modal proof].[421]

Here Dharmakīrti explains how the intrinsic capacity to prove the thesis is correctly located in the evidence that is characterized by the three modes, and the capacity to directly produce the recollection of the three modes resides in the proof statement expressing the three modes. [324]

THE THESIS: THE IMPLIED CONTENT OF THE PROOF STATEMENT

Now having briefly explained the topic of proof statements, let us discuss the nature of a thesis implied by a proof statement. In general, the word *thesis* or "what is to be established" (also known as the *probandum*) can have numerous senses. For example, the term can refer to the *purpose* for which a person strives, or to what they *act upon*, or to the *goal* of their activity. Here, in this particular context of the science of reasoning, the term *thesis* refers to the following. Consider someone who, on the basis of taking the presence of smoke rising on a mountain pass as the evidence, infers the presence there of fire. Here the smoke is the evidence or reason, and given that what is being proven is the presence of fire on that smoky pass, the existence of fire on that pass is the "thesis to be proven." Since such a thesis is to be understood on the basis of evidence, it is "that which is to be understood." And since it is being inferred by means of evidence, it is the "thing to be inferred by evidence." It is the "validly inferred object," since it is the object inferred through ascertaining the three modes of valid evidence. Since it is taken as a position by a party in debate it is the "position"; and since it is taken as a proposition by a party in debate it is the "proposition." These terms are all synonymous.

In view of the above, the presence of fire is referred to as the "predicate of the thesis," since it is the particular or universal feature of the thesis "the existence of fire on that pass." The smoky pass is the "locus of debate" in that it is the locus in reference to which the proponent and opponent discuss whether there is fire. [325] It is also referred to as the "inferential locus," since the smoky pass is the locus where the presence of fire is being inferred through inferential cognition of the thesis. Those two expressions, the "locus of debate" and the "inferential locus," are synonymous. All this is from the perspective of the terminology used within the texts of the science of inferential reasoning.

The *thesis* refers to something that is to be understood on the basis of some appropriate evidence. It can be of two types: a correct thesis, such as "sound is impermanent," and a fallacious thesis, such as "sound is permanent." What is known as a *correct thesis* is one that must be realized by valid inferential cognition in dependence on correct reasoning or evidence. Within this category of a correctly reasoned thesis, there is a "thesis for one's own purpose" in the context of establishing a thesis by way of employing evidence by oneself, and a correctly reasoned thesis in the context of the purpose of others is known as a "thesis for the purpose of others." Here, in the following, we will discuss the notion of a thesis to be proven for the sake of others.

The definition of a thesis for the purpose of others is: that which satisfies the five features: (1) proper form, (2) only, (3) intended, (4) himself, (5) not refuted. The phrase "proper form" indicates that the object to be proven on that occasion is actually something to be established; it must be something that has not already been ascertained by the valid cognition of the opponent. The word "only" indicates that the disputants take it as something to be inferred and not as evidence. The word "intended" indicates that, regardless of whether a disputant has verbally expressed it, it must be something that he accepts as the thesis to be inferred on the basis of that evidence. The word "himself" indicates that the disputant on that occasion must himself accept that object to be the thesis to be proven. [326]

The phrase "not refuted" indicates that it must not be refuted by valid cognition. The *Compendium of Valid Cognition* says:

> [A valid thesis] is one that is intended by [the disputant] himself
> as something to be stated in its proper form only; it is not refuted.[422]

Each feature stated here in the *Compendium of Valid Cognition* has the purpose of both excluding something and including its opposite. The phrase "its proper form" in the definition of a proof statement for the purpose of others excludes and includes as follows. In terms of exclusion, anything that has already been established by an opponent targeted by a specific argument is excluded from being a correct thesis of that specific argument. In terms of what is affirmed or included, if something is a correct thesis of a specific argument for the purpose of others, then it must not have been already established by the opponent targeted by that specific argument. This is the significance of the phrase "its proper form." Devendrabuddhi's *Commentary on Difficult Points in the Exposition of Valid Cognition* says: "The phrase 'its proper form' indicates [the thesis] is not established."[423]

The word "only" (Skt., *eva*) in the verse has the purpose of excluding and including the following. In terms of exclusion, evidence posited as the evidence for a specific argument and evidence with an unestablished example are excluded from being the correct thesis of that specific argument. In terms of what is included, if something is a correct thesis of a specific argument for the purpose of others, [327] then it cannot at that time be posited as evidence. This is the significance of the word "only" here. *Commentary on Difficult Points in the Exposition of Valid Cognition* says: "The phrase 'through its proper form only' indicates that it is not the evidence."[424]

The word "intended" in the verse has the purpose of excluding and including the following. In terms of exclusion, it excludes the notion that if something is a correct thesis of a specific argument, it must necessarily be explicitly stated by a correct proof statement of that specific argument. In terms of what is included, that very thesis is implicitly expressed by the proof statement. This is the significance of the term "intended" here.

Commentary on Difficult Points in the Exposition of Valid Cognition says: "The word 'intended' means that [the thesis] becomes an object through intention."[425]

The word "himself" in the verse has the purpose of excluding and including the following. In terms of exclusion, it excludes the notion that every scriptural claim accepted by the opponent is a correct thesis of that specific argument. In terms of what is included, if something is a correct thesis of a specific argument for the purpose of others, then it must be accepted as the thesis by the proponent. This is the significance of the term "himself" here. *Commentary on Difficult Points in the Exposition of Valid Cognition* says: "The word 'himself' indicates that it is something the person considers as the thesis to be established by the stated proof."[426] [328]

The phrase "not refuted" in the verse has the purpose of excluding and including the following. In terms of exclusion, if something is refuted by valid cognition, then it is excluded from being a correct thesis for the purpose of others. In terms of what is included, if something is a correct thesis, then it must be established by means of valid cognition. This is the significance of the phrase "not refuted" here. *Commentary on Difficult Points in the Exposition of Valid Cognition* says: "It is not undermined by the two types of valid cognition: perception and inferential cognition."[427]

In brief, to be a correct thesis for the purpose of others: (1) it must not yet be established by the appropriate opponent's valid cognition, (2) it must not be something that is presently being used as a proof, (3) the disputants must have the intention for it to be inferred, (4) the disputants must accept it to be the thesis, and (5) it must not be refuted by valid cognition.

There are two types of theses: an explicit thesis and an implicit thesis. For example, in the argument proving that sound is impermanent as evidenced by being produced, the explicit thesis is "sound is impermanent," and the implicit thesis is "sound is not permanent." [329]

The Inferential Evidence: The Explicit Content of a Proof Statement

Inferential evidence is the actual means of ascertaining the thesis itself, as explained above. Since it is the only means of ascertaining a hidden phenomenon, it is crucially important. The cognition of perceptible things, as when a visual consciousness sees a form or an auditory consciousness hears a sound, does not depend on the use of evidence. However, the cognition of hidden phenomena, such as a physical form being impermanent and momentary, must depend on reliable inferential evidence characterized by the three modes. The three modes of correct evidence are also presented in Dharmakīrti's *Exposition of Valid Cognition*:

> [Correct] evidence is an attribute of the subject of the thesis,
> and it is pervaded by the other part [of the thesis—the predicate].
> Such evidence is only of three kinds due to the necessary relation of
> unaccompanied nonoccurrence.[428]
> Fallacious evidence is other than these three.[429]

This stanza presents the three modes, the defining characteristics of correct inferential evidence. It also states that correct inferential evidence is only of three types: (1) *effect* evidence, (2) *nature* evidence, and (3) evidence consisting in *nonperception*. It further states that the grounds for this classification lie in the fact that only two kinds of relations—intrinsic relations and causal relations—pertain to the invariable relation of unaccompanied nonoccurrence. [330] Finally, it states how the inverse of these three types of evidence is fallacious evidence.

In general, the definition of evidence is: that which is adduced as evidence. When applied to a specific case, the definition of evidence in the proof that sound is impermanent is: that which is adduced as evidence in the proof that sound is impermanent. The phrase "that which is adduced as evidence" means the same as "that which is posited as a reason." For example, the existence of smoke is posited as a reason in the proof that there is fire on the mountain pass. The terms *proof, evidence, reason, inferential evidence*, and *cause of understanding* are all synonymous.[430] There are two types of evidence: correct evidence and fallacious evidence. The terms *correct evidence, correct reason*, and *correct inferential evidence* are all synonymous. The definition of correct evidence is: that which is characterized by the three modes. When applied to a specific case, the definition of correct evidence in the proof that sound is impermanent as evidenced by being produced is: that which is characterized by the three modes proving that sound is impermanent as evidenced by being produced. For example, "produced" is the evidence in the case of arguing "Sound is impermanent because it is produced."

THE THREE MODES: THE CRITERIA OF CORRECT EVIDENCE

In the context of the science of inferential evidence, any evidence that is capable of proving a thesis must satisfy the three modes. *Introduction to Entering into Valid Reasoning* says:

> Inferential evidence is characterized by the three modes. What are these three modes? [331] That [the evidence] is an attribute of the thesis's subject, that it is ascertained to be present only in the set of homologous cases, and that it is also ascertained to be only absent in the set of heterogeneous cases.[431]

According to this source, the three modes are: (1) being an attribute of the subject of the thesis (Skt., *pakṣadharma*), (2) the positive pervasion (*anvaya*), and (3) the negative pervasion (*vyatireka*). Sometimes the three modes are also referred to, respectively, as the first mode, the second mode, and the third mode. For example, in the case of arguing that sound is impermanent, "produced" is an attribute of the thesis's subject, "sound," and it stands in a relation of both positive pervasion and negative pervasion with respect to the predicate, "impermanent." Each of the three modes may be defined in their respective order as follows, in relation to a specific case.

Applied to a specific case, the evidence being an attribute of the subject of the thesis in the proof that sound is impermanent as evidenced by being produced is defined as the ascertainment by valid cognition that, in accord with the way the proof has been stated, the evidence ("produced") is present in "sound," which is the subject of the thesis sought to be proven. When a person has ascertained by valid cognition that sound is produced and then doubts whether sound is impermanent, for that person the evidence is established as an attribute of the subject of the thesis. In this case, sound is the subject of the thesis of interest in the proof that sound is impermanent as evidenced by being produced.

The positive pervasion in the proof that sound is impermanent as evidenced by being produced is defined as the ascertainment by valid cognition that, in accord with the way the proof has been stated, the evidence is present in *only* the set of homologous cases owing to its relation to the predicate of the thesis to be proven. [332] The negative pervasion in the proof that sound is impermanent as evidenced by being produced is defined as the ascertainment by valid cognition that, in accord with the way the proof has been stated, the evidence is *only* absent in the set of heterogeneous cases owing to its relation to the predicate of the thesis to be proven.

With regard to establishing the positive pervasion in a proof that sound is impermanent as evidenced by being produced, the opponent must ascertain by valid cognition the relation between "produced" and "impermanent," and that "produced" is present only in something that is impermanent. Likewise, with regard to establishing the negative pervasion in a proof that sound is impermanent as evidenced by being produced, the opponent must ascertain by valid cognition the relation between "produced" and "impermanent," and that "produced" is not present in anything that is permanent.

When some given evidence is present only in the set of homologous cases in a specific proof, it necessarily precludes its presence in the set of heterogeneous cases; therefore the positive pervasion and the negative pervasion are mutually inclusive. Dharmakīrti taught that the positive and negative pervasions of a specific argument can be realized either explicitly or implicitly by a single cognition.

To establish both the positive and the negative pervasions of a given proof statement, not only does this need to be preceded by three valid cognitions, one must also ascertain the relation between the evidence and the predicate. The three valid cognitions being referred to here are: (1) valid ascertainment that the predicate of the probandum (thesis) and the predicate of the negandum (antithesis) are directly contradictory, (2) valid ascertainment of the evidence itself, and (3) valid cognition that negates the evidence's presence in the predicate of the negandum. This may be explained by applying it to an example: "The subject, sound, is impermanent because it is produced; an example is a pot." The three valid cognitions that must go before establishing the evidence's positive pervasion in a proof that sound is impermanent as evidenced by being produced are the following: [333] valid ascertainment that "permanent" and "impermanent"—which are, respectively, the predicate of the negandum and predicate of the probandum for that evidence—are directly contradictory; valid ascertainment of the evidence itself, namely, "produced"; and valid cognition that negates the evidence, "produced," to be present in the predicate of the negandum, "permanent." The same pattern applies to the other instances of proof statements.

The term *ascertainment* is included in the definitions of the three modes for the following reasons. Taking the example of reasoning that "Sound is impermanent as evidenced by being produced," the inclusion of the term *ascertainment* as part of the definition of the first mode—the "attribute of the subject of the thesis" of that argument—precludes "produced" from being unestablished evidence in the proof that sound is impermanent. Likewise, the inclusion of the term *ascertainment* as part of the definition of the second mode—the positive pervasion of that argument—precludes "produced" from being contradictory evidence in the proof that sound is impermanent. And including the term *ascertainment* as part of the definition of the third mode—the negative pervasion of that argument—precludes "produced" from being inconclusive evidence in the proof that sound is impermanent. The *Exposition of Valid Cognition Autocommentary* says:

> Therefore [Dignāga] said that the ascertainment of inferential evidence in terms of the three modes prevents unestablished, contradictory, and inconclusive evidence.
>
> When there is no relation, there can be no ascertainment of the positive and negative pervasions. Therefore, indicating this very point, he stated the ascertainment [of the three modes]. By ascertaining the positive pervasion, the *contradictory* and those in its similar category are excluded; and by ascertaining the negative pervasion, the *inconclusive* and those in its similar category, such as remainder evidence (Skt., *śeṣavat*), are excluded.[432] [334]

In general, the reason one can set forth a reliable presentation of evidence on the basis of establishing invariable logical relations is because the thesis is being proven on the basis of correct evidence that stands in the relation of unaccompanied nonoccurrence with the predicate. This means that when the pervaded is present, the pervader too is necessarily present, and when the pervader is absent, the pervaded too is necessarily absent. Similarly, it is in dependence on a specific cause that an effect comes into being, and when the cause is absent, the effect too is necessarily absent.

For example, on a smoky mountain pass in the east, one can infer the presence of fire when there is smoke. This is because smoke is related causally to fire. Similarly, for example, on a lake without fire at night, one can infer the absence of smoke when there is no fire. The main point here is that there is a cause-and-effect relation between smoke and fire. Likewise, on the basis of the presence of a juniper one can prove the presence of a tree, and similarly, one can prove the absence of a juniper on the basis of the absence of a tree because there exists an intrinsic relation between a tree and a juniper. If there is no such relation, one cannot establish the positive and negative pervasions. For example, you cannot prove that a person has horses with the evidence that he has cows; similarly, you cannot prove that a person has no cows with the evidence that he does not have horses. This is because there is no relation of unaccompanied nonoccurrence between cows and horses whereby one would not occur without the other. *Exposition of Valid Cognition* says: [335]

> If that were not so, then by excluding one thing,
> why would there be the exclusion of another?
> By saying "That person has no horses,"
> must he thereby also have no cows?
>
> Likewise, how could the presence of one thing
> follow from the presence of another?
> By saying "This person has cows,"
> must he thereby also have horses?[433]

These are the components of the inference "The subject, sound, is impermanent because it is produced":

Subject	sound
Predicate of the probandum	impermanent
Evidence	produced
Probandum (thesis)	Sound is impermanent.

20

Categories of Correct Evidence

Correct evidence in a general sense is of three types: (1) correct effect-evidence, (2) correct nature-evidence, and (3) correct evidence consisting in nonperception. From the perspective of the predicate of the probandum, correct evidence is of two types: (1) correct evidence for a negation and (2) correct evidence for an affirmation. From the perspective of how the proof functions, correct evidence is of two types: (1) correct evidence proving the expression [336] and (2) correct evidence proving the things themselves. From the perspective of its thesis, correct evidence is of three types: (1) correct evidence based on empirical fact, (2) correct evidence based on popular convention, and (3) correct evidence based on trustworthy testimony. From the perspective of how it is present in homologous cases, correct evidence is of two types: (1) correct evidence that is pervasively present in the set of homologous cases and (2) correct evidence that is present in a subset of the set of homologous cases. From the perspective of the respondent, correct evidence is of two kinds: (1) correct evidence for oneself and (2) correct evidence for the purpose of others. All of these types of evidence, categorized in various ways, are subsumed within the three types of correct inferential evidence—effect, nature, and nonperception. Dharmakīrti's *Drop of Reasoning* says: "There are only three types of evidence that are characterized by the three modes: nonperception, nature, and effect."[434]

Furthermore, correct evidence is confined to effect, nature, and nonperception because relations are confined to causal and intrinsic relations. In general, all forms of correct evidence fall into one of two categories: correct evidence for a negation and correct evidence for an affirmation. In the case of evidence for a negation, regardless of the kind of relation there is between the evidence and the predicate, insofar as it is negating evidence, it is evidence consisting in nonperception. In evidence used to affirm something, the relation between evidence and predicate must be either a causal relation or an intrinsic relation. The kind of evidence that establishes a causal relation are effects, and the kind of evidence that establishes the intrinsic relation is a thing's nature. [337]

Correct Effect-Evidence: An Effect Used as Evidence

The definition of correct effect-evidence is: that which is characterized by the three modes pertaining to an effect. An instance of something that is correct effect-evidence in a specific proof is defined as follows. It is correct evidence for an affirmation in a specific proof—and in the proof using this evidence, there is something that is both the evidence's cause and the object taken as the explicit predicate of the probandum. Here the actual predicate of the probandum's effect is characterized by the three modes and is posited as the evidence; it is thus called *correct effect-evidence*. For example, when smoke is posited as evidence in the proof of the existence of fire on a smoky mountain pass, that smoke is the effect of the fire that is the predicate of the probandum, and it is characterized by the three modes in the proof of the existence of fire located on the mountain pass. Thus it is said to be "characterized by the three effect modes."

There are five types of correct effect-evidence: (1) correct effect-evidence proving the actual cause, (2) correct effect-evidence proving the preceding cause, (3) correct effect-evidence proving a general cause, (4) correct effect-evidence proving a particular cause, and (5) correct effect-evidence inferring the cause's attributes. These are types of effect evidence that prove, respectively, the actual cause, the preceding cause, the mere existence of a cause, the existence of an additional cause, and the cause's attributes. [338]

First, regarding correct effect-evidence proving the actual cause, the *Sūtra of the Ten Dharmas* says:

> Just as fire is known from smoke
> and water from waterfowl . . .[435]

As is illustrated by this citation, we can posit, for example, "On a smoky mountain pass, there is fire because there is smoke," and, "In a place in the east where there are waterfowl, there is water because waterfowl have been hovering about for a long time." *Exposition of Valid Cognition* says:

> An effect is inferential evidence for a cause
> in terms of that number of essential properties
> without which it could not occur.[436]

Second, regarding correct effect-evidence proving a preceding cause, we can posit the following example: "The awareness of a newborn infant is preceded by a prior awareness because it is an awareness." *Exposition of Valid Cognition* says:

> When taking birth,
> respiration, the senses, and awareness
> do not arise from the body alone
> without relying on something similar to them.[437]

Third, regarding correct effect-evidence proving a general cause, we can posit, for example, "The aggregates that are characterized by suffering must have their own causes because they arise episodically." *Exposition of Valid Cognition* says: [339]

> Because it is episodic [i.e., only occurs sometimes],
> suffering is established to be caused.[438]

Fourth, regarding correct effect-evidence proving a particular cause, we can posit, for example, "A perceptual sense consciousness apprehending a form has another condition apart from its dominant condition and immediately preceding condition because it does not arise through the mere fulfillment of its dominant condition and immediately preceding condition and it does nonetheless arise occasionally." *Exposition of Valid Cognition* says:

> Since sense consciousnesses do not always occur
> when those causes are present,
> it can be inferred that there is some other cause.[439]

Fifth, regarding correct effect-evidence inferring the cause's attributes, we can posit, for example, "In a lump of brown sugar in the mouth, the causal complex that produced the preceding taste of the sugar has the ability to cause the present form of the sugar because there is now the taste of sugar." *Exposition of Valid Cognition* says:

> By means of an inference of the cause's attributes,
> through taste one cognizes the [other qualities] such as form
> that depend on the same causal complex,
> just as changes occurring in firewood are inferred from
> smoke.[440]

Here the "cause" refers to the preceding moment of the substantial cause of the taste; and the "attributes" [340] refers to capacity of the causal complex that includes the earlier taste to generate the subsequent instance of form. Since it establishes that such a capacity is present in the lump of brown sugar in one's mouth, it is called "effect evidence inferring the cause's attributes." Also, (a) the earlier taste of sugar is the direct substantial cause of the later taste and is the direct cooperative condition of the later sugar's form, and (b) the earlier sugar's form is the direct substantial cause of the later sugar's form and is the direct cooperative condition of the later sugar's taste. Therefore both the taste and the form of the sugar are established as having a relation dependent on a single, direct causal complex.

In general, as part of the presentations of effect evidence in the epistemological texts, there are extensive discussions about the validation of the causal relation between fire and smoke. This is not simply to help one understand the causal relation between fire and smoke. The aim of such a presentation is to use the relationship between fire and smoke as an illustration so that one can understand the nature of causal relation in the case of all instances of cause and effects, whether they are part of the external world or they are part of the internal world of experience. [341]

Correct Nature-Evidence: A Thing's Nature Used as Evidence

The definition of correct nature-evidence is: that which is characterized by the three modes in terms of intrinsic nature. Applied to a specific case, the definition of correct nature-evidence in a specific proof is as follows. Some *x* is correct evidence for an affirmation in a specific proof, and whatever is held to be the explicit predicate of the probandum in that proof is necessarily of the same nature as *x*. We can posit, for example, "Sound is impermanent because it is produced." *Kanakavarṇa's Past Practice* says: "All phenomena that have arisen even slightly are subject to cessation within a short while."[441] *Exposition of Valid Cognition* says:

> A natural property is evidence for another natural property
> that is invariably concomitant with the mere presence [of the
> property used as evidence].[442]

There are two types of correct nature-evidence: correct nature-evidence involving qualification and correct nature-evidence free of qualification. The definition of correct nature-evidence involving qualification in a specific proof is: [342] it is correct nature-evidence in a specific proof where the term expressing it implies an agent as its qualifying feature. We can posit, for example, "The sound of a conch is impermanent because it arises from effort." The definition of correct nature-evidence free of qualification in a specific proof is: it is correct nature-evidence in a specific proof where the term expressing it does not imply an agent as its qualifying feature. We can posit, for example, "Sound is impermanent because it is a real thing." Moreover, the words "arises from effort" imply a living being as the agent, while the words "real thing" imply no agent at all. *Exposition of Valid Cognition* says:

> [When adduced as evidence], a natural property, whether
> depending on a particular contingent condition or on its
> own,
> is stated to prove the probandum. For example, either "the fact
> of being an effect"
> or "existence" are stated to prove [inherent] disintegration.[443]

This discusses nature evidence that is stated in order to prove the probandum. In the case of those involving qualification, the very words that express the evidence depend upon implying its particular agent as a qualifying feature, such as stating that sound is an effect as the evidence to prove that sound [inherently] disintegrates. Evidence that stands alone does not depend upon such implication, such as positing a functioning thing's existence as the evidence for its disintegration. [343]

Buddhist epistemological texts explain that a *produced* thing is intrinsically connected to its *disintegration* from the very moment it comes into being. For a produced thing to disintegrate requires no other cause than the cause that gave rise to it. Rather, the very cause of its existence guarantees its disintegration. It is along these lines that Buddhist epistemological treatises present the topic of nature evidence extensively in order to establish all conditioned things as impermanent, to explain the theory of exclusion (*apoha*) in a detailed manner, and to identify the intrinsic relation and its meaning. In brief, *nature evidence* refers to evidence for an affirmation where the evidence and the predicate have the same intrinsic nature.

Correct Evidence Consisting in Nonperception

The definition of correct evidence consisting in nonperception is: that which is characterized by the three modes pertaining to nonperception.[444] When applied to a specific case, the definition of correct evidence consisting in nonperception in a specific proof is: it is correct evidence in a specific proof where there occurs a common locus of a negation and the object held as the explicit predicate of the probandum in the proof of that probandum by the evidence. We can posit, for example, "On a lake without fire at night, there is no smoke because there is no fire." [344] The *Sūtra Teaching the Nonarising of All Phenomena* says:

> Having seen this fact, the Tathāgata gave the following teaching. A person should not judge another. Were a person to judge another, that person would be harmed. Only I and others who are like me can judge other people.[445]

Exposition of Valid Cognition says:

> In all cases, negations are
> effected through nonperceptions

> because, for those who speak of [existence as] established by
> valid cognitions,
> the opposite [establishes nonexistence] by implication.[446]

There are two types of correct evidence consisting in nonperception: correct evidence consisting in nonperception of the imperceptible and correct evidence consisting in nonperception of the perceptible. In the case of evidence consisting in the nonperception of the imperceptible, the referent of the predicate of the negandum, even though it exists, is not perceptible to the opponent. In the case of evidence consisting in the nonperception of the perceptible, if that object existed, it would be perceptible to the opponent. Furthermore, each of these has two types: evidence consisting in the nonperception of a related object, where the negation of an object related to the predicate of the negandum is adduced as evidence; and evidence consisting in the perception of something contradictory, where the perception of something contradictory to the predicate of the negandum is adduced as evidence. [345]

Correct Evidence Consisting in the Nonperception of the Imperceptible

The definition of correct evidence consisting in the nonperception of the imperceptible in a specific proof is: it is correct evidence consisting in nonperception in a specific proof where it is the case both that, in the proof with this as its evidence, the referent of the predicate of the negandum exists in general, and that, for the person who has ascertained that this evidence is an attribute of the subject in that proof, the referent of the predicate of the negandum is imperceptible. We can posit, for example, "Regarding the space in front of oneself, it is not reasonable for a person for whom a ghost is an imperceptible thing to claim 'There is a ghost,' because a person for whom a ghost is an imperceptible thing has not perceived a ghost with valid cognition." *Exposition of Valid Cognition* says:

> The nonoccurrence of valid cognitions
> results in nonengagement with the nonexistent.[447]

In general, things are imperceptible because of their spatial location, their temporal location, their nature, and so on. In the present context, since ghosts are by their very nature very subtle, they remain imperceptible to some people and are not ascertained by their valid cognition. So it would be inappropriate for such a person to make any claim about the presence or absence of ghosts. In the same manner, it would be unreasonable for ordinary beings to claim that "All things exist in such-and-such a way" [346] while their valid cognition does not ascertain such a truth. Similarly, it would be inappropriate for persons to make allegations about such-and-such faults in others while they have not ascertained through valid cognition the presence of those faults in others. In this and other ways, it is shown that it is not right to engage in acts of exaggeration or denial while one does not see through one's own valid cognition what is actually the case.

Correct Evidence Consisting in the Nonperception of the Perceptible

The definition of correct evidence consisting in nonperception of the perceptible in a specific proof is: it is correct evidence consisting in nonperception in a specific proof where, if the referent of the predicate of the negandum in that proof existed on the basis of negation, it would appear to the opponent. There are two types: correct evidence consisting in the nonperception of something perceptible that is related and correct evidence consisting in the perception of something perceptible that is contradictory. *Exposition of Valid Cognition* says:

> Some [instances of nonperception] result in knowledge of
> actual absence
> on the basis of specific types of evidence.
>
> There are four kinds of nonperception that prove the absence
> of something:
> those that prove (1) the perceptible presence of something
> *contradictory* [to the negandum] or (2) the perceptible *effect*
> [of something contradictory]
> and those that disprove the presence of a perceptible (3) cause
> or (4) nature.[448] [347]

The definition of correct evidence consisting in the nonperception of something perceptible that is related is: correct evidence consisting in the nonperception of something perceptible where the negation of something related to the predicate of the negandum in that proof is adduced as the evidence. There are four types: (1) nonperception of a perceptible cause, (2) nonperception of a perceptible pervader, (3) nonperception of a perceptible nature, and (4) nonperception of a perceptible immediate effect. These are posited in terms of the adducing as evidence the negation of the cause, pervader, essential nature, or immediate effect of the predicate of the negandum.

Illustrations of these four, respectively, are as follows. First is "there is no fire" in the case of positing "On a lake without fire at night, there is no smoke because there is no fire." Here the negation of fire—the cause of the smoke, which is the predicate of the negandum—is adduced as evidence, and it is correct evidence; thus it is called "correct evidence consisting in the nonperception of a perceptible cause."

The second is illustrated by "there are no trees" in the case of positing "On a rocky cliff without trees, there are no ashoka trees because there are no trees." Here "the presence of a tree" is a pervader of "the presence of an ashoka tree," which is the predicate of the negandum; the negation of the presence of trees is adduced as evidence, and it is correct evidence; thus it is called "correct evidence consisting in nonperception of a perceptible pervader."

The third is illustrated by "a pot is not perceived" in the case of positing "In a place where a pot is not perceived by valid cognition, a pot does not exist because a pot is not perceived by valid cognition." Here the negation of a pot perceived by valid cognition—the essential nature of an existent pot, which is the predicate of the negandum—is adduced as evidence, and it is correct evidence; [348] thus it is called "correct evidence consisting in the nonperception of a perceptible nature."

Fourth is "there is no smoke that is its immediate effect" in the case of positing "In a roofless enclosure without smoke, the immediate, unobstructed cause[449] of smoke does not exist because there is no smoke that is its immediate effect." The negation of smoke—the immediate effect of its immediate, unobstructed cause, which is the predicate of the negandum—is adduced as evidence, and it is correct evidence; thus it is called "correct evidence consisting in the nonperception of a perceptible immediate effect."

The definition of correct evidence consisting in the perception of something perceptible that is contradictory is: correct evidence consisting in the nonperception of something perceptible where the perception of something that contradicts the predicate of the negandum in that proof is adduced as evidence. There are two types: correct evidence consisting in the perception of something contradictory in the sense of being contrary and correct evidence consisting in the perception of something contradictory in the sense of being mutually exclusive. The distinction between these is expressed in *Exposition of Valid Cognition*:

> A contradictory quality, whether different or not different,
> can be used [in a nonperception inference];
> for example, "fire" in proving the absence of snow [different nature and contrary]
> and "existence" to refute arising [same nature and mutually exclusive].[450]

In the case of correct evidence consisting in the perception of something contradictory in the sense of being contrary, [349] the evidence and the predicate of the negandum must be different substances, as in the case of positing fire as the evidence to prove the absence of snow. In the case of correct evidence consisting in the perception of something contradictory in the sense of being mutually exclusive, the evidence and the predicate of the negandum cannot be different substances, as in the case of pointing out that effects such as a sprout already exist as evidence to prove that the sprout does not need to be produced again.

The definition of correct evidence consisting in the perception of something contradictory in the sense of being contrary is: it is correct evidence consisting in the perception of something contradictory in a specific proof where, owing to being contradictory in the sense of being contrary, it cannot coexist with the predicate of the negandum in that proof. The definition of correct evidence consisting in the perception of something contradictory in the sense of being mutually exclusive in a specific proof is: it is correct evidence consisting in the perception of something contradictory where, owing to being contradictory in the sense of mutually excluding each other, it cannot coexist with the predicate of the negandum in that proof.

Correct evidence consisting in the perception of something contradictory in the sense of being contrary is of six types: (1) perception of something that contradicts the *cause* of the predicate of the negandum, (2) perception of something that contradicts the *pervader* of the predicate of the negandum, (3) perception of something that contradicts the *nature* of the predicate of the negandum, (4) perception of the *effect of something that contradicts* the predicate of the negandum, (5) perception of something that contradicts the *effect* of the predicate of the negandum, (6) perception of the *effect of something that contradicts the cause* of the predicate of the negandum. An example of the first type, perception of something that contradicts the cause, would be "completely pervaded by a very powerful fire" in the case of positing "In a place in the east completely pervaded by a very powerful fire, there is an absence of persistently bristling hair as an effect of cold because of being in a place completely pervaded by a very powerful fire." Once it has been shown that fire and tangible cold[451] are the contradictor and the contradicted object, [350] we can take tangible cold's nature, tangible cold's pervaded object, tangible cold's effect, and tangible cold's cause as what are contradicted by fire. Also, we can take fire, fire's pervaded object, and fire's effect as what contradicts cold. Tangible cold's *nature* refers to tangible coolness; its *pervaded object* refers to the coldness of snow; its *effect* refers to continuously bristling hair as an effect of cold; its *cause* refers to the immediate unobstructed cause of tangible cold. Also, fire's *pervaded object* refers to fire in the east; its *effect* refers to strongly billowing smoke. From this, the formulation of the other types of correct evidence consisting in the perception of something contradictory (types 2 to 6) may be understood.

Suppose someone counters, "Since there is no tenable distinction between the cause and effect of fire in terms of their contribution to contradicting tangible cold, the cause of fire too would end up contradicting the cold."

To respond, we say that it is not owing to its own sheer presence that strongly billowing smoke, an effect of fire, contradicts tangible cold. Rather, it contradicts the cold owing to the particular circumstances of the place permeated by the heat of the fire, the time coincident with smoke billowing forth, and its unfaltering association with fire's endowed capacity to counter cold. As for fire's cause being unable to contradict tangible cold, it is as in the case of firewood being unable to counter tangible cold. *Exposition of Valid Cognition* says:

We assert that, in the case of an effect that contradicts [the probandum],
[this use of evidence] is contingent on place, time, and so on;
otherwise, [the evidence] is misleading,
like ash [as evidence] for it not being cold.[452] [351]

Second, correct evidence consisting in the perception of something contradictory in the sense of being mutually exclusive is of two types: correct evidence that, through definiteness, negates dependence and correct evidence that, through dependence, negates definiteness. We may posit illustrations of these two, respectively, as follows. The first is "A pot's disintegration does not depend on other causes and conditions that arise later than itself because it definitely disintegrates owing to its own mere existence." The second is "A red cloth does not definitely have that color inherently, from its inception, because it comes to have that color in dependence on causes that arise later than itself."

Thus one needs to understand the following points. Correct evidence consisting in the nonperception of the *imperceptible* merely negates veridical conventions about the definite existence of the referent of the predicate of the negandum in the subject of interest. It does not prove the actual absence of the referent of the predicate of the negandum. To prove that the referent of the predicate of the negandum does not exist on the basis of negation, one needs to employ correct evidence consisting in the nonperception of the *perceptible*. In the explanation of correct evidence involving something perceptible, it is shown that tangible cold is contradicted by fire. This is shown so that one may understand how distorted cognitions are undermined by cultivating the correct cognitions that are their opposites. For example, the awareness cognizing impermanence undermines grasping at permanence, and the forbearance one has cultivated undermines anger. [352]

Categorizing Correct Evidence in Terms of the Predicate of the Probandum

When correct evidence is categorized in terms of the predicate of the probandum, there are two categories: correct evidence for a negation and correct evidence for an affirmation. The definition of correct evidence for a negation in a specific proof is: it is characterized by the three modes in that proof—and whatever is the explicit predicate of the probandum in that proof must be a negation. An example is the evidence "produced" in the case of positing "Sound is empty of being permanent because it is produced." The definition of correct evidence for an affirmation in a specific proof is: it is characterized by the three modes in that proof—and whatever is the explicit predicate of the probandum in that proof must be an affirmation. An example is the evidence "capable of performing a function" in the case of positing "Sound is a real thing because it is capable of performing a function." Evidence consisting in nonperception and evidence for a negation are synonymous. Effect evidence and nature evidence must both be evidence for an affirmation. The *Exposition of Valid Cognition Autocommentary* says: "Of those, while two are evidence for the affirmation of real things, one is evidence for a negation."[453] [353]

Categorizing Correct Evidence in Terms of How the Proof Functions

When correct evidence is categorized in terms of how the proof functions, there are two categories: correct evidence proving the expression and correct evidence proving the thing itself. The definition of correct evidence proving the expression is: it is characterized by the three modes in a specific proof—and whatever is the explicit predicate of the probandum in that proof must be a definiendum. An example is the evidence "produced" in the case of positing "Sound is impermanent because it is produced." The definition of correct evidence proving the thing itself is: it is characterized by the three modes in a specific proof—and whatever is the explicit predicate of the probandum in that proof must be a definition. An example is "produced" in the case of positing "Sound is momentary because it is produced." The first is called *evidence proving the expression* because on the basis of the subject, sound, it proves the expression, "impermanent," which is the definiendum of "momentary." The second is called *evidence proving the thing itself* because on the basis of the subject, sound, it proves the thing itself, "momentary," which is the definition of "impermanent." [354]

Categorizing Correct Evidence in Terms of the Thesis

When correct evidence is categorized in terms of the thesis, there is a threefold division: (1) correct evidence based on empirical fact, (2) correct evidence based on popular convention, and (3) correct evidence based on trustworthy testimony. In general, objects of cognition are of two types: an object established based on empirical fact and an object established based on popular convention. In the first case, for example, "hot and burning" exists as fire's reality, exists as fire's nature, and actually exists in fire; whoever touches it will not experience anything other than a hot and burning sensation. So such a thing is called "an object established based on empirical fact." In the second case, for example, fire being known as "the purifier" and as "that which has a crown tuft" is not established based on empirical fact. The expression "the purifier" is a convention used widely in ancient times to refer also to a pig, and "that which has a crown tuft" is a convention also used to refer to a peacock. So such a thing is an object established by popular convention, and it is called "a language-based popular convention" or "an object established by popular convention."

It is the subjective that constitutes popular convention, and this includes both concepts and verbal expressions. Instances of these are, respectively: an ordinary thought that wishes to apply a linguistic convention such as the word "moon" to the rabbit bearer; motivated by that, one then uses the word "moon" to refer to the rabbit bearer. The way in which popular conventions are established and become well known in the world [355] is through someone first applying whatever name he wishes to something previously unnamed so as to establish the use of a popular convention; for example, the initial application of the linguistic convention "pot" to something flat-based and bulbous.

As for taking the trustworthy words of an authority or a scriptural passage as testimony, this is for proving a thesis that is very hidden. An object that is hidden from oneself, if it cannot be proven by evidence based on empirical fact, must be proved by scripture. Also, when positing a scriptural passage as evidence, it is not enough for it merely to be a scripture; it must be a passage that passes a threefold analysis. Just as, for obvious flaws in a piece of gold, one analyzes the color and so forth after burning it in fire, what is expressed in some scriptural passage must not be contradicted by analysis based on direct perception. For slightly hidden flaws in a piece of gold, one analyzes the internal faults by cutting and dividing it; likewise, what is expressed in some scriptural passage must not be undermined by empirical analysis through fact-based valid cognition. For those faults and qualities of gold that are very difficult to discover, one analyzes the value through rubbing it with a touchstone; so too, the words of the scripture must not be contravened by internal contradictions in different parts of the text, direct and indirect contradictions, and so forth. When it is cleared through this threefold analysis, a scriptural passage is worthy to be accepted. Thus to be considered authoritative, a scriptural passage must meet several criteria.

In brief, a suitable scripture passage: (1) is not refuted by valid direct perception with regard to teaching about directly perceptible things, (2) is not refuted by valid inference based on empirical fact [356] with regard to what it teaches about slightly hidden things that are not directly perceptible, and (3) is not invalidated by contradictions between earlier and later assertions in the scripture, direct and indirect contradictions, and so forth with regard to teaching about very hidden, imperceptible things. In the epistemological texts, the manner of clearing a scriptural passage through the threefold analysis refers to something like this.

The definition of correct evidence based on empirical fact is: it is characterized by the three modes in a specific proof—and the thesis in that proof is established by empirical fact. An example is the evidence "produced" in the case of positing "Sound is impermanent because it is produced—for example, a pot." Dharmottara's *Explanatory Commentary on the Ascertainment of Valid Cognition* says: "It is because its three modes are established by valid cognition that evidence based on empirical fact engages its object. One does not seek for some additional evidence in a scriptural source."[454]

The definition of correct evidence based on popular convention is: it is characterized by the three modes in a specific proof—and the thesis in that proof must be a language-based popular convention. An example is "it exists as a conceptual object" in the case of positing "The rabbit bearer is suitable to be called 'the moon' because it exists as a conceptual object." It is merely due to a popular assertion or wish that the rabbit bearer is considered suitable to be called "the moon." It is not established by empirical fact. *Exposition of Valid Cognition* says: [357]

> Words depend on linguistic conventions,
> and a linguistic convention in turn depends on mere intentions.
> [The conceptual objects] established by words are not nonestablished;
> [Dignāga] spoke of this as "language-based popular convention."
>
> Because that kind of [cognition about the possibility of any] convention is inferential,
> [Dignāga's discussion of "language-based popular convention"] shows that, in the case of what is established by [empirical] inference,
> there is no contradictory, invalidating (Skt., *viruddhāvyabhicārin*) [evidence based on mere convention].[455]

Just as it is possible to apply the word "moon" to the rabbit bearer, so too it is possible to use the term "eye" for an ear. Asvabhāva's *Explanation of the Compendium of the Mahāyāna* explains: "It is said that terms are not restricted in their reference. For example, the Drāvidians call an ear an 'eye,' cooked rice a 'thief,' lips 'the separated,' and so on."[456] Furthermore, "moon" arbitrarily refers to the rabbit bearer; in treatises on perfumery, the word "moon" refers to camphor; in treatises on medicine, the word "moon" refers to mercury; in treatises on poetry, the word "moon" refers to the face of a maiden. Thus no matter what the object, it is possible to take that as the subject of predication and call it "moon."

Correct evidence based on empirical fact and correct evidence based on popular convention are differentiated according to whether their thesis is established by virtue of an empirical fact or by virtue of a linguistic convention, [358] which is explained in Dharmottara's *Explanatory Commentary on the Ascertainment of Valid Cognition* as follows: "The classification of objects is as follows: the object of a popular convention is something that comes from a linguistic convention; the object of an empirical inference is an actual thing."[457]

The definition of correct evidence based on trustworthy testimony is: that which is characterized by the three modes in a specific proof and whose thesis is very hidden. An example is the evidence "It is a scripture that has passed the threefold analysis" in the case of positing "The scriptural quote 'Generosity brings wealth, and ethical conduct brings happiness' is nondeceptive with regard to the meaning of what it teaches because it is a scripture that has passed the threefold analysis." Also, the thesis—that the gaining of material resources comes from generosity—is very hidden, and to realize it most people definitely have to rely on a scripture that has passed the threefold analysis. The subject of this argument is the kind discussed in *Exposition of Valid Cognition*:

> A statement that is coherent, presenting suitable methods,
> and speaking of some human aim . . .[458]

The same text says:

> [That statement's] trustworthiness consists in its not being contradicted
> by direct perception or by
> the two types of inferential cognition
> regarding perceptible and imperceptible objects.[459]

This identifies the evidence for that argument. [359]

Categorizing Correct Evidence in Terms of How It Is Present in Homologous Cases

When correct evidence is categorized from the perspective of how it is present in the set of homologous cases, there is a twofold division: correct evidence that is pervasively present in the set of homologous cases and correct evidence that is present in a subset of the set of homologous cases. The definition of correct evidence that is pervasively present in the set of homologous cases is: it is characterized by the three modes in a specific proof—and it is entirely present in all homologous cases of that proof. An example is "produced" in the case of positing "Sound is impermanent because it is produced." The definition of correct evidence that is present in a subset of homologous cases is: it is characterized by the three modes in a specific proof—and it is present in some homologous cases of that proof. An example is "arises from effort" in the case of positing "Sound is impermanent because it arises from effort." Also, if some subject bears the predicate of the thesis "impermanent," then it is pervaded by also bearing the attribute "produced"; thus "produced" is evidence that is pervasively present in the set of homologous cases of the argument proving that thesis. In cases of subjects that bear the predicate "impermanent," there are two alternatives: those that arise from effort and those that do not; thus "arises from effort" is evidence that is present in a subset of homologous cases for the argument proving that thesis. [360]

Categorizing Correct Evidence in Terms of the Respondent

When correct evidence is categorized in terms of the respondent, there are two divisions: correct evidence for oneself and correct evidence for others. The definition of correct evidence for oneself is: it is characterized by the three modes in a specific proof—and it is not set forth to an opponent. For example, without a proponent setting forth the reasoning "Sound is impermanent because it is produced," "produced" becomes correct evidence in the proof that sound is impermanent just by oneself thinking about the evidence for why sound is impermanent. The definition of correct evidence suitable for others is: it is characterized by the three modes in a specific proof—and it is set forth to an opponent. For example, "produced" becomes correct evidence in the proof that sound is impermanent in dependence on a proponent (other than oneself) positing "Sound is impermanent because it is produced." In brief, although both evidence for oneself and evidence for others are similar in being correct evidence in a specific proof, they are different in terms of having, or not having, a genuine proponent who posits the evidence. [361]

21

Fallacious Inferential Evidence

FALLACIOUS INFERENTIAL EVIDENCE is explained so that one may know the ways one can deviate from correct evidence. Based on the explanation above that correct evidence is characterized by the three modes, one can understand, conversely, that fallacious evidence is not characterized by the three modes. *Exposition of Valid Cognition* says:

> Fallacious inferential evidence is other than these three.[460]

In general, inferential evidence either is or is not established to be an attribute of the subject. In cases where the evidence is established to be an attribute of the subject, the positive and negative pervasion between the evidence and the predicate is either ascertained accurately or ascertained in a distorted manner. In the first case, the evidence is correct evidence, whereas in the second case it is *contradictory evidence.* If neither a correct pervasion nor an inverse pervasion is ascertained, then the evidence is *inconclusive evidence.* When the evidence is not established as an attribute of the subject, it is *unestablished evidence.*

The definition of fallacious evidence is: it is posited as evidence in a specific proof and it is not characterized by the three modes. Fallacious evidence is of three types: (1) contradictory inferential evidence, (2) inconclusive inferential evidence, and (3) unestablished inferential evidence. [362]

CONTRADICTORY INFERENTIAL EVIDENCE

Evidence that is established as an attribute of the subject in the given proof but that contradicts the predicate of the thesis mainly by way of the positive pervasion being unestablished is called *contradictory evidence.* The definition of contradictory evidence in a specific proof is: it is an attribute of the subject in a specific proof—and the positive and negative pervasions in that proof are ascertained in a distorted manner by the person who has ascertained it as an attribute of the subject. Here the phrase "attribute of the subject" precludes it from being unestablished evidence, and the phrase "the positive and negative pervasions in that proof are ascertained in a distorted manner by the person who has ascertained it as an attribute of the subject" precludes it from being inconclusive evidence. An example is "produced" in the case of positing "Sound is permanent because of being produced."

Inconclusive Inferential Evidence

Evidence that is established as the attribute of the subject in the given proof but where primarily the negative pervasion is not established or not ascertained is called *inconclusive evidence*. The definition of inconclusive evidence in a specific proof is: that which is an attribute of the subject in a specific proof—and where the positive and negative pervasions in that proof are neither ascertained accurately nor ascertained in a distorted manner by the person for whom it is the attribute of the subject. [363] There are two types: uncommon inconclusive evidence and common inconclusive evidence.

The definition of *uncommon inconclusive evidence* in a specific proof is: it is inconclusive evidence in a specific proof—and it is neither ascertained to be present in the set of homologous cases in that proof nor ascertained to be present in the set of heterogeneous cases in that proof by the person for whom it is the attribute of the subject in that proof. An example is "audible" in the case of positing "Sound is impermanent because it is audible." Now here, we may ask, although the evidence "audible" is not ascertained to be present in the set of heterogeneous cases—that is, things that are permanent—why is it not ascertained to be present in the set of homologous cases, things that are impermanent? This is because the person for whom the evidence "audible" is an attribute of the subject in the proof that sound is impermanent will have already ascertained the relation between the word *sound* and its meaning "audible." So if that person ascertains that the evidence "audible" is present in that proof's set of homologous cases, the impermanent, then he or she must already know that "sound" also is present in that proof's set of homologous cases, the impermanent, and this would entail the fault of already cognizing that sound is impermanent prior to employing the proof. In brief, from among the three—the proof's subject under discussion, the set of homologous cases, and the set of heterogeneous cases—the evidence "audible" is known to be present within the subject under discussion, "sound," but since it is not ascertained to be present within the other two, it is called *uncommon inconclusive evidence*. [364]

The definition of *common inconclusive evidence* in a specific proof is: it is inconclusive evidence in a specific proof—and it is either ascertained to be present in the set of homologous cases in that proof or ascertained to be present in the set of heterogeneous cases in that proof by the person for whom it is an attribute of the subject in that proof. For example, "an object of knowledge" in the case of positing "Sound is impermanent because it is an object of knowledge" for someone who knows that "an object of knowledge" is an attribute of the subject in a proof that sound is impermanent and has ascertained that "an object of knowledge" is present in that proof's set of homologous cases (the impermanent) but does not ascertain that it is present in the set of heterogeneous cases (the permanent). Also, for this proof's opponent, "an object of knowledge" is inconclusive evidence that has been ascertained to be present "in common" in both the subject under discussion, "sound," and the set of homologous cases, the impermanent. Therefore such reason is called *common inconclusive evidence*.

There are two types of common inconclusive evidence: actual inconclusive evidence and inconclusive evidence with remainder. The definition of *actual inconclusive evidence* in a specific proof is: it is inconclusive evidence in a specific proof—and it is ascertained to be present in *both* the set of homologous cases and the set of heterogeneous cases of that proof by a person who knows that it is an attribute of the subject of that proof. An example is "an object of knowledge" in the case of positing "Sound is permanent because it is an object of knowledge" for someone who knows that "an object of knowledge" is an attribute of the subject in the proof that sound is permanent, and who ascertains that it is present in both the set of homologous cases (the permanent) and the set of heterogeneous cases (the impermanent). [365]

The definition of *inconclusive evidence with remainder* in a specific proof is: it is common inconclusive evidence in a specific proof—and, by a person who knows that it is an attribute of the subject, it has been ascertained to be present in that proof's set of homologous cases and then doubted whether it is present or absent in the set of heterogeneous cases; or it has been ascertained to be present in that proof's set of heterogeneous cases and then doubted whether it is present or absent in the set of homologous cases. Since, for the person who knows that it is an attribute of the subject, there remains a doubt about whether that evidence is present in the set of homologous cases or whether it is present in the set of heterogeneous cases, it is called *with remainder.* Furthermore, for example, "an object of knowledge" in the case of positing "Sound is impermanent because it is an object of knowledge"—for an opponent who knows that "an object of knowledge" is an attribute of the subject in the proof that sound is impermanent and who ascertains "an object of knowledge" to be present only in that proof's set of homologous cases (the impermanent) and does not ascertain it to be present in the set of heterogeneous cases (the permanent)—is known as *with correct remainder.* Also, for example, "an object of knowledge" in the case of positing "Sound is impermanent because it is an object of knowledge"—for an opponent who knows that "an object of knowledge" is an attribute of the subject in the proof that sound is impermanent and who ascertains "an object of knowledge" to be present only in that proof's set of heterogeneous cases (the permanent) and does not ascertain it to be present in the set of homologous cases (the impermanent)—is known as *with contradictory remainder.* [366]

Unestablished Inferential Evidence

Evidence that is not established as an attribute of the subject is called *unestablished evidence.* The definition of unestablished evidence is: that which is posited as evidence in a specific proof—and it is not established as an attribute of the subject in that proof. There are three types: evidence that is (1) unestablished in relation to the fact, (2) unestablished in relation to the mind, and (3) unestablished in relation to the disputant.

The first, evidence that is unestablished in relation to the fact, can be subdivided into seven types: (1) unestablished because the evidence lacks existence, as in the case of positing "This being is suffering because of having been pierced by a rabbit's horn"; (2) unestablished because the subject lacks existence, as in the case of positing "A rabbit's horn is impermanent because it is produced"; (3) unestablished because the evidence and the predicate are not distinct, as in the case of positing "Sound is impermanent because it is impermanent"; (4) unestablished because the evidence and the subject under discussion are not distinct, as in the case of positing "Sound is impermanent because it is sound"; (5) unestablished because the predicate and the subject under discussion are not distinct, as in the case of positing "Sound is sound because it is produced"; (6) unestablished because the evidence does not exist in relation to the subject in the way the reasoning formulates it, [367] as in the case of positing "Sound is impermanent because it is an object apprehended by an eye consciousness"; (7) unestablished because the evidence does not exist in one area of the subject, as in the case of positing "A shrub has a mind because at night it curls up its leaves and sleeps."

The second, evidence that is unestablished in relation to the mind, can be subdivided into four types: (1) unestablished because of having doubts about the reality of the evidence, as in the case of positing to a person who does not validly ascertain microbes[461] "Sound is impermanent because it is an object of valid cognition of a microbe"; (2) unestablished because of having doubts about the reality of the subject, as in the case of positing to an opponent for whom microbes are not perceptible objects "A microbe is impermanent because of being produced"; (3) unestablished because of having doubts about the relation between the evidence and the subject under discussion, as in the case of positing to a person who does not know where the peacock is "In the middle of the three valleys is a peacock because there is the sound of the peacock's call"; (4) unestablished because of not having a nondeficient subject of interest, as in the case of positing to an opponent who has already established and not forgotten that sound is impermanent "Sound is impermanent because it is produced."

The third, evidence that is unestablished in relation to the disputant, is subdivided into three types: (1) unestablished in relation to the proponent, as in the case of a proponent who has not validly established that sound is impermanent but who posits to an opponent "Sound is impermanent because it is produced"; [368] (2) unestablished in relation to the opponent, as in the case of a proponent positing to an opponent who has no interest in knowing whether sound is impermanent "Sound is impermanent because it is produced"; (3) unestablished in relation to both the proponent and the opponent, as in the case of a proponent who has not validly established that sound is impermanent positing to an opponent who has no interest in knowing whether sound is impermanent "Sound is impermanent because it is produced."

22

Dignāga's *Drum of a Wheel of Reasons*

To FACILITATE AN UNDERSTANDING of the very important categories of inferential evidence, Dignāga composed a text presenting nine specific types of evidence where the evidence is established as the attribute of the subject. We explain this presentation briefly here. First, to identify the nine types of inferential evidence that are the attribute of the subject, Dignāga's *Drum of a Wheel of Reasons* says:

> Object of knowledge, produced, and impermanent,
> produced, audible, and arisen from effort,
> impermanent, arisen from effort, and nonphysical.[462]

This above passage identifies each of the nine examples of inferential evidence appearing in the *Drum of a Wheel of Reasons*. [369] Then:

> Permanent, impermanent, and arisen from effort,
> permanent, permanent, and permanent,
> not arisen from effort, impermanent, and permanent—
> these state what is to be proved.[463]

This above passage indicates the nine predicates of the theses. Next:

> In this magical wheel of nine types of evidence,
> their specific examples are as follows:
> (1) space and a pot, and (2) a pot and space,
> (3) like a pot and lightning or space,
> (4) space and a pot, and (5) space and a pot,
> (6) like space and a pot or lightning,

(7) like lightning or space and a pot,
(8) like a pot or lightning and space,
(9) like space, atoms, movement, and a pot.[464]

These lines present the concordant examples and the discordant examples for each of the nine proof statements. [370]

Thus, among the nine examples of inferential evidence that are attributes of the subject as set forth in the *Drum of a Wheel of Reasons*, the evidence under discussion may apply to the set of homologous cases in one of three ways: as pervasively present in the set of homologous cases, as absent in the set of heterogeneous cases, or as present in a subset of the set of homologous cases. Each of these subdivisions is further divided according to the set of heterogeneous cases: as pervasively present in it, as absent in it, or as present in a subset of it. When the first three are mapped onto the second three, there are nine combinations altogether. These nine may be expressed as follows: (1) as pervasively present in both the set of homologous cases and the set of heterogeneous cases, (2) as pervasively present in the set of homologous cases and absent in the set of heterogeneous cases, (3) as pervasively present in the set of homologous cases and present in a subset of the set of heterogeneous cases, (4) as absent in the set of homologous cases and pervasively present in the set of heterogeneous cases, (5) as absent in both the set of homologous cases and the set of heterogeneous cases, (6) as absent in the set of homologous cases and present in a subset of the set of heterogeneous cases, (7) as present in a subset of the set of homologous cases and pervasively present in the set of heterogeneous cases, (8) as present in a subset of the set of homologous cases and absent in the set of heterogeneous cases, (9) as present in both a subset of the set of homologous cases and a subset of the set of heterogeneous cases. The *Drum of a Wheel of Reasons* says:

In the set of homologous cases:
present, absent, and both;
in the set of heterogeneous cases: likewise.
There are three ways of combining each set of three.[465] [371]

In the case of the homologous cases, "pervasively present in the set of homologous cases" means that if something bears the predicate of the probandum, then it necessarily also bears the evidence as a property. "Absent in the set of homologous cases" means that if something bears the predicate of the probandum, then it necessarily does not bear the evidence as a property. "Present in a subset of the set of homologous cases" means that if something bears the predicate of the probandum, then it does not necessarily bear the evidence as a property, nor does it necessarily not bear the evidence as a property. Likewise, for the set of heterogeneous cases, "pervasively present in the set of heterogeneous cases" means that if something bears the predicate of the negandum, then it necessarily also bears the evidence as a property. "Absent in the set of heterogeneous cases" means that if something bears the predicate of the negandum, then it necessarily does not bear the evidence as a property. "Present in a subset of the set of heterogeneous cases" means that if something bears the predicate of the negandum, then it does not necessarily bear the evidence as a property, nor does it necessarily not bear the evidence as a property.

Among the nine examples of inferential evidence, the second and eighth are correct evidence; the fourth and sixth are contradictory evidence; the first, third, seventh, and ninth are common inconclusive evidence; and the fifth is uncommon inconclusive evidence. The *Drum of a Wheel of Reasons* says:

> The upper and lower two are correct ones;
> on the two sides [right and left] are two contradictory ones;
> the four corners are common inconclusive ones;
> and in the center is the uncommon one.[466] [372]

Among these nine examples of inferential evidence, the evidence presented at the top of the middle column and at the bottom of the middle column are the two correct instances of evidence. [373] The evidence presented on the left of the middle line and on the right of the middle line are the two instances of contradictory evidence. The evidence presented at the four corners of the chart are the four forms of common inconclusive evidence. And the evidence presented in the center of the chart is uncommon inconclusive evidence.

The twenty-four examples (stated for these nine instances of evidence) should be understood in the following manner. The pair—space and a pot—are stated as examples in the proof "Sound is permanent because

Inferential Evidence in Dignāga's *Drum of a Wheel of Reasons*[467]		
1. Actual Inconclusive "Sound is permanent because it is an object of knowledge." *Concordant example*: space *Discordant example*: a pot	*2. Correct Evidence* "Sound is impermanent because it is produced." *Concordant example*: a pot *Discordant example*: space	*3. Actual Inconclusive* "The sound of a conch is arisen from effort because it is impermanent." *Concordant example*: a pot *Discordant examples*: lightning and space
4. Contradictory Evidence "The sound of a conch is permanent because it is produced." *Concordant example*: space *Discordant example*: a pot	*5. Uncommon Inconclusive* "Sound is permanent because it is audible." *Concordant example*: space *Discordant example*: a pot	*6. Contradictory Evidence* "The sound of a conch is permanent because it is arisen from effort." *Concordant example*: space *Discordant examples*: a pot and lightning
7. Actual Inconclusive "The sound of a conch is not arisen from effort because it is impermanent." *Concordant examples*: lightning and space *Discordant example*: a pot	*8. Correct Evidence* "The sound of a conch is impermanent because it is arisen from effort." *Concordant example*: a pot *Discordant example*: space	*9. Actual Inconclusive* "The sound of a conch is permanent because it is not a tangible object." *Concordant examples*: space and atoms *Discordant examples*: movement and a pot

it is an object of knowledge" because they are, respectively, examples of something pervasively present in the set of homologous cases and of its opposite, something pervasively present in the set of heterogeneous cases. Space is an example of something pervasively present in the set of homologous cases of that proof, and a pot is an example of something pervasively present in the set of heterogeneous cases of that proof. The way in which they serve as examples is this. Space is a common locus of both "object of knowledge" and "permanent," whereas a pot is a common locus of "object of knowledge" and "impermanent."

In the last type of evidence in the wheel, the four things—space, atoms, movement, and a pot—are stated as examples in the proof "The sound of a conch is permanent because it is not a tangible object," as illustrations of something present in a subset of the set of homologous cases and in a subset of the set of heterogeneous cases. This is because space and atoms [374] are examples of something present in a subset of the set of homologous cases of that proof. And movement and a pot are examples of something present in a subset of the set of heterogeneous cases of that proof. Here is how they serve as these examples. Space is cited for someone adhering to the non-Buddhist Vaiśeṣika school as an example of something that "If permanent, it is not necessarily a tangible object." Atoms are cited for that school as an example of something that "If permanent, it is not necessarily not a tangible object." Movement is cited for that school as an example of something that "If impermanent, it is not necessarily a tangible object." And a pot is cited as an example of something that "If impermanent, it is not necessarily not a tangible object." In accordance with this explanation of how the examples function with respect to the first and the last types of evidence in the wheel, one should understand how the examples function for those that are in between.

As for the purpose of Dignāga's presentation of the nine types of inferential evidence that are attributes of the subject, Dharmakīrti's *Exposition of Valid Cognition* says:

> The two types of correct evidence and the two contradictory ones
> are to establish nature and effect evidence;
> the "unique" and the "general" are to counter the disputes;
> the rest are to preclude [the evidence's presence in the set of heterogeneous cases].[468]

The two correct types of evidence—effect and nature—are discussed for the purpose of understanding that the invariable relation of unaccompanied nonoccurrence that exists between the evidence and the predicate in the case of both effect evidence and nature evidence is not something established merely through asserting it to be so. For if such an invariable relation were established purely on the basis of assertion, then the two instances of contradictory evidence would also turn out to be correct evidence; yet those two are entirely erroneous [375] with respect to the pervasion between the evidence and the predicate. Therefore the two instances of contradictory evidence are included for the purpose of understanding that ascertaining the positive and negative pervasion is contingent on establishing the invariable relation. As an implication of showing that the negative pervasions of effect evidence and nature evidence are not established merely by way of assertion, it is easy to realize that the nonperception of a cause and the nonperception of a pervader too are forms of correct evidence. Thus four of the instances of evidence that are established as the attribute of the subject—the two instances of contradictory evidence on the two sides (4 and 6) and the two instances of correct evidence at the top and the bottom (2 and 8)—are included in the wheel of reasons to show that correct evidence is subsumed within the three categories—effect, nature, and nonperception. This is what Dharmakīrti explains.

To refute correct evidence having only the negative pervasion or only the positive pervasion, it is proved that if one of the two is established (the positive or the negative pervasion), then the other must be established also. And since this is established on the basis of having established that the evidence is an attribute of the subject, two instances of evidence established as the attribute of the subject—namely, where "audible" and "an object of knowledge" are stated as evidence (in 1 and 5)—are included in the wheel of reasons so as to establish that there are exactly three modes that characterize correct evidence.

For were the negative pervasion alone sufficient for correct evidence, then "audible" would be correct evidence proving that sound is permanent. But in all cases alleged to be instances of correct evidence that have only the negative pervasion, there is in fact no establishment of the negative pervasion. This is because those examples are similar to an example in which "audible" is posited as evidence for proving that sound is impermanent. It is for the purpose of demonstrating this point that "audible" is included in the wheel of reasons. [376]

Likewise, were a positive pervasion alone sufficient for correct evidence, then "an object of knowledge" would be correct evidence for proving that sound is permanent. But in all cases alleged to be instances of correct evidence that have only the positive pervasion, there is in fact no establishment of the positive pervasion. This is because those examples are similar to an example in which "object of knowledge" is posited as evidence for proving that sound is impermanent. It is for the purpose of demonstrating this point that "object of knowledge" is included in the wheel of reasons. This is what Dharmakīrti explains.

Dignāga included in the *Drum of the Wheel of Reasons* the remaining three instances of evidence established as the attribute of the subject (3, 7, and 9) for the purpose of understanding that evidence that is merely present in the set of homologous cases, or overwhelmingly present, or just present and absent in equal portions, would be unsuitable as a correct proof of the thesis. If evidence that is merely present in the set of homologous cases were taken to be a correct proof, then impermanence would be correct evidence proving that the sound of a conch is arisen from effort. Thus "the sound of a conch is arisen from effort because it is impermanent" is stated for the purpose of understanding this fault. If evidence that is overwhelmingly present in the set of homologous cases were taken to be a correct proof, then impermanence would also be correct evidence proving that the sound of a conch is not arisen from effort. Thus "the sound of a conch is not arisen from effort because it is impermanent" is stated for the purpose of understanding this fault. If evidence that is just present and absent from the set of homologous cases in equal portions were taken to be a correct proof, then something that is not a tangible object would be correct evidence proving that sound is permanent. Thus "the sound of a conch is permanent because it is not a tangible object" is stated for the purpose of understanding this fault. [377]

In brief, this ninefold classification of evidence that is the attribute of the subject is presented for the purpose of distinguishing between correct and fallacious evidence by way of showing the following: (1) with regard to a given subject, what is excluded from the set of heterogeneous cases in its entirety is the thesis; (2) the definition of correct evidence is necessarily characterized by the three modes; and (3) correct evidence so defined is confined to three types—effect, nature, and nonperception.

This concludes the discussion of proof statements and inferential evidence, which are included as part of the presentation of the ways that the mind engages objects. [379]

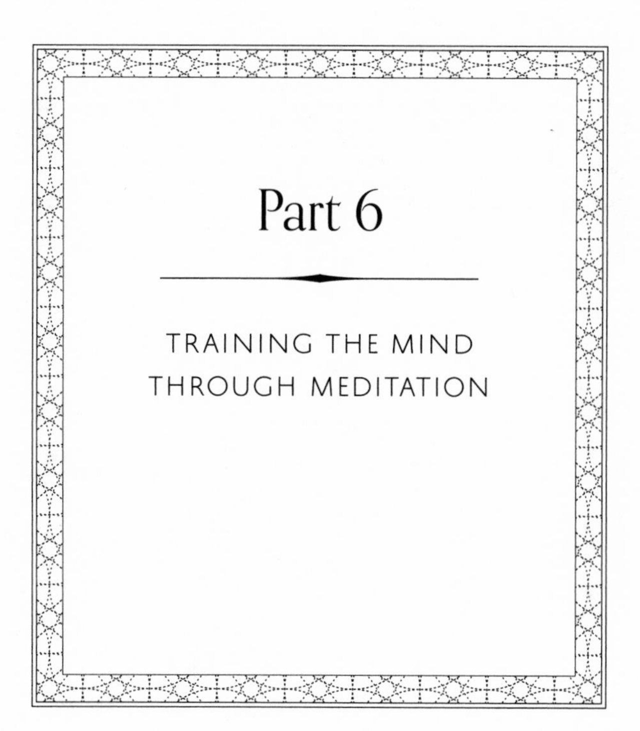

Part 6

TRAINING THE MIND THROUGH MEDITATION

The Purpose of Analysis

Throughout this volume, a consistent but often subtle theme is the underlying purpose for all these rigorous and detailed analyses of the mind, cognition, concept formation, and reasoning. In this final part, that purpose comes to the fore: namely, that all that has been presented thus far is concerned with transforming the mind and with the changes in behavior that induces. A key element in that process of transformation is meditation, and in keeping with our authors' attention to that issue, much of my introduction will present some key aspects of meditation. To begin, however, I will follow our authors in surveying the Buddhist theories of psychological and behavioral change, since the Buddhist understanding of that process is precisely what helps us to appreciate the transformative role played by meditative practices on the Buddhist path.

Counteracting *Kleśas*

By now, the term *kleśa* is likely familiar: it refers to the mental afflictions that cause suffering. The Buddhist approach to eliminating suffering thus focuses on how we work with our mental afflictions such as ignorance, craving, and aversion so as to reduce and eventually eliminate them. Through their analysis of that process, Buddhist theorists have arrived at an intriguing insight: it is possible to counteract mental afflictions in ways that temporarily prevent their occurrence, but to uproot them completely, one must address the underlying cognitive mechanism that makes the afflictions possible. We can characterize these two approaches by introducing two terms implied by our authors' analysis, namely *suppressing* and *undermining.* These two approaches rest on two different ways that a mental state can be incompatible with another. When one mental state (such as aversion) is *suppressing* another (such as craving), this is due to the incompatibility of these two states. Aversion, for example, is focused on an object that is construed in some fashion as negative or unattractive, while craving is focused on an object that is construed as positive or attractive. These two afflictions cannot exist simultaneously in a mental state focused on a single object because they require incompatible construals of that object as either attractive or unattractive. In contrast, when a mental state such as wisdom *undermines* an affliction such as craving, it has eliminated the cognitive mechanisms that are required for craving to arise. Thus, when an affliction is *suppressed*, it is prevented from arising temporarily because another, incompatible mental state is present. But when a mental state is *undermined,* the very conditions for its occurrence have been eliminated, and if that undermining is completely effective in a lasting way, then any mental affliction that depends on those mechanisms can never occur again.

Buddhist contemplative practice tends to be eminently practical, and since the process of completely uprooting the mechanisms that underlie the mental afflictions is a long-term and arduous task, it is tremendously useful to have some techniques to suppress particular mental afflictions, even if the techniques are essentially a "band-aid" that cannot undermine the problem altogether. In this regard, one set of well-known practices focus on compassion (Skt., *karuṇā*) and loving kindness (*maitrī*; Pali, *mettā*). While one is in a mental state with compassion or loving kindness, it is not possible for any of the afflictions connected to aversion—such as anger, hatred, and resentment—to arise. Thus, for practitioners who have mindfully observed a predominance of aversive afflictions in their mental life, meditations that cultivate loving kindness and compassion can be an effective—if temporary—balm that reduces their aversion.

Techniques to suppress negative mental states are noteworthy, and if cultivated effectively, Buddhist traditions acknowledge that they can help one to manage one's afflictions. However, the task of completely undermining mental afflictions such as aversion, such that they simply cannot ever arise again, requires a different approach. To understand what is involved, it is helpful to examine the role that a distorted form of attention, as a product of ignorance, plays in engendering mental afflictions.

IGNORANCE AND INAPPROPRIATE ATTENTION

One of the most important insights offered by our authors' account of mental and behavioral transformation is the notion that afflictions *cognitively distort* their objects. The technical term for this type of distortion is *inappropriate attention* (Skt., *ayoniśomanasikāra*), which occurs within the overall context of our conditioned experiences. As mentioned previously, each moment of experience is conditioned by the preceding moment (and that preceding moment, too, is likewise conditioned), and the various appetites, expectations, and fears that condition an experience are relevant to the way that we attend to the particular object presented to us in that moment. This inheritance of conditioning falls under the general rubric of *karma*, and it is in itself reinforced by the behaviors—and afflictive mental states—that follow from that highly conditioned experience. Thus, if I have somehow acquired the conditioning to be especially fond of strawberries, when I encounter a bowl of strawberries, my highly conditioned cognitive system immediately imposes a series of largely unconscious expectations and worries, or "hopes and fears" (Tib., *re dogs*), about the strawberries themselves. My past conditioning prompts me to draw on a fundamental tendency to exaggerate and essentialize my past experiences, which in this case manifests as the latent attitude that "all strawberries are delicious." Combined with an overall yearning to interpret any experience in relation an unchanging, absolute "me" that craves pleasure and fears pain, I exaggerate and yearn for the scrumptious qualities of the strawberries as somehow truly satisfying all my needs, or perhaps I fear the loss of my precious strawberries, which clearly deserve considerable protection, even if it requires an aggressive confrontation with someone who has eaten too many of them.

Perhaps it is already obvious that the basic mechanism of inappropriate attention is a compulsion to attribute qualities, entirely good or entirely bad, to the current object of one's concern. In this way, both craving and aversion—and all the many mental afflictions, such as hatred and addiction, that follow from them—are rooted in a fundamentally distorted experience that involves exaggerating the qualities that make us attracted or averse. In actuality, a strawberry could never satisfy my deepest needs, even though my craving may seem to present it that way in experience. Likewise, some focus of aversion—such as an unpleasant coworker—may seem to be completely devoid of any redeeming qualities, but that too is not the case. This very tendency to distort reality—the tendency to filter all experiences through the exaggerations of craving and aversion—is precisely what Buddhists mean by the inappropriate attention that arises from ignorance.

Fortunately, as we learned in part 1, the mental afflictions, including ignorance, are not essential qualities of the mind, and as a result, ignorance (along with the inappropriate attention that it causes) can be completely eliminated. Moreover, since the cognitive distortions of ignorance lie at the root of all the other mental afflictions, one can eradicate all of the mental afflictions by eliminating ignorance. And how does one do so? Through contemplative practices, including meditation, that cultivate wisdom.

The Practice of Wisdom

As noted in the essay introducing part 1, wisdom eradicates ignorance by "seeing the way things truly are"; in other words, it undermines the cognitive distortions that constitute ignorance. This immediately raises two issues: first, what exactly is the way things truly are? And second, what kind of "seeing" (Skt., *darśana*) is involved here? The first question is highly complex, and an adequate answer would require a lengthy account of the various Buddhist philosophical positions concerning the nature of reality and the way reality is actively misperceived due to ignorance. Fortunately, subsequent volumes in this series will offer detailed accounts of the relevant Buddhist philosophies, and to understand the contemplative process of cultivating wisdom, we do not need to get into those details. Instead, it is the second question about the notion of "seeing" that is most relevant here. In short, the metaphor of "seeing" points to a crucial aspect of wisdom's transformative power. For it to be effective, wisdom cannot be simply a matter of intellectual understanding. It must instead become experiential.

The distinction between a mere intellectual understanding and an experiential one largely comes down to the difference between conceptual and perceptual cognition. Our authors have already discussed this difference in detail, and without reiterating their analysis, we can simply note that perceptual experience has a clarity or phenomenal intensity that involves a visceral level of response, such that thinking of eating an apple and actually doing so are manifestly distinct. We have also learned that conceptual cognitions always involve some degree of distortion (for example, the thought of an "apple" imputes a false sameness to all apples), whereas perception is free of those distortions. Thus, from an epistemological perspective, wisdom must become perceptual, if it is to be a truly undistorted encounter with reality. This epistemological issue is important, but in the context of understanding the contemplative process of cultivating wisdom, a perhaps more crucial issue is the role that perceptual experience plays in psychological and behavioral change. In short, Buddhist theorists maintain that, without the visceral impact of an actual experience, merely thinking about some transformative content will be much less effective in creating those changes.

Our authors discuss the process of cultivating wisdom through two interrelated schemas: the "three stages of wisdom" and the notion of "view, meditation, and conduct." The three stages of wisdom outline a contemplative process that begins with learning some conceptual content, for example, the Buddhist analysis of personal identity. This act of learning is literally called "listening" (Skt., *śruti*), since traditionally, even written texts (including philosophical ones) are transmitted through oral discourse and commentary. This also points to the contemplative aspects of this process, inasmuch as the guidance for proper listening includes instructions about motivation, attention, affective orientation, and so on. Once one has a clear understanding of the material learned from the text and teacher, the "wisdom arisen from learning" is in place. One then applies rational analysis to that material, primarily employing the type of inferential reasoning discussed in part 5. At this stage, one is critically reflecting on the meaning of the material that one has learned, and once one reaches a point of clarity and certainty about that meaning, one has achieved the "wisdom arisen from critical reflection." Finally, in order to bring one's critical understanding to the point of a visceral, nonconceptual experience, one engages in meditative practices that eventually end in such an experience, which is itself the "wisdom arisen from meditation."

The contemplative process outlined by the three stages of wisdom fits into a larger context of contemplative practice, and the notion of "view, meditation, and conduct" points to some key features of that larger context. The *view* (Skt., *dṛṣṭi*) concerns especially the intellectual understanding and experiential knowledge that constitutes the wisdom that uproots ignorance, but the view can also be understood to include other theoretical and experiential knowledge concerning other topics, such as compassion, that are crucial to the Buddhist path. The term *conduct* (*samācāra*) refers to the contemplative lifestyle required for effective practice, and one of its main features is the importance of ethical behavior. In short, an unethical life is one filled with mental afflictions, and since mental afflictions necessarily cause mental disturbance, they induce chaotic mind-states that are not suitable to contemplative practice. Finally, *meditation* (*bhāvanā*) refers not only to practices for cultivating the aforementioned wisdom arisen from meditation but also to the wide variety of meditative practices that Buddhist contemplatives use to advance on their spiritual path. This raises the overall question of what is meant by "meditation," and our authors address this issue at length. Let us now look at some key aspects of their account.

Meditation

The main technical term for meditation in Buddhist Sanskrit texts is *bhāvanā*. This term comes from the verbal root *bhū* ("to be" or "to become") in its causal form, and it literally means "causing to become" or "making be." The Tibetan translation of this term is *sgom*, and it emphasizes a particular interpretation of *bhāvanā* where meditation involves a process of familiarization. As a general translation, "cultivation" may best capture the range of meanings suggested by *bhāvanā*, such that meditation could involve "cultivating" a particular mental state such as compassion or "cultivating" knowledge of or familiarity with a particular object or topic, such as impermanence.

A key rubric, discussed at length by our authors, is the distinction between cultivating calm abiding (*śamatha*) and cultivating special insight (*vipaśyanā*). These two are said to lie at the core of all Buddhist meditative practices, in part because the process that culminates in the emergence of meditative wisdom requires the integration of both. *Calm abiding* most literally refers to a particular achievement, such that one can sustain a completely stable meditative state with a type of mental and physical "fluency" (*praśrabdhi*) that enables one to effortlessly remain undistracted while sustaining great mental clarity. *Special insight* refers to a state in which wisdom is fully present, especially the wisdom that uproots ignorance. In usage, both calm abiding and special insight commonly refer to meditative practices that are meant to culminate in these achievements. Thus one can say that one is "practicing calm abiding," in the sense that one is engaged in a practice that is meant to culminate in actual calm abiding. Many sources maintain that calm abiding is usually cultivated first, with special insight arising later. But in all cases, the two must eventually be combined. As one traditional metaphor puts it, calm abiding is like the shade around an oil lamp that keeps it stable and unperturbed by the winds of distraction, and special insight constitutes the intensity of the flame itself. Without both, one will not be able to see what needs to be seen.

The rubric of calm abiding and special insight relate to another pair of terms: *placement meditation* and *analytical meditation*. When engaged in the practice of placement meditation (Tib., *'jog sgom*), one is not employing discursive content or analysis. In contrast, when engaged in analytical meditation (*'dpyad sgom*), discursive content or analysis is an explicit feature of the practice. As our authors note, calm abiding will generally fall under the rubric of placement meditation. Special insight is cultivated through an analytical process, so it will generally be characterized as analytical. However, the integration of calm abiding and special insight is a special case, where the intense focus of the analysis in special insight itself engenders the fluency and undistracted stability of calm abiding. The relationship between calm abiding and special insight points to a more general feature of many other meditative practices, such as the cultivation of compassion. Specifically, an analytical meditation—where, for example, one follows a script of discursive contemplations that are intended to induce compassion—is often followed in the same session by placement meditation, such that one lets go of the discursive content and remains nondiscursively in the state induced by the analytical meditation. And when that state degrades or dissipates, one again returns to the analytical meditation. In this way, a practitioner might alternate between analytical meditation and placement meditation. This approach characterizes much of what is described as meditation in traditional manuals.

Mindfulness, Meta-awareness, and Regulating Attention

Although we speak of meditation with the single term *bhāvanā*, the large variety of practices encompassed by that term can differ so significantly that one may wonder whether they share any universal features. In this regard, our authors helpfully turn to the central role played by mindfulness and meta-awareness in any meditative practice, especially as a means to regulate attention. To appreciate their presentation, however, it is important to unpack both of these terms.

These days, the term *mindfulness* is widespread, and secularized mindfulness programs now show up in numerous contexts, such as schools and workplaces, that are often outside a Buddhist cultural milieu. Both within Buddhism and more broadly in the contemporary world, *mindfulness* can describe a range of practices that vary considerably in technique and goals. In the more general usage found both in some Buddhist texts and in contemporary contexts, the Sanskrit term *smṛti* connects to meanings that range from its literal sense of "remembering" to meanings such as "moment-by-moment nonjudgmental awareness." Our authors, however, are using this term in a much more restricted, technical sense. Drawing on the account of mental factors in part 2, *mindfulness* here refers specifically to a mental factor that supports stable attention.

More specifically, in this technical context, mindfulness is the mental factor that prevents distraction. In other words, it prevents the mind from losing track of an object or dropping out of a target state. This loss of focus often occurs due to "attentional capture," where an unintended object or stimulus catches one's attention, and one involuntarily loses track of one's intended object. Sometimes, however, one simply loses focus on the object without being drawn to a new object, as when one dozes off in the middle of a meditation session. The role of the mental factor mindfulness is to prevent the loss of one's meditative focus in any of these ways. In this usage, one can think of the original term *smṛti* and its Tibetan translation *dran pa*, which literally mean "memory," as a kind of metaphor. When one loses track of an object, it is as if one is "forgetting" it. Thus, the mental factor that prevents one from "forgetting" or losing the object can be metaphorically called "remembering." Unfortunately, the usual English translation of this term as "mindfulness" does not convey this nuance.

The second term cited by our authors is *samprajanya* in Sanskrit, and we have translated this as "meta-awareness." Readers familiar with this term in its Pali context (where it is rendered as *sampajañña*) may be surprised by this translation, since in those contexts it is usually rendered as "clear comprehension," where one of its main functions is to recognize key features of the meditative object. However, in the Sanskrit sources cited by our authors, and especially as these sources are interpreted in Tibet, *samprajanya* plays a different role. Specifically, this term refers to the aspect of a meditative awareness that *monitors* the quality of one's attention, along with other mental and physical aspects of an ongoing meditative experience. For example, if one were stabilizing one's attention on the breath, *samprajanya* is what enables one to notice that one has become distracted, such that instead of attending to the sensations of breathing, one is now thinking about a beach vacation. In other words, *samprajanya* is what enables one to notice that mindfulness (in the technical sense described above) has been lost. For this reason, *samprajanya* is clearly a form of what cognitive scientists call meta-awareness.

The technical terms *mindfulness* and *meta-awareness* point to a process that is essential to meditation practice: namely, the regulation of attention. Our authors explore this topic in their lengthy account of calm abiding, but the issues raised there apply more broadly to all other practices. In brief, the regulation of attention is conceptualized especially in terms of the two main ways that a meditative state degrades: excitation (Skt., *auddhatya*) and laxity (*laya*). These two features stand along a spectrum of what psychological science refers to as "arousal," such that the strongest form of excitation amounts to a form of high arousal, and the deepest form of laxity is a form of low arousal. High-arousal states of excitation involve mental scattering and instability, whereby the mind is constantly losing track of its focal object or target state and is instead attending to irrelevant objects (sensory stimuli, memories, and so on) that attract the mind. In low-arousal states involving laxity, the mind loses the clarity or intensity required to sustain focus on an object or task, and the meditation degrades as a result.

In this model of attention regulation within meditation, meta-awareness is what detects excitation and laxity, and advanced practitioners can notice their presence before the meditation actually degrades. In other words, as the capacity for meta-awareness improves, practitioners can detect subtle degrees of excitation and laxity, and they can do so even while maintaining their focus on a meditative object. In most contexts, excitation and laxity are closely tied to the clarity and stability of the meditative state, and a general goal of practice is thus to strike the right balance between these two features. Too much clarity or intensity tends to produce excitation and the mental scattering that follows. Too much stability can lead to laxity and an increasingly dull mental state that eventually can even transition into sleep. Especially for beginners, many practices involve learning how to cultivate the kind of meta-awareness that enables them to notice the imbalance of stability and clarity, along with the potential for excitation and laxity, in a way that does not completely interrupt the meditative state.

Some styles of practice even attempt to cultivate a form of meta-awareness that persists without any explicit focus on any object. In those practices, the goal is to sustain "mindfulness of mere nondistraction" (Tib., *ma yengs tsam gyi dran pa*), such that one drops focus on any object and, with meta-awareness still present, one remains in a state free of any attentional capture whatsoever. This style of practice, known as *objectless calm abiding*, is radically different from those that seek to cultivate object-oriented focus, yet the basic features of mindfulness, meta-awareness, and attentional regulation are in many ways the same.

Examples of Meditative Practices

Our authors conclude this part with a presentation of two different examples of contemplative practices: the *applications of mindfulness* (Skt., *smṛtyupasthāna*) and the cultivation of equanimity toward the *eight worldly concerns* (Tib., *'jig rten chos brgyad*). The latter practice is an excellent example in the *mind training* (Tib., *blo sbyong*) style that is highly discursive and thus relies primarily on analytical meditation. As with other parts of this volume, here again the work of Śāntideva is especially influential, with its emphasis on pithy arguments and aphorisms that point out the absurdity or pointlessness of our usual attitudes toward mundane issues such as praise and blame, fame and infamy, and so on. In many ways, our authors' account of this practice actually reproduces the practice itself, and as one reads the text, it may be useful to see how the arguments and aphorisms affect one's frame of mind.

The key point here is to recognize that the target of such a practice is our ordinary, unreflective way of going through the day under the delusion of attempting to protect or satisfy a sense of self that is, in fact, nonexistent. Likewise, even in such a highly discursive practice that can almost be realized just by reading the text, one may also be able to see how mindfulness and meta-awareness—along with the regulation of attention—must be in place for one to attend to the contemplations with enough focus for them to transform one's experience. Allowing the contemplation on the eight worldly concerns to have an impact enables one to see how this is a genuine meditative practice, even if it does not conform to the stereotype of meditation as sitting quietly in deep, inward serenity.

The other practice presented by our authors, the applications of mindfulness, is sometimes translated as the "foundations of mindfulness," which might be a more familiar term to some readers. The practice as described, however, is quite different from contemporary notions of mindfulness, and this points both to the complexity of the term *mindfulness* and to the general diversity of meditative practices. Our authors describe the applications of mindfulness in terms of eliminating four false conceptions: that the body is pure; that sensations are truly pleasant; that experience is stable; and that there is, in relation to our mind-body components, some form of absolute self. This way of interpreting mindfulness practice draws on classical accounts in the Abhidharma literature. In contrast, many contemporary accounts emphasize the notion of attending purposefully to the present moment without judgment or reactivity. This too is an authentic practice, but in relation to the account presented by our authors, it is much closer to a calm abiding or *śamatha* practice, especially according to the styles found in the nondual Tibetan traditions such as Mahāmudrā and Dzokchen.

All this is to say that, as this part of the volume amply indicates, when we use terms such as *meditation* or *mindfulness* in the singular, we may very easily miss the tremendous diversity that characterizes these Buddhist contemplative practices.

John Dunne

Further Reading in English

Jay Garfield discusses key aspects of the transformative process of training the mind in chapter 6 of *Engaging Buddhism: Why It Matters to Philosophy* (New York: Oxford University Press, 2015).

For the overall structure of the Buddhist path and the role that contemplative practices play therein, see chapters 3 and 7 of Rupert Gethin, *The Foundations of Buddhism* (New York: Oxford University Press, 1998).

A detailed examination of the process of counteracting ignorance from a Dharmakīrtian perspective is presented by Vincent Eltschinger in "Ignorance, Epistemology and Soteriology—part I," *Journal of the International Association of Buddhist Studies* 32 (2009): 39–83. See also Eltschinger's *Self, No-Self, and Salvation: Dharmakīrti's Critique of the Notions of Self and Person* (Vienna: Austrian Academy of Sciences Press, 2013).

For an account of mindfulness practice in early Buddhism, see Anālayo, *Satipaṭṭhāna: The Direct Path to Realization* (Birmingham, UK: Windhorse Publications, 2003).

For a discussion of meta-awareness in later styles of mindfulness, see John D. Dunne, Evan Thompson, and Jonathan Schooler, "Mindful Meta-Awareness: Sustained and Non-Propositional," *Current Opinion in Psychology* 28 (August 2019): 307–11, https://doi.org/10.1016/j.copsyc.2019.07.003.

Several features of contemplative practices discussed in this part have been the subject of scientific studies. For an accessible account of that scientific work, including reflections on some of the Buddhist theoretical models discussed here, see Daniel Goleman and Richard J. Davidson. *Altered Traits: Science Reveals How Meditation Changes Your Mind, Brain, and Body* (New York: Avery, 2017).

23

How the Mind Is Trained

The Causes and Conditions that Produce Mental Afflictions

THUS FAR WE HAVE completed our presentation of the mind—the cognizing subject, how the mind engages its object, the reasoning that may be used by the mind to determine its object, and so on. Now we will discuss methods of training the mind aimed at goals such as enhancing the mind's ability to determine its object and, while the mind is engaging with it, making it remain single-pointedly focused on that object.

The methods for training the mind can be subsumed into two categories: (1) gradually reducing the force of harmful mental states, mental afflictions such as attachment and so on, and (2) gradually increasing and making habitual beneficial mental states such as love, compassion, and wisdom. When we presented the mental factors, we identified afflictive mental states as consisting of both root and secondary mental afflictions. To gradually eliminate these afflictions by means of their antidotes, it is not enough to merely understand the definition and function of those mental afflictions. [380] We also need to understand the causes and conditions that give rise to each of these afflictions. Because this is of pivotal importance, many classical Buddhist texts provide detailed explanations of the causes and conditions of the mental afflictions as well as of the faults and defects of each. Drawing on these sources, we offer a brief explanation here.

In general, Buddhist texts speak of numerous causes of afflictions. Asaṅga's *Levels of Yogic Deeds* mentions six principal causes:

> What are the causes of the mental afflictions? There are six types of causes from which a mental affliction can arise: a basis, a focal object, social distraction, explanation, habituation, and

> attention. A *basis* refers to the latent seed from which it arises. A *focal object* refers to an object that appears that is conducive to mental afflictions. *Social distraction* refers to going along with unwise beings. *Explanation* refers to listening to unwise teachings. *Habituation* refers to arising through the force of prior familiarity. *Attention* refers to arising from inappropriate attention.[469]

First, a *basis* refers to a seed or latent potential. It is to have within one's continuum a seed or causal basis that easily gives rise to mental afflictions whenever suitable conditions are encountered—just as, for example, when one has failed to remove the underlying cause for an illness, even a slightly inappropriate diet or behavior can lead to a flare up of that illness. [381] Second, an object or *focal object* refers to something that appears attractive or unattractive and that gives rise to a mental affliction when encountered. Third, *social distraction* refers to going along with bad companions, those who are nonvirtuous. Fourth, *explanation* refers to listening to [and studying] the opposite of the Dharma—such as treatises on warfare or treatises on sensual desire. Fifth, *habituation* refers to having repeatedly acquainted oneself with mental afflictions in the past. Sixth, *attention* refers to inappropriate attention that elaborates more and more about objects of attachment, hatred, and so forth. An example of this is to think repeatedly about someone or something attractive or unattractive, as in thinking that an object of attachment such as a suit of clothing has a delightful color, a beautiful shape, and an exquisite texture; or thinking that an object of anger such as an enemy has done such-and-such harm to oneself in the past and will certainly do likewise in the future.

In the *Compendium of Knowledge*, however, three principal causes of mental afflictions are mentioned: "Mental afflictions arise owing to their latent seeds not having been abandoned, owing to the appearance of things that are conducive to mental affliction, and owing to the activation of inappropriate attention."[470] (1) Not having abandoned the latent seeds of mental afflictions refers to not having abandoned the power of the underlying cause—the afflictive subtle proclivities. [382] (2) The appearance of things that are conducive to mental affliction refers to the power of the object—the proximity of an object that is suitable to act as an objective condition for the arising of attachment and so on. (3) The activation of inappropriate attention refers to the power of the preparatory activity—the presence of inappropriate attention as an immediately preceding condition. A statement similar to this one in the *Compendium of Knowledge* is found also in the *Treasury of Knowledge* and its autocommentary. For example, the *Treasury of Knowledge Autocommentary* says:

> As for their cause, mental afflictions arise owing to three causes—"not having abandoned the subtle proclivities, proximity of the object, and inappropriate attention." For example, in the case of attachment to pleasure, the subtle proclivity for attachment to pleasure has not been abandoned and goes unnoticed; objects conducive to entanglement in attachment to pleasure appear; and there is inappropriate attention to them. In this way, attachment to pleasure arises. Respectively, these are the power of the cause, the object, and the preparatory activity. One should understand that other mental afflictions too arise in the same way.[471]

The mental afflictions rest on ignorance, which is the root of all faults and defects, as stated in Nāgārjuna's *Fundamental Treatise on the Middle Way*:

> Karma and afflictions arise from conceptualizations;
> these arise from the tendency to proliferate.[472] [383]

Contaminated actions that bring forth various types of suffering arise from mental afflictions. Mental afflictions arise from the conceptualizations of inappropriate attention, which is attention directed toward attractive and unattractive objects that exaggerates their modes of being. These conceptualizations of inappropriate attention arise from long habituation to the proliferating tendency of ignorance that is confused about the way things actually exist.

Here the word *ignorance* does not refer to mere not knowing or to a lack of knowing; it refers to the opposite of the wisdom that knows. It is like, for example, the way that the words *unfriendly* and *untrue* do not merely mean "not friendly" and "not true" but rather mean the opposite of *friendly* and *true*. The *Treasury of Knowledge* says: "Ignorance—like unfriendly, untrue, and so on—is itself a factor, one that is the opposite of knowing."[473] This is explained in the *Treasury of Knowledge Autocommentary* as follows:

> When the word *friendly* is negated, it indicates its opposite, someone unfriendly; it does not indicate someone who is just other than a friend nor the mere absence of a friend. Something factual is called *true,* and the term opposite to it is *untrue*. Also, *unrighteous*, *worthless*, *improper*, and so on are the opposite of *righteous* and so on. Likewise, one must understand *ignorance* to be itself a factor, one that is the opposite of knowing.[474] [384]

There are in general various types of ignorance: ignorance that pertains to confusion about gross everyday facts, such as the way to get somewhere; moral confusion about what to adopt or to abandon, as in the case of ignorance with regard to karmic causes and effects; or ignorance that pertains to distorted confusion about the way that things exist. The antidote to these types of confusion is simply the relevant knowledge that understands how things actually are.

The reason ignorance itself is the root of all defects and faults is that the mental afflictions, such as attachment, as well as all defects without exception—such as birth, aging, sickness, and death—arise from ignorance. First there arises the ignorance grasping at a self, which is confused about its object's mode of existence. This ignorance obscures the ultimate nature of reality. Then, on this basis, attachment and aversion arise because of the conceptualizations of inappropriate attention that superimpose notions such as "friend" or "enemy." What is called "conceptualization of inappropriate attention" superimposes onto its object additional features that do not accord with the way the thing exists. For example, when attachment arises, the person who is the object of attachment is seen as wonderful in every respect, in a way that exceeds how they actually are. And when aversion arises, the person who is the object of aversion is seen as unpleasant in every respect, in a way that exceeds how they actually are. This is how the conceptualizations of inappropriate attention exaggerate features of their objects.

Thus the conceptualizations of inappropriate attention are induced by ignorance that is confused about reality; on that basis, attachment and aversion arise; and from attachment and aversion arise the various forms of suffering. [385] Therefore deluded ignorance itself is the root of all mental afflictions and suffering. Accordingly, just as when the roots of a tree are cut away, its branches and leaves wither, so when ignorance is extinguished, all the mental afflictions cease. This is explained as follows in the *Sūtra Teaching the Tathāgata's Inconceivable Secret*: "Śāntimati, it is like this. For example, if the root of a tree is cut away, all its parts, such as twigs, leaves, and branches, wither. Śāntimati, like that, if the view of the perishable collection is completely pacified, all mental afflictions will be completely pacified."[475] The *Fundamental Treatise on the Middle Way* says:

> By extinguishing karma and afflictions, there is liberation;
> karma and afflictions arise from conceptualizations;
> these arise from the tendency to proliferate;
> the tendency to proliferate, however, ceases in emptiness.[476]

Also, *Four Hundred Stanzas* says:

> Just as the tactile sense faculty pervades the whole body,
> delusion abides in all [the mental afflictions].
> Therefore, by destroying delusion,
> all mental afflictions are destroyed.[477] [386]

Here "delusion" must be understood to mean "ignorance grasping at a self." Now, when someone speaks sweetly to us, we think something such as, "They are speaking well of me." The sense of "me" that appears hovering deep in the mind is held to be an autonomous self that is the experiencer. Owing to that delusion, one creates biased categories of self and other, such that one is attached to self and averse to other; one is attached to the helpful, averse to the harmful, and confused about the neutral. In this way, all defects and faults of the three doors (i.e., body, speech, and mind) arise. *Exposition of Valid Cognition* says:

> One thinks "I" about the one who sees the self,
> and there is constant infatuation with it.
> Owing to that infatuation, one craves pleasures,
> and that craving obscures the flaws [in what one craves].[478]

The same text says:

> When there is "self," there is the idea of "other";
> from this distinction of self and others, grasping and aversion arise;
> and in connection with these,
> all the faults [of saṃsāra] occur.[479]

Also the *Entering the Middle Way Autocommentary* says:

> Here "afflictions" refers to attachment and so on, while "faults" refers to birth, aging, sickness, death, misery, and so on. [387] All these without exception arise from the view of the perishable collection. It says in a sūtra that all mental afflictions "have the view of the perishable collection as their root, have the view of the perishable collection as their cause, have the view of the perishable collection as their source." It is taught that the mental afflictions have the view of the perishable collection as their very cause. So for those who have not abandoned the view of the perishable collection, it follows that conditioning actions will ensue and that suffering such as birth and so on will arise. Therefore, without exception, all [afflictions and faults] have the view of the perishable collection as their cause.[480]

As explained above, a mind with attachment focused on an attractive object conceives it as attractive beyond its actual reality, and a mind with aversion focused on an unattractive object conceives it as unattractive beyond its actual reality. And when attachment and so on arise, they do so on the basis of focusing on the experiencer (the self) and the experienced (the aggregates, food, clothing, housing, close friends, and so on).

Generally, with respect to desire or aspiration, one type, such as an aspiration pursuing an activity beneficial to others, is not distorted, and another type is afflictive, such as a distorted desire pursuing an object of attachment. Likewise, on the reverse side, in the case of a mental process that is avoiding something undesirable, one type, such as hatred, is afflictive, and another type is not afflictive, [388] such as a mental process that, recognizing the faults of suffering and mental afflictions, rejects them. It is important to be cognizant of such differences.

Vasubandhu's *Treasury of Knowledge* explains how the other root afflictions as well as the secondary mental afflictions arise from ignorance in the following way. From *confusion* about how things exist comes *doubt* that wonders whether the object exists in such a way or not. From erroneous learning and thinking induced by doubt comes *wrong view*, thinking, "That is not how the object exists." From this wrong view comes the *view of the perishable collection*, viewing the aggregates as the self or as belonging to the self. Owing to the view of the perishable collection, the *view holding to an extreme* then views the self as permanent or as annihilated. Owing to the view holding to either of these extremes, the *view holding ethics and vows to be supreme*, the belief that purification and liberation will be attained by means of them, arises. From the belief that purification and liberation will be attained by holding ethics and vows to be supreme, the *view holding views to be supreme* arises. From the view holding views to be supreme comes attachment to one's own views, haughtiness, and pride, as well as hatred that despises others' views. The *Treasury of Knowledge* says:

> From confusion comes doubt; from that
> wrong view; from that the view of the perishable collection;
> from that the view holding to an extreme; from that
> the view holding ethics and vows as supreme;
> from that pride and attachment to one's own view
> and hatred of others' views, in such an order.[481]

In general, there is no certainty that these afflictions will arise in this specific sequence; [389] however, since it is possible for the mental afflictions to arise in such an order upon misperceiving the object's way of existing, such a sequence is presented.

As for the arising of the secondary afflictions from the root afflictions, from attachment comes shamelessness, excitation, avarice, pretense, and arrogance. With regard to concealment, some assert that it arises from attachment, others say that it arises from delusion, and yet others maintain that it arises from both. From ignorance comes dullness, afflictive sleep, and nonembarrassment. From doubt comes nonvirtuous regret. From anger comes rage, jealousy, resentment, and violence. From holding views to be supreme comes spite. From wrong view comes guile. The *Treasury of Knowledge* says:

> Due to attachment there arises
> shamelessness, excitation, and avarice;
> concealment is debated; from ignorance comes
> dullness, sleep, and nonembarrassment;
> from doubt, regret; rage and jealousy
> arise from anger as their cause.

And:

> Pretense and arrogance
> arise from attachment;
> resentment and violence arise from anger;
> spite comes from holding views to be supreme;
> guile arises from wrong views.[482] [390]

How Some Mental Factors Are Mutually Incompatible

Earlier it was explained that there are various kinds of mental factors with specific functions. Likewise, the causes and conditions that give rise to the afflictive mental factors and such have also been explained. Now let us explain briefly how some types of mental factors, such as anger and tolerance, are mutually incompatible. In general, when Buddhist texts explain how different states of mind are mutually incompatible, numerous antidotes to specific mental states are discussed, such as the following: compassion as the antidote to ill will; tolerance as the antidote to anger; rejoicing as the antidote to jealousy; attention focused on the unattractive as an antidote to desirous attachment; breath meditation such as focusing on the outbreath and inbreath as an antidote to obsessive thinking; contemplation of the taxonomies of aggregates, elements, and bases as an antidote to pride; learning as an antidote to not understanding; mental uplift as an antidote to laxity; contemplation of the causes of disheartenment and disenchantment (such as death, impermanence, and suffering) as an antidote to excitation; and so forth.

This mutual incompatibility between mental factors—this way of functioning as the counteractor and the counteracted—may be illustrated by looking at physical phenomena. In one type of illustration, the two opposite factors—light and darkness, heat and cold, and so on—act *directly* as the counteractor and the counteracted. Another type concerns effects, [391] such as hairs bristling, where the opposing factors of heat and cold and so on become the counteractor and the counteracted in an *indirect* manner. We must also distinguish two different categories of mental phenomena. One type is where the counteractor and the counteracted fall into the aspiration-type category of mental states, such as compassion and hatred. Another type is where the counteractor and the counteracted fall into the intelligence-type category of mental states, such as wisdom and ignorance.[483]

First, regarding the intelligence-type category, *Four Hundred Stanzas* says:

> The consciousness that is the seed of saṃsāra
> has objects as its sphere of activity.
> By seeing the absence of self in objects,
> the seed of saṃsāra will completely cease.[484]

This indicates that ignorance must be reversed by seeing that its way of apprehending is distorted. *Exposition of Valid Cognition* says:

> Because ascertainment and a mind of false superimposition
> are related as counteractor and counteracted . . .[485]

This states that, for example, an ascertaining consciousness that ascertains blue and a false superimposition that holds blue as not blue are the counteractor and the counteracted in the sense of having directly contradictory ways of holding their object. The same text also says:

> If its object is not refuted,
> then that [mental state] cannot be abandoned.[486]

A mental state that is a counteractor effectively counteracts whichever mental state is opposite to it and directly counteracted by it in the sense of [392] refuting that mental state's conceived object or way of grasping. For example, the cognition ascertaining the aggregates to be impermanent is a counteractor in that it refutes the conceived object of the mind holding the aggregates to be permanent. And the cognition realizing that a coiled rope in a dark corner is not a snake is a counteractor in that it knows that the conceived object of the mind holding this coiled rope to be a snake does not exist.

However, two cognitions or mental states do not become counteractor and counteracted merely because of having directly contradictory ways of holding a single focal object. Although attachment and hatred directed toward a specific object, such as a person, have contradictory ways of apprehending that object—as attractive and as repellent—those two minds do not constitute the counteractor and the counteracted. This can be understood from *Exposition of Valid Cognition*, which says:

> Attachment and anger, although
> different from each other, do not counteract each other.[487]

So while it is the case that whenever either attachment or hatred becomes manifest, the other cannot become manifest, it is not that those two are counteractor and counteracted; it is simply that when one becomes manifest, the conditions for the other's arising cannot be satisfied. [393]

The mental states included within the aspiration-type, such as love and hatred, are counteractor and counteracted in that they have contradictory modes of apprehension. For when directed toward the same focal object, one of them has the intention to benefit others, whereas the other has the intention to harm others. They are not counteractor and counteracted in the sense of the mind that is the counteractor realizing that the conceived object of the mind that is the counteracted does not exist.

In the case of tolerance as the antidote to hatred, it is not a matter of directly countering it by first generating the opposite perspective of the hateful mind, such as fondness. Instead, the antidote to hatred is applied through cultivating the means to ensure that the causes and conditions of hatred do not arise, and even when they do, that they do not maintain their continuum. *Engaging in the Bodhisattva's Deeds* says:

> When the undesired happens
> and the desired is obstructed,

> anger gets its fuel—discontent—
> and thereby inflamed, it destroys me.
>
> Seeing this, I will eradicate
> the fuel for this enemy of mine.[488]

Discontent arises owing to certain conditions, such as the occurrence of what one doesn't want for oneself or one's close friends and the obstruction of what one wants for oneself and others. This discontent, which is like fuel, gives rise to a mind of hatred that destroys oneself. Therefore it says here that we must eliminate the discontent itself that is the fuel of our inner enemy, hatred. Furthermore, as *Engaging in the Bodhisattva's Deeds* instructs us, we need to investigate the nature and function of the counteractor and reflect again and again on [394] clear explanations about how to apply the antidotes to the mind that is the opposite of tolerance, especially through contemplating how the perpetrator of harm can help our cultivation of tolerance and so on.

In this way, too, we should understand the rest of the aspiration-type mental states that relate to each other as the counteractor and the counteracted, such as generosity and avarice, ethical conduct and unethical conduct, diligence and laziness, and so on.

In brief, with respect to the manner in which mental states relate to each other as the counteractor and the counteracted, both in the case of intelligence-type and aspiration-type mental states, there are the following two kinds. One is like the example of heat and cold—when you increase the heat, the force of cold that is its opposite is reduced by a corresponding amount. Similarly, when you increase the thought of helping others, the thought of harming others is stopped or reduced by a corresponding degree. There is also another kind where in the case of a counteractor and its counteracted, the elimination of the opposite is simultaneous. When a light is lit, for example, the darkness to be dispelled vanishes at the same time.

This said, the early Buddhist masters on the whole maintain that, since all the mental factors explained within the category of mental afflictions are ultimately rooted in the delusion of self-grasping that distortedly apprehends reality or the view of the perishable collection that grasps at I and mine, cultivating specific antidotes, such as loving kindness against hatred and rejoicing in others' happiness against jealousy, will not be able to eliminate these afflictions permanently. Rather, they only help to temporarily subdue them. For something to be an antidote that functions to eliminate all mental afflictions, it must definitely be the wisdom that realizes how things exist. This point will be discussed in greater detail in the philosophy volumes of the present series. [395]

Three Stages of Wisdom: Learning, Reflection, and Meditation

Analyzing an object by means of correct reasoning will gradually produce three levels of wisdom generated through learning, critical reflection, and meditative cultivation in that order. Learning about a topic through listening to someone else teach it is mere *learning*. The wisdom properly comprehending what has been heard is the wisdom arisen from learning. This is generated through the causal capacity of someone else. Then carefully analyzing, using scripture and reasoning, what one has understood from that learning is *critical reflection*. The insight arisen from ascertaining what this analysis reveals is the wisdom arisen from critical reflection. This is generated through one's own causal capacity. *Meditative cultivation* refers to bringing the subject to mind again and again in accordance with what one has ascertained through learning and reflection. It is through such meditative cultivation by way of repeated familiarity that the wisdom arisen from meditative cultivation is generated. There are two types of such meditative cultivation: analytical meditation involves rationally analyzing and bringing the ascertained meaning to mind again and again; and stabilizing meditation, which is without such discursive analysis, consists of keeping the mind single-pointedly focused on the chosen object. Buddhist texts explain these three types of wisdom—arisen from learning, reflection, and meditative cultivation—as the crucial means for the mind to enhance its capacity to engage reality. [396]

In the beginning it is extremely important to generate the wisdom arisen from learning. Its beneficial qualities are stated as follows in Āryaśūra's *Garland of Birth Stories*:

> Learning is the lamp that dispels the darkness of delusion;
> it is the best wealth, one that cannot be stolen by thieves and others;
> it is the weapon that destroys the enemy, confusion;
> it is the supreme friend that reveals personal instructions, the best course to take;
> it is the unwavering dear friend even when one suffers misfortune;
> it is the medicine for the sickness of sorrow with no side effects;
> it is the destroyer of legions of great faults;
> it is also the supreme treasure of glory and renown;
> it is the best gift when meeting with wise people;
> when in an assembly, it brings pleasure to the learned ones.[489]

Learning is like the lamp that dispels the darkness of delusion because, for example, even knowing only one letter of the alphabet, such as *a*, removes the obscuration of not knowing *a* and produces an experience of understanding that knows it; then knowing all twenty-six letters of the alphabet removes the obscuration of not knowing them and produces an experience of understanding that knows them. Likewise, whatever learning we accomplish, just that purifies more and more the darkness of ignorance and increases the dawning of wisdom more and more. [397] Learning is the best wealth, one that cannot be stolen by thieves and others. Ordinary wealth can be stolen by thieves or plundered by enemies, but it is impossible for the good qualities of learning to be stolen. It is the weapon that destroys the enemy, confusion; indeed, by relying on learning, all the enemies, the mental afflictions, can be completely uprooted. Learning is also the supreme friend, expert in giving faultless instruction to oneself. When it is time to do something, then through having previously engaged in learning, one knows whether that action is suitable or unsuitable, how to do what is suitable or stop what is unsuitable, and their respective benefits and faults. Likewise, learning is the medicine for the sickness of mental afflictions and the army destroying legions of enemies that are the faults. Also, it is the supreme treasure, glory, and renown. It is the best gift when meeting with wise people. When in an assembly, it brings pleasure to the learned ones.

Buddhist texts explain the definitions of learning, reflection, and meditation in the following way. "Wisdom arisen from learning" is a cognition differentiating phenomena that is generated on the basis of another source, such as a teacher or a valid scripture. For example, in dependence on hearing or reading a text teaching that sound is impermanent, an understanding that arrives at knowing a general statement—namely, that sound is impermanent—is such a level of understanding. "Wisdom arisen from critical reflection" is a cognition that has reached ascertainment through rationally analyzing the meaning of what has been learned. An example is an inferential cognition realizing that sound is impermanent. "Wisdom arisen from meditative cultivation" is an awareness supported by special pliancy that has arisen from [398] repeatedly and undistractedly habituating or familiarizing the mind, through either stabilizing or analytical meditation as appropriate, with the meaning experienced as a result of learning and reflection. Examples of this are both calm abiding and special insight. The *Treasury of Knowledge Autocommentary*, for example, says: "It is explained, 'Wisdom arisen from learning is ascertainment that arises from trustworthy valid scriptures, wisdom arisen from reflection is ascertainment that arises from definitive analysis with reasoning, and wisdom arisen from meditative cultivation is ascertainment that arises from concentration.' This seems to faultlessly state their definitions."[490]

The text also describes "causes of calm abiding, firm apprehension, and equanimity"; these, respectively, mean the following. When special insight is stronger and calm abiding is weaker, the mind fluctuates like an oil lamp in the wind; thus the meditation object is not seen clearly. At that time, attention focuses on the cause or sign of calm abiding, the method for drawing the mind inward. In the opposite situation, it is like being asleep, thus ultimate reality is not seen clearly. At that time, attention should be focused on firm apprehension's sign, the method for uplifting the mind. When both calm abiding and special insight operate equally of their own accord, attention should be brought to equanimity's sign. [399]

In general, mere wisdom arisen from learning is not certain to bring about a change in one's physical or verbal behavior. If, however, one is able to attain mastery of the relevant subject through critical reflection and meditation, and especially gain experiential understanding through meditative cultivation, this can greatly transform one's habitual ways of thinking as well as one's physical and verbal behavior. Therefore both critical reflection and meditative cultivation are ways for a person to imbue their mindstream with cognitions that are correct with regard to understanding the meaning of what one has heard. In this way, one must tether one's behavior of body, speech, and mind to what one has understood at the time of hearing, as if bound by a rope. Thus, given the crucial importance of critical reflection and meditative cultivation with regard to positive transformation of a person's physical, verbal, and mental conduct, the Buddhist texts provide detailed presentations of the wisdom arisen from reflection (induced by powerful reasoning) as well as the meditative practices that constitute the method for cultivating the wisdom arisen from meditation. [400]

The Importance of Mindfulness and Meta-Awareness

Mindfulness is the mental factor that prevents one from forgetting the object and its aspects that are to be adopted or abandoned. Meta-awareness is the mental factor that analyzes one's conduct of body, speech, and mind from moment to moment and brings about an understanding of whether it is virtuous or nonvirtuous. Both of these mental factors are very important—not only at the time of learning and reflecting about what one needs to know—but also in everyday life. By maintaining continual awareness of each point to be adopted or abandoned without confusing them, mindfulness helps retain them without forgetting. Then, having understood which points are to be adopted or abandoned, meta-awareness acts as a spy at the time of engaging in any conduct and thinks "I will do such-and-such." When mindfulness and meta-awareness have both been developed, then one behaves skillfully without error—doing what is to be adopted and not doing what is to be abandoned.

To generate such mindfulness and meta-awareness, one must always practice heedfulness [which protects the mind from afflictive mental factors and promotes virtue]. To give rise to heedfulness, one must consider the benefits of practicing heedfulness and the faults of not doing so. How does this work? If one gives rise to heedfulness, then mindfulness and meta-awareness will be generated. In dependence on that, one will adopt good qualities and abandon faults without error. [401]

In the context of learning, reflecting, and so on, one can explain the practice of mindfulness and meta-awareness as follows. Whatever the mind is engaged in, one needs a method both to prevent it from wandering away from the main object of focus and to recognize whether it is wandering away or not. The first is the function of mindfulness and the second is the function of meta-awareness. Vasubandhu's *Explanation of the Ornament for the Mahāyāna Sūtras* says: "Mindfulness and meta-awareness enable the mind to focus because the former prevents the mind from wandering away from its object while the latter makes one clearly aware if it is wandering."[491] If we have mindfulness and meta-awareness in our mindstreams, then not only do we study the textual traditions without distortion and understand their meaning without forgetting, but we also give rise to good qualities that we did not have before, and those already arisen are further improved. Also, our behavior of body, speech, and mind will not fall under the power of the contrary side, the mental afflictions.

Furthermore, we will remain immune to the faults of nonvirtuous tendencies if we are able to sustain mindfulness and meta-awareness, as is mentioned in the *Great Play Sūtra*:

> Having skillfully applied mindfulness
> and having well cultivated wisdom,
> I will act with meta-awareness. [402]
> Evil mind, what will you do now?[492]

This indicates the following: one applies mindfulness focused on the proper object of one's activity, while analyzing it accurately with wisdom differentiating good and bad; and using the spy of meta-awareness that recognizes whether the mind has wandered away from that meaningful activity to be done, then evil minds—mental afflictions such as laziness, distraction, and so on—cannot do any harm. Conversely, if we do not apply mindfulness, meta-awareness, and heedfulness, then although we may appear to know many points at the time of learning and reflecting, when we enter into actual practice, what was studied earlier does not become an object of our mindfulness or recollection. This is like pouring water into a pot that has a leak. *Engaging in the Bodhisattva's Deeds* says:

> In one whose mind lacks meta-awareness,
> what has been learned, contemplated, and meditated on

does not remain in mindful recollection,
just like water in a leaky pot.[493]

If mindfulness and meta-awareness degenerate, the mind will be disturbed by mental afflictions. In that case, whatever activity of listening, reflecting, and studying that one may engage in will not give rise to any greatly beneficial result owing to a weak capacity to accomplish that activity. [403] It is like people whose physical elements are so disturbed by sickness that they are too weak to do any useful work. *Engaging in the Bodhisattva's Deeds* says:

Just as people disturbed by illness
have no power to do any useful work,
those whose minds are disturbed by confusion
have no power to do any useful work.[494]

In brief, without mindfulness and meta-awareness, previously present good qualities deteriorate, and new ones do not arise. Nāgārjuna's *Friendly Letter* says:

With mindfulness impaired, all qualities diminish.[495]

Having pointed out the benefits of applying mindfulness and meta-awareness and the faults of not applying them, we now discuss why, in general, a person who is a beginner must guard his or her mind in the following way. When someone with a physical wound is in a throng of excited people whose behavior is uncontrolled, then he or she is careful of that wound and protects it. Likewise, when you are in a mob of careless people who give license to attachment and hatred, then you must always guard the wound of your mind. With mindfulness you remember what is to be adopted and abandoned according to the circumstances without forgetting, and with meta-awareness you scrutinize whether a specific action is suitable [404] or unsuitable. If you do not guard it in this way, you will thereby miss the point of learning and reflecting, and you will cut the root of present and future happiness for yourself and others. *Engaging in the Bodhisattva's Deeds* says:

> Just as I take care of a bodily wound and cover it
> when amid a rough and jostling crowd,
> so too, when amid a crowd of careless people,
> I must always protect the wound of my mind.[496]

How do we apply mindfulness? In a worldly situation, if a man's sword falls from his hand when fighting an enemy, he immediately picks it up out of fear that the other may kill him. Likewise, if we lose our mindfulness—which ensures that we do not forget the object and its aspects that are to be practiced or abandoned when struggling with mental afflictions such as heedlessness—then we must immediately take up mindfulness again out of fear of falling into a wrong course of action. *Engaging in the Bodhisattva's Deeds* says:

> If someone drops his sword in battle
> he immediately picks it up out of fear;
> likewise, if I lose my weapon of mindfulness,
> then recollecting the fears of hell, I must quickly retrieve it.[497]

Thus this text explains from numerous perspectives the great importance of applying mindfulness. As for how to apply meta-awareness, *Engaging in the Bodhisattva's Deeds* says:

> At the outset, first recognizing that
> my mind has faults, [405]
> I should remain controlled,
> like a piece of wood.[498]

At the outset of engaging in any action, one should first discern the motivation for engaging in that action, the purpose of doing it, whether it is appropriate or inappropriate, and so on. Then, if one recognizes through meta-awareness that faulty motivations and so forth have arisen, one must be able to hold one's mind stable, like a piece of wood, without allowing it to be moved by the defilements. This is how one relies on meta-awareness at the start of an action. Next, in all contexts, whether walking, moving, lying down, sitting, speaking, thinking, and so on, one mindfully recollects what is appropriate and inappropriate, and with meta-awareness acting as a spy, one examines again and again whether one's activities of body, speech, and mind are concordant or discordant [with the appropriate activities that one is mindfully recollecting]. This is how meta-awareness protects one when one is about to engage in any action of body, speech, and mind. *Engaging in the Bodhisattva's Deeds* says:

> To observe again and again
> the state of one's body and mind:
> just this alone, in brief,
> is the definition of guarding meta-awareness.[499]

In brief, how does someone wanting victory over the mental afflictions—such as ignorance, heedlessness, and distraction—apply mindfulness and meta-awareness? Here *Engaging in the Bodhisattva's Deeds* says: [406]

> Those who engage in practice should be as intent
> as a frightened man carrying a jar full of mustard oil
> while being surveilled by someone brandishing a sword
> and threatening to kill him if he spills any.[500]

Suppose a person has been ordered by a cruel king to walk along a path carrying a jar filled with mustard oil while accompanied by a man wielding a sword who has been told, "If he spills even one drop of oil, you must kill him instantly!" Such a person will try very hard indeed to be vigilant, out of fear for his life. When a person engages in any actions of body, speech, and mind, he or she must be vigilant like this, maintaining mindfulness and meta-awareness with regard to what is suitable or unsuitable to engage in. One can infer from their behavior what degree of seriousness and alertness such a person has adopted at that time. As for the causes, definitions, functions, and classifications of both mindfulness and meta-awareness, these should be understood from the earlier section where the mental factors were explained. [407]

View, Meditation, and Conduct

In general, phenomena that are categorized as mental arise through causal conditioning. Hence, the following consequences ensue: when the causes and conditions producing them either increase or decrease, the resultant mental states will also increase or decrease accordingly. In proportion to the degree of habituation to a mental state, that state arises more easily. In this way, the features of that mental state will gradually become part of one's habitual nature.

Buddhist texts contain specific discussions about how, on the basis of learning, reflection, and meditative cultivation—or alternatively, through view, meditation, and conduct—a gradual transformation of the mind can take place. In particular, one can gain conviction about how, through targeted mental training, the expressions or behavior of both one's internal mind and one's external body and speech can be transformed. An important feature of this explanation of how inner transformation within one's mind brings about external changes in one's conduct of body and speech is how Buddhist texts distinguish among view, meditation, and conduct. First, we need to change the faulty *views* we might have. Then, on the basis of such methods as analytical and stabilizing *meditation*, we need to cultivate familiarity with those perspectives connected with the correct view. In this way, so it is explained, our *conduct* of body and speech will change so that they proceed in a virtuous direction. [408]

Concerning view, meditation, and conduct, in general meditation involves familiarizing oneself again and again with the chosen object that is the focus of one's meditation. Dharmamitra's *Clarifying Words* commentary on the Perfection of Wisdom says: "To meditate is to make the mind acquire the object's qualities or understand its reality."[501] As we saw above, there are two types: *analytical meditation*, which is to meditate by analyzing with the wisdom of fine investigation, and *stabilizing meditation*, which is to place the mind single-pointedly on the object without analyzing. In general calm abiding is a type of stabilizing meditation, and special insight is a type of analytical meditation. However, not all single-pointed meditation is calm abiding, and not all analytical meditation is special insight. As explained extensively below, stabilizing meditation is to meditate while placing the mind single-pointedly without distraction on a single basis or object of meditation, and analytical meditation is to meditate while contemplating many scriptural citations, reasons, examples, and so on concerning the deeper meaning of the object of meditation.

There are, however, many other forms of meditation. For example, in the context of meditating on impermanence and emptiness, one meditates in relation to the object as cognized. One may also meditate by generating a particular subjective state, as in the cultivation of loving kindness and compassion. Other forms of meditation involve contemplating the characteristics or aspects that resemble states more advanced than one's own, as when meditating on higher qualities that one has not yet attained. Alternatively, one can speak of meditating on a feature of an object, as when one meditates on all conditioned things as impermanent, [409] or of meditating on a subjective feature, as when one cultivates compassion by generating one's mind into the state of compassion.

As for the object of meditation, Asaṅga's *Śrāvaka Levels* says: "What are the objects [of meditation]? There are four types of objects: pervasive objects, objects that purify behavior, objects of mastery, and objects that purify mental afflictions."[502] As for a detailed explanation of these four types of meditation objects, these need to be learned from the great texts themselves.[503] In this same text of Asaṅga, it also states that one may categorize meditation practice in terms of the degree of acuity of the meditator's faculties and also in terms of the behavior or temperament from the standpoint of afflictive mental states. In terms of habituation to mental afflictions, there are seven types of temperaments: (1) attachment predominates, (2) aversion predominates, (3) delusion predominates, (4) pride predominates, (5) distorted thinking predominates, (6) equally balanced temperament, (7) less afflictive temperament.[504] Also, in terms of actual engagement in practice, the same text explains that there are persons who engage by way of faith and those who do so by way of things themselves [410]—that is, those who rationally analyze the way things exist. One should understand these points by referring to detailed scriptural sources such as Asaṅga's *Śravaka Levels*, Vasubandhu's *Treasury of Knowledge* and its *Autocommentary*, Śāntideva's *Compendium of Training*, and Kamalaśīla's *Stages of Meditation*; these texts contain extensive explanations about how reliance on specific types of meditation objects by these different types of persons will have widely varying impact on the transformation of their minds. Thus these texts explain, in their presentations on meditation, types of people who meditate, what are the objects of meditation, how to engage in meditation, and how meditation gradually brings about mental transformation. A brief explanation of these will be offered below in the context of how to cultivate calm abiding and special insight.

Seven Types of Temperament

Asaṅga's *Śrāvaka Levels* clearly distinguishes the seven different temperaments of persons on the basis of evidence. Here we provide a brief summary.

First, in one with a predominance of attachment, the text mentions the following evidence. In such persons, even small things producing attachment cause a huge entanglement of desire to arise, and the attachment that has arisen [411] remains within the mindstream for a long time. Such a person becomes completely habituated to attachment. He becomes overpowered by the things to which he is attached. He has faculties that are graceful, gentle, and without harshness, and by nature he does not inflict injury on others. He finds it hard to separate himself from persons or things. He is not easily disheartened. He wishes for what is inferior. He accomplishes actions in a firm and stable manner and keeps vows in a firm and stable manner. He experiences long-lasting delight and has a nature that is greedy for goods and attributes, and so he experiences hankering. He is frequently happy; he is frequently joyful; he does not develop full-blown rage; his face has a radiant hue, a smiling expression, and so on. Regarding these attributes, *Śrāvaka Levels* states: "What are the signs of a person in whom attachment predominates? For such a person, if even a small and inferior thing triggering attachment produces huge and dense entanglements of attachment to arise, what need is there to mention those things that are mediocre and good?"[505]

Second, in one with a predominance of aversion, the text mentions the following evidence. Even slight triggers for anger cause frequent entanglements of anger to arise, and as in the explanation of attachment, that entanglement remains within the mindstream for a long time. [412] Such a person becomes completely habituated to aversion. He becomes overpowered by the conditions giving rise to anger and has faculties that are ungraceful, ungentle, harsh, and so on—contrary to those stated above in the case of attachment. He finds it easy to separate himself from persons and things. He is easily disheartened. He makes senseless remarks and lacks a sense of shame. He frequently lacks aspiration. He accomplishes actions in a firm and stable manner, and he keeps vows in a firm and stable manner. He often has mental discomfort; he is frequently disturbed; he lacks tolerance and experiences short-lived delight. He has an unharmonious nature; he is frequently angry; he has a nature that is difficult to rectify; his thinking is fierce and wrathful; his speech is spiteful and abusive; he generates ill will even when only a minor fault or disagreement is mentioned; he wallows in such ill will; he instigates quarrels; he develops full-blown rage; his face has a darkened hue; he is frequently jealous and angry with regard to others' possessions and so on. Regarding these attributes, *Śrāvaka Levels* says: "What are the signs of a person in whom aversion predominates? For such a person, if even a slight reason for anger toward a thing triggering anger causes a dense entanglement of anger to arise frequently, what need is there to mention those reasons that are middling and great?"[506] [413]

Third, in one with a predominance of delusion, the text mentions the following evidence. Even a small thing producing bewilderment causes frequent dense entanglements of delusion to arise, which remain within the mindstream for a long time. Such a person becomes completely habituated to delusion and so on, as in the case of attachment and aversion. He has faculties that are slow, foolish, and weak. His physical conduct is lax; his verbal conduct is lax. His thinking is unskillful; he makes unskillful remarks; he does unskillful deeds. He is extremely lazy; he lacks the ability to accomplish things; his speech is meaningless; his mind is inattentive; he is forgetful. He does not practice meta-awareness, and he apprehends things in a distorted manner. He finds it hard to separate himself from people and things. He is not easily disheartened. He wishes for what is inferior; he is stupid; he staggers around; he gestures with his hands; he is unable to understand the meaning of good and bad; he is carried away by circumstances; he is dependent on others. Regarding these attributes, *Śrāvaka Levels* says: "What are the signs of a person in whom delusion predominates? In the case of such a person, if even a small thing producing confusion causes a dense entanglement of delusion to arise frequently, what need is there to mention those things that are middling and great?"[507] [414]

Fourth, in one with a predominance of pride, the text mentions the following evidence. Even a small object producing pride causes dense entanglements of pride to arise frequently, and pride remains within the mindstream for a long time. Such a person becomes completely habituated to pride, as explained in the case of attachment, aversion, and so on. He has faculties that are excited, lofty like a reed, and haughty. He makes great effort to adorn his body. He gives orders without showing respect. He is haughty; he does not bow down; he is not inclined to put his palms together and pay homage; he considers only himself; he praises himself;

he criticizes others. He wants gain; he wants respect; he wants fame. His thinking is muddled. He finds it hard to separate himself from persons and things. He is not easily disheartened and wishes for greatness. He has little compassion and has a strong view of the self. He develops very fierce rage and is quarrelsome. Regarding these attributes, *Śrāvaka Levels* says: "What are the signs of a person in whom pride predominates? For such a person, if even a small thing producing haughtiness causes a dense entanglement of pride to arise frequently, what need is there to mention those things that are middling and great?"[508] [415]

Fifth, in one with a predominance of distorted thinking, the text mentions the following evidence. Even a small thing producing distorted thinking causes a dense entanglement of distorted thinking to arise frequently, and it remains in the mindstream for a long time. Such a person becomes completely habituated to distorted thinking, as explained in the case of attachment, aversion, and so on. He has faculties that are unstable, fluctuating, changeable, and disturbed. He finds it hard to separate from people and things. He is not easily disheartened. He likes proliferation; he enjoys proliferation; he often oscillates; he has many doubts; he has schemes. He is unstable in keeping vows; he is uncertain in keeping vows; his actions are unstable and uncertain; he is frequently anxious; he is forgetful. He dislikes solitude and is frequently distracted and wholly taken up with attachment wanting various mundane things. He is clever; he is not lazy; he has the ability to accomplish his aims. Regarding these attributes, *Śrāvaka Levels* says: "What are the signs of a person in whom distorted thinking predominates? In the case of such a person, if even a small thing producing distorted thinking causes a dense entanglement of distorted thinking to arise frequently, what need is there to mention those things that are middling and great?"[509] [416]

Thinking that the sixth, an equally balanced temperament, and the seventh, a less afflictive temperament, are both types that are easy to understand, the *Śrāvaka Levels* simply states: "Whoever is equally balanced is one having an equally balanced temperament, and whoever has a nature with few mental afflictions is one with a less afflictive temperament."[510] Other than this, the text does not identify the specific signs of these two types of persons.

In brief, the divisions of persons by temperament are posited mainly on the basis of mental inclination and general temperament. Thus individuals are differentiated in terms of their everyday temperament into seven different types: (1) one whose nature is very kindly or prone to attachment, (2) one whose nature is rough and easily angered by small conditions, thus short-tempered, (3) one who has naturally slow intelligence and is prone to forgetfulness and laziness, (4) one who is arrogant, has a haughty nature, and strongly holds his own views to be the best, (5) one who has a nature of obsessive thinking and suspicion, (6) one who is equally balanced whereby none of those mental states manifestly predominates, (7) and one who has a disposition that is naturally less afflictive.[511] [417]

24

Calm Abiding

How to Cultivate the Single-Pointed Mind of Calm Abiding

As explained above, Buddhist texts distinguish among many categories of "gross" and "subtle" states of mind. They speak of how one can go beyond the gross states of mind of the desire realm and attain higher states of mind, such as the states of mind encompassing the meditative absorption (*dhyāna*) of the form and formless realms. Achieving these higher states mainly arises through attaining a single-pointed mind focused on one's chosen meditation object in dependence on a method that combines mindfulness and meta-awareness. Ultimately this depends on achieving calm abiding, where the mind remains single-pointedly focused. Hence, we now present a brief explanation of the topic of calm abiding (*śamatha*).

In general, the purpose of cultivating calm abiding is to gather one's scattered mind inward and make the mind serviceable so that one can focus, as one wishes, on any virtuous object. Afflictions such as attachment arise and lead to all sorts of suffering because one's mind is scattered. When scattered, the mind's clear and knowing capacity dissipates toward various objects, thus creating the conditions that prevent it from being channeled in a focused manner. When calm abiding is attained, the mind's capacities become concentrated and highly efficient with respect to whatever higher quality one might seek to cultivate. It is stated, therefore, that it is important to cultivate a stable calm abiding at the outset. [418]

When the mind's capacities can be channeled single-pointedly, then the mind focusing on its chosen object, whatever that object might be, will be very powerful. In general, the mere gathering of the mind inward also occurs during a state of deep sleep, but in such cases, other than the simple fact of the mind not being distracted outwardly, there is neither clarity nor vibrancy. In contrast, the gathering of the mind inward that occurs as a result of calm abiding meditation possesses *stability*, in that the mind remains firmly focused on its meditation object, and *clarity*, in that there is the clear appearance of its object. Such a mind is also characterized by *vibrancy*, in that it is lucid and alert. Therefore, although both sleep and concentration (*samādhi*) are similar in that they both gather the mind inward, there differ enormously in their meaning.

Ensuring the Prerequisites of Calm Abiding

Although many prerequisites giving rise to calm abiding are mentioned in the texts, Kamalaśīla's *Stages of Meditation* explains the principal causes of calm abiding to be as follows: "What are the prerequisites of calm abiding? Staying in a suitable area, having few desires, being content, giving up many activities, maintaining pure ethical conduct, and abandoning thoughts of desire and so on."[512] [419] Here the text speaks of a total of six essential prerequisites that give rise to calm abiding, helping to produce it where it has not yet arisen and, where it has arisen, averting its degeneration and enhancing it.

Among these prerequisites: (1) "Staying in a suitable area" refers to being in an environment that has these five attributes: ease in obtaining food, clothing, and so on; a good area with no harmful living beings such as predators and robbers; a good piece of ground with hygienic water and no causes of sickness; good companions who observe ethical conduct and engage in similar practices; and conditions comfortable for yogic practice without many people about in the daytime and little noise at night. (2) "Having few desires" means not craving for more or better external objects of desire, such as food and clothing. (3) "Being content" means being satisfied with just the bare necessities of food and clothing. If one fails to maintain a state of having few desires and being content, then one will be distracted toward outer objects of desire, as well as toward accumulating and protecting one's wealth and possessions, owing to which concentration will not arise. (4) "Giving up many activities" means that if one does not give up social distractions, one will engage in purposeless activities and waste one's time chatting and so on. (5) "Maintaining pure ethical conduct" means pacifying subtle inner distraction, which depends on abandoning gross outward distractions. If gross distorted thinking predominates, then the mind will not rest naturally; therefore the mind is made peaceful by means of ethical conduct that refrains from faulty behavior of body and speech. [420] (6) "Abandoning thoughts of desire and so on" is to contemplate, "Since all outer things, whether attractive or unattractive, are subject to disintegration and do not remain long, it is certain that I will be separated from them. So why am I totally attached to them?"

When one has gathered the prerequisites of calm abiding, such as a place with the five characteristics and so forth, one should sit on a comfortable seat in the bodily posture of Vairocana, which has seven or eight features:[513]

(1) On a seat slightly elevated at the back, it is good to keep the legs in either a full-lotus position or a half-lotus position. The purpose is to help produce pliancy, because the body must be well disciplined in the yogic practice of sitting cross-legged; also, the position enables one to sit for a long time without tiring the body. The back of the right hand is placed on the palm of the left hand; the two elbows are placed comfortably and directed slightly away from the sides of the body so that the empty area between will allow air to move back and forth.

(2) The eyes should be neither wide open nor completely shut and should be left as if directed toward the tip of one's nose. It is said that if the eyes are just looking at the edges of the nose, it more easily eliminates laxity and excitation. Also, it is taught that if the eyes are wide open, the mind comes under the sway of the eye consciousness, thus increasing the risk of distraction. And if the eyes are completely shut, then an appearance of redness arises that obstructs the clarity of the meditation object; it also increases the risk of sleep and mental dullness. [421]

(3) The body should be straight, without leaning back too far nor bending forward too much. It is said that holding it upright, balanced between leaning back and bending forward, helps prevent the occurrence of sleep and confusion. Also, when the body is straight, the channels are straight, and based on that, the winds flowing in the channels flow straight, which makes the mind more serviceable.

(4) The shoulders should be straight and evenly placed. (5) The head should be neither raised nor lowered but kept facing forward without turning to the side. (6) The teeth and lips should remain naturally in their usual position. (7) The tongue should be raised slightly to touch the top teeth. When the tip of the tongue is drawn up behind the teeth, it helps to

prevent the mouth from becoming dry, and in deep meditative equipoise it helps prevent drooling.

(8) As for counting the breath, one settles into a behavioral state—such as the bodily posture of Vairocana with its seven features—that is conducive to mental absorption, and one checks one's mind. If it turns out that one's mind is being swayed by a motivation that is nonvirtuous, such as attachment or aversion, then it will be difficult to have a good meditation. So, for the time being, one needs to develop a neutral mind. A neutral mind is neither virtuous nor nonvirtuous; it is said to be like a piece of stainless white cloth. A neutral mind can be developed through counting the breath in the following way. Breathing through the nostrils, gently exhale in a way that is not noisy, forceful, or unsteady, and think "breathing out." In the same manner, inhale, and think "breathing in." Thereby think "one round." Then keep counting in rounds from "two" up to "seven," "nine," "eleven," "fifteen," or "twenty-one," and so on, in the same way. Without using a rosary to count, [422] keep note of the out-breaths and in-breaths with an unfluctuating mind. The texts explain that, by doing this practice, attachment and so on will be calmed, and one's mind will enter a neutral state.

There is also the practice of "clearing stale air in nine rounds," which is as follows. Through both nostrils, breathe out while imagining that all nonvirtuous mental states—such as attachment, aversion, and ill will—appear as black smoke that, upon exhaling, vanish into the realm of space. Through the right nostril, breathe in while imagining that all types of virtuous mental states, such as love and compassion, appear as white light that, upon inhaling, dissolve into one's heart and manifest as a special virtuous mind. Through the left nostril, breathe out while imagining the same as in the above case of breathing out. Visualizing in this way, breathe in through the right nostril and breathe out through the left three times; then breathe in through the left nostril and breathe out through the right three times; and finally breathe in through both nostrils and breathe out through both three times. Although it is said that one should resolve to do the clearing nine-round breathing exercise, breath exercises such as this one are just enhancement practices, so they are not absolutely necessary. If one's mind is not presently arising with afflictions such as attachment, and one is easily able to induce virtuous states to mind, then there is no need for exercises like counting the breath. This is the reason for the "seven-point" or "eight-point" Vairocana posture.

Likewise, the Guhyasamāja explanatory tantra, *Vajra Garland*, says:

> The adept, having sat on a comfortable seat,
> fixes both eyes on the tip of the nose [423]
> while keeping the nose in line with the navel,
> the top of the shoulders level, the tongue against the palate,
> and the teeth and lips placed comfortably.[514]

Also, Kamalaśīla's *Stages of Meditation* says:

> On a very comfortable seat, it is good to sit in the full-lotus posture of Vairocana or the half-lotus posture. The eyes, neither too widely open nor too tightly closed, should be directed toward the tip of the nose. The body should be held upright, neither bending forward nor leaning back, and the mental attention directed inward. The shoulders should be kept level, and the head should face forward without tilting up or down or turning to either side. The nose should be in line with the navel. The teeth and the lips should remain as usual, with the tongue resting against the upper palate. The inflow and outflow of the breath should be smooth without being noisy, forceful, or unsteady. In any case, inhale and exhale naturally, gently, and imperceptibly.[515] [424]

The Kālacakra tantra texts give a slightly different instruction on the bodily posture with five features in the context of gathering the winds into the central channel. They are to: (1) keep the spine straight upright, (2) place the legs in the vajra lotus posture, (3) place both hands in the vajra-fist gesture[516] crossed at the belly, or alternatively, place the left hand underneath and the right hand on top of it with the thumbs touching to form the gesture of meditative equipoise, or place both hands in the vajra-fist gesture on the thighs, (4) keep the tongue against the palate, (5) keep both eyes looking upward. This way of sitting is different from the one described above, which shows there are alternative postures for particular circumstances.

Objects of Calm Abiding Meditation

In general, anything can be used as a meditation object of calm abiding—a flower, a tree, and so on. Thus any suitable object one is inclined to is acceptable. If one takes as a meditation object the form of a deity, for example, then the body of the chosen deity is visualized as being one arm span in front of oneself, about the size of a thumb tip, very dense, emanating light, radiant, and at the level of one's forehead, or heart, or navel, in accordance with the meditator's inclination and comfort. When one is able to observe that form clearly, [425] visualizing it as very small makes it easier for the mind to gather inward and tone down. It is taught that meditating on it as very dense can help the mind become less scattered and prevent excitation, while meditating on it as emanating light can help prevent laxity. Since it is difficult to have a clear meditation object initially, if just a rough part of the basic meditation object, such as the head, the hands, the legs, and so on, appear to the mind, then one must hold that with mindfulness without losing it and use meta-awareness to check that mindfulness does not lose the object.

In the above manner, if one practices by engaging in many short sessions and maintains continuity of the meditation object, it is said that the object will become clearer and clearer. It is also said that holding it firmly prevents laxity; not being distracted prevents excitation; practicing in many sessions allows concentration to arise; and keeping the sessions short inclines one to engage in meditation again and again. One must meditate without altering the basic meditation object until calm abiding is established because changing the basic meditation object becomes an obstacle to establishing calm abiding. Āryaśūra's *Compendium of the Perfections* says:

> One should stabilize the mind's attention
> by firmly placing it on just a single object;
> going through many meditation objects
> will only disturb the mind with afflictions.[517] [426]

Similarly, Atiśa's *Lamp for the Path to Enlightenment* says:

> Focusing on any single meditation object,
> the mind is placed in virtue.[518]

Here we should unpack the specification "single." Initially one must establish calm abiding by focusing on a single meditation object. Later, however, it is suitable to focus on many. Kamalaśīla's *Stages of Meditation* says: "When attention has been mastered, one focuses more broadly in terms of categories such as the aggregates, the elements, and so on. Thus various types of meditation objects are spoken of in the *Unraveling the Intention Sūtra* and so on in terms of different meditative objects for yogis, such as the eighteen aspects of emptiness."[519]

The criterion of the mind having first "obtained" or "found" the meditation object that one is focusing on can be understood as follows. Suppose one takes the body of a deity as the meditation object. Initially, one visualizes the head, the two arms, the trunk of the body, and the two legs, several times in sequence. After that, when one focuses on the whole body in general, if some coarse image of the limbs—perhaps just half—can appear in one's mind, then even if what is appearing lacks radiant clarity, one should be content with simply that much and focus the mind on it. If one is not content with that and does not hold one's mind on it but instead keeps visualizing it again and again seeking further clarity, [427] then the meditation object will indeed become slightly clearer; however, not only will one *not* gain the meditative concentration of the mind's stability aspect, but this will also *prevent* one from gaining it. Conversely, even though the meditation object may not be very clear, if one holds that mind on the meditation object itself with only half the features, then one will quickly achieve concentration; and since this enhances clarity, one will easily accomplish the mind's clarity aspect also.

If some parts of the deity's body appear very clearly, then focus on those, but if they are unclear, then focus on the entire body in general. It might happen that a red one appears when one wants to meditate on a yellow one, or a standing one appears when one wants to meditate on a seated one, or two of them appear when one wants to meditate on one, or a very small one appears when one wants to meditate on a large one, and so on. If this happens, it would be totally unsuitable to pursue these appearances; instead, one should use only that which is the original object of focus as the object of one's meditation.

THE NATURE OF CALM ABIDING

The definition of calm abiding is: a concentration arisen from meditative cultivation that, through being conjoined with a special pliancy, engages its object exactly as desired. The *Unraveling the Intention Sūtra* says:

> One stays alone in solitude. Then, having correctly settled the mind within, one focuses attention on those phenomena that one has reflected upon. The mind that is paying attention is continuously attending inwardly; thus this is *attention*. [428] In one who focuses inwardly in this way and abides like that repeatedly, bodily pliancy and mental pliancy arise; when that happens, it is called *calm abiding*. In this way bodhisattvas seek calm abiding.[520]

To explain the term *calm abiding*, when the outward wandering of the mind has been *calmed* and it *abides* single-pointedly on the inner object of meditation, then it is called *calm abiding*. The identification and analysis of *pliancy* is explained below.

THE EIGHT MENTAL APPLICATIONS AS ANTIDOTES TO THE FIVE FAULTS

The instructions on developing calm abiding found in Maitreya's *Distinguishing the Middle from the Extremes* teach how to meditate by way of relying on the eight applications that are the antidotes countering the five faults. Concerning this method, first, the five faults are: (1) *laziness* in the preparatory stage, such that when one is first cultivating concentration, one does not want to begin doing it, or although one has begun, one is not able to maintain continuity; (2) *forgetting the instructions* that apply when one's mindfulness has lost the meditation object; (3) *laxity and excitation* during the actual session when the meditation object is being retained by mindfulness; (4) *not applying the antidotes* when laxity or excitation arise; (5) *applying the antidotes* when neither laxity nor excitation has arisen. *Distinguishing the Middle from the Extremes* says: [429]

> Countering the five faults arises from its cause,
> which is to rely on the eight applications.

> Laziness, forgetting the instructions,
> laxity and excitation,
> not applying, and applying [the antidotes]:
> these are said to be the five faults.[521]

Of the eight applications that are the antidotes countering those five faults, four are antidotes to laziness and four are antidotes to the other four faults. The first four are (1) *diligence*, which is the direct antidote to laziness; (2) the cause of that, which is *aspiration* seeking to attain concentration; (3) the cause of that, which is *faith* that sees the good qualities of concentration; (4) the result of diligence, which is *pliancy*. *Distinguishing the Middle from the Extremes* says:

> The basis, that which is based on it,
> the cause, and its result.[522]

Here it should be understood that "basis" refers to aspiration, "that which is based on it" refers to diligence, the "cause" of these two is faith that sees the good qualities of concentration, and pliancy is the "result."

The four antidotes to the other four faults are: (1) *Mindfulness*, which does not forget the meditative object or its aspects, is the antidote to forgetting the instructions. (2) *Meta-awareness*, which recognizes when laxity or excitation has arisen, is the antidote to laxity and excitation. (3) *The intention to counter laxity and excitation* is the antidote to the fault of not applying an antidote when laxity or excitation have arisen. [430] (4) *Meditative equanimity* that functions effortlessly and spontaneously is the antidote to the fault of applying an antidote when neither laxity nor excitation has arisen. *Distinguishing the Middle from the Extremes* says:

> Not forgetting the object of meditation,
> noticing laxity and excitation,
> applying [the antidotes] to counter them,
> and when it is calm, resting naturally.[523]

Since not applying an antidote is a fault when either laxity or excitation arises, as the antidote to this, one must rely on an adjusting attitude that consists in applying the antidote. For example, if one thinks that an enemy might be coming and receives a message confirming it, then one immediately prepares to fight. Likewise, when meta-awareness recognizes that laxity or excitation is arising, one does not ignore it and let it be; instead one applies the adjusting attitude that is its antidote. Someone with acute intelligence becomes aware of laxity or excitation just before one of these faults arises, a person of middling intelligence becomes aware as soon as the fault arises, and a person of inferior intelligence becomes aware only after a fault has already fully arisen.

When laxity and excitation are absent, applying an antidote is a fault. Once one has mastered the way to sustain concentration, the mind abides single-pointedly on the meditation object without laxity or excitation. At that time, since stability is impaired by using meta-awareness to investigate whether laxity or excitation has arisen, one does not employ meta-awareness. Instead, as an antidote to this fault, one applies equanimity, the adjustment that acts as an antidote. [431]

The concentration to be attained must have two qualities: a vibrancy of clarity and a single-pointed abiding on the meditation object. Abiding single-pointedly on the meditation object requires both a method to prevent distraction and an understanding of whether distraction is occurring. For the first one must rely on mindfulness, and for the second on meta-awareness.

Once meta-awareness has identified laxity and excitation, mindfulness holds the basic meditation object without interruption, and meta-awareness monitors whether laxity and excitation are arising. Meta-awareness of this kind is needed occasionally, not all the time. What is needed constantly is mindfulness. Mindfulness joins the mind to the meditation object thus. Once the meditation object is visualized, mindfulness produces a strong way of holding it such that, without forgetting the meditation object, the mind holds it without distraction. And without analyzing any new object whatsoever, the mind remains serenely placed exactly on the meditation object previously visualized. *Engaging in the Bodhisattva's Deeds* says:

> If the elephant of my mind is bound completely
> by the rope of mindfulness,
> all fears will disappear,
> and all virtues will be obtained.[524]

If the crazed elephant that is one's mind is completely bound to the virtuous meditation object by mindfulness, then all fears will disappear and all virtues that bring benefit to oneself and others will be obtained as if delivered into one's hands. [432]

The definition of mindfulness is: a mental factor that, focusing on an object to which there has been previous familiarization, prevents by its own power the forgetting of it. It has three distinguishing qualities: (1) the distinguishing quality of its object—the object is a familiar thing; (2) the distinguishing quality of its aspect—it focuses on that object without forgetting it; and (3) the distinguishing quality of its function—it prevents the mind from being distracted from the object and serves as the basis for the mind to remain continuously on that object. How mindfulness functions to keep the mind from wandering from the meditation object has already been explained in our presentation of mental factors with a determinate object in part 2.

Engaging in the Bodhisattva's Deeds explains that the cause of meta-awareness is mindfulness that does not forget the mind's object of meditation:

> When mindfulness stands guard
> at the gate of the mind,
> then meta-awareness will come,
> and even if it leaves, it will return.[525]

When mindfulness stands at the gate of the mind in order to guard it from the mental afflictions and so on, then meta-awareness, having been caused by that mindfulness, arrives knowing correctly what is to be done or not done. And even if meta-awareness is lost, it will arise again.

The definition of meta-awareness is: an intelligence-type mental factor that monitors again and again which activities that should or should not be done are being engaged in through one's own body, speech, and mind. [433] Śīlapālita's *Commentary on the Finer Points of Discipline* says: "Meta-awareness is intelligence."[526]

The function of meta-awareness is to monitor whether one's body, speech, and mind are engaging or not engaging in what should or should not be done. The *Great Final Nirvāṇa Sūtra* says: "Householder, if one's mind is not guarded, then one's body and speech will not be guarded; but if one's mind is guarded, then one's body and speech will also be guarded."[527] In the case of meta-awareness, one should understand that there is meta-awareness that monitors before laxity or excitation have arisen, meta-awareness that monitors after laxity or excitation have arisen, and so on.

The effect of meta-awareness is that one's body, speech, and mind become peaceful, for they no longer go in the wrong direction under the power of mental afflictions. The benefits of relying on mindfulness and meta-awareness, and the manner of relying on them, can be understood from the slightly longer explanation given earlier in part 6.

Laxity and Excitation

The main impediment to the arising of faultless concentration within the mental continuum is laxity and excitation. Thus it is very important to identify them both. [434] In general, the stability of the mind is obstructed by the mind scattering outward toward distractions, but excitation is said to be the main hindrance due to its intense, continuous scattering toward an object of desire. The clarity of the mind is obstructed by laxity. When clarity does not manifest, it is said to become as if covered by obscurations.

Laxity

The definition of laxity is: a mental factor that is either virtuous or "neutral without defilement," wherein the mind maintains stability but has lost the vibrancy of clarity and which arises owing to heaviness of body and mind from dullness, sleepiness, and so on. Kamalaśīla's *Stages of Meditation* says: "When one sees that the mind is lax or suspects that it will become lax owing to being overwhelmed by dullness or sleepiness . . ." And it also says, "When the mind does not see the meditation object clearly, like a blind person or like someone in the dark or with eyes shut, then one should know that laxity is present."[528] After having stopped the scattering of the mind, if there is a lack of clarity even while abiding single-pointedly on the meditation object, it is like having fallen asleep on one's cushion even while sitting upright. If the mind dwells in such oblivion, it has come under the sway of dullness, and it causes laxity.

There is a difference between dullness and laxity. *Dullness* is associated with delusion and is either nonvirtuous or "neutral with defilement"; it has the function of accompanying all mental afflictions and secondary afflictions [435] and makes the body and mind unserviceable. *Laxity* is a mental factor that does not apprehend the meditation object clearly or firmly because the mind loosened its hold on the meditation object when dull and sleepy. Laxity is the effect of dullness, and dullness is the cause of laxity. Vasubandhu's *Treasury of Knowledge Autocommentary* says: "What is dullness? It is either the heaviness of the body, or the heaviness of the mind, or the body's unserviceability, or the mind's unserviceability."[529] The *Unraveling the Intention Sūtra* says: "If you become afflicted by any secondary mental afflictions of meditative absorption, or by experiencing a taste of meditative absorption, or by laxity owing to dullness and sleepiness, it is a case of internal mental distraction."[530]

Laxity may be gross or subtle. In gross laxity, although the mind stays on the basic object of meditation, either there is no lucidity or the meditation object is very unclear, as if a fog had settled on the mind. Both lucidity and vibrancy are absent at such a time. In the case of subtle laxity, although clarity, lucidity, and stability are present, the clarity lacks vibrancy, due to which the mind becomes stupefied.

Laxity may also be categorized into subtle, gross, and medium. Gross laxity is when, while mindfulness apprehends the object of meditation, stability is present but lucidity and vibrancy are absent. [436] Medium laxity is when stability and lucidity are present but vibrancy is absent. Subtle laxity is when stability and lucidity are present but there is a slight loss of vibrancy, which is said to be the main obstacle to producing flawless meditation. "Lack of vibrancy" refers to a mind that, while having stability, has come under the sway of laxity because of excessive loosening. If stability is firm on that occasion, it is said to be a cause of subtle laxity.

A mind that possesses the vibrancy of clarity is one that abides brightly alert on the object of meditation. A mind that lacks the vibrancy of clarity, even though it seems to be focused alertly on the meditation object, is said to be sluggish. "Clarity" here is not the clarity of the object; rather, it refers to the clarity of the subject, one's own mind.

To prevent the arising of subtle laxity, one must hold the meditation object firmly. When the hold is loosened, a more relaxed mind arises and subtle laxity returns. Subtle laxity and flawless concentration are similar in having both lucidity and clarity. Because the only difference between them is in terms of whether a slight degree of vibrancy of focus is missing, it is said to be difficult to distinguish between those two states.

Excitation

The second impediment to the arising of concentration, excitation, involves the mind scattering toward objects of desire. For example, when there is attachment to food, mental scattering in that direction occurs after having thought about eating some food. This is a case of both scattering and excitation. Excitation belongs to the class of attachment, [437] so mental scattering owing to other mental afflictions and scattering toward a virtuous object are instances of scattering but not of excitation. Therefore, although excitation is necessarily an instance of scattering, not all scattering is excitation.

Similarly, there is a difference between excitation and distraction. Excitation is scattering toward an attractive object and is included only within attachment. Distraction, on the other hand, involves scattering toward a variety of objects and may belong to any of the three poisons (attachment, hatred, or delusion). As for the distinction between scattering and distraction, scattering makes the mind wander freely and widely, and it may be virtuous, nonvirtuous, or neutral. Distraction, within the class of secondary mental afflictions, is only nonvirtuous.

Excitation may be gross or subtle. Gross excitation is scattering that involves having lost the basic object of meditation. Subtle excitation is scattering that does not involve losing the meditation object. For example, just as water flows beneath the ice, although the object of meditation is not lost and is still held by the mind, one part of the mind has the mental factor of distraction to which the attractive aspect of an external object is just about to appear. For the time being, that distraction does not cause the mind to fluctuate outward away from the meditative focus, but it does indirectly induce fluctuations.

The object of excitation is something attractive or pleasant. As for its aspect, since it makes the mind disquiet and scattered outward, it belongs to the class of attachment; thus it engages its object by way of craving. Its function is to hinder the mind from abiding on the meditation object. When the mind is focused inwardly on the meditation object, [438] excitation—being attached to forms, sounds, and so on—draws the mind helplessly toward those things and causes distraction. Candragomin's *Praise of Confession* says:

> Just as when trying to focus within calm abiding,
> turning your mind to that object again and again,
> the noose of afflictions pulls your mind away
> helplessly with the rope of attachment to things.[531]

Causes and Elimination

As for the causes that give rise to the two impediments to concentration—laxity and excitation—*Levels of Yogic Deeds* says:

> What are the signs of laxity? Not guarding the doors of the senses; not eating in moderation; not making an effort to practice rather than sleeping during the early and later parts of the night; abiding without meta-awareness; deluded behavior; oversleeping; not knowing the proper method; having excessively slack aspiration, diligence, intention, and analysis; giving only partial attention to calm abiding without accustoming yourself to it and fully mastering it; letting your mind abide as if in darkness; and not delighting in focusing on the object of meditation.[532]

Here one should understand "signs of laxity" to mean the causes of laxity. [439] Again, *Levels of Yogic Deeds* says:

> What are the signs of excitation? In addition to the four listed above for the signs of laxity—not guarding the doors of the senses and so on—there are desirous behavior; having a disquieting manner; lacking a sense of disenchantment; not knowing the proper method; and as before, having excessively tight aspiration and so on; not accustoming yourself to diligence; engaging in only partial meditation without firmly holding the object and fully mastering the practice; being distracted by thinking about relatives; and being captivated by any sort of exciting topic.[533]

Here, as in the case of laxity above, one must understand "signs of excitation" to mean the causes of excitation.

The way to eliminate laxity is as follows. In general excitation arises when the mind is too aroused and laxity arises when it is too deflated. Therefore, if the mind enters into subtle laxity while practicing concentration, one should tighten one's focus in meditation. If that approach does not remove the subtle laxity, then one should think about things that make the mind joyful, such as contemplating the good qualities of calm abiding, as a way to uplift the mind, and then engage in meditation again. If that too does not remove it, then one should end the session and go to a spacious area or to a place that is brightly illuminated, or walk in the hills and so on to look at distant views and breathe in fresh pure air, or wash one's face in cold water, and such like. [440]

The way to eliminate excitation is as follows. If the mind is distracted even while continuously not losing the basic object of meditation, this is subtle excitation. Since it is a fault of holding too tightly, one must loosen the hold a little. If that does not stop the distraction, then it is medium excitation. In that case, first one sets the meditation object aside. Then, since excitation arises from a tightened and aroused mind, one must think about impermanence, suffering, and so on, as a way to lower the state of mind, and engage in the practice again after the mind has slackened. If even that does not stop the distraction, then it is gross excitation. In that case, one must rely on the instructions on the outflow and inflow of the breath, identified as a way to firmly contain excitation. When exhaling, think of the first breath as *out-breath*, and when inhaling, think of it as *in-breath*. When one counts using that method—in one's mind, not on a rosary—thinking *one* and so on, excitation will be eliminated. If even that does not stop it, then it is very gross excitation. One should conclude the session for the time being and rest and then take up the practice again later.

When the mind has been uplifted and its way of holding tightened through developing vibrant clarity, it is freed from the fault of laxity, but the danger of excitation increases. Conversely, when the mind has been considerably slackened and deflated, scattering and excitation is reduced to that extent, but the danger of laxity increases. Now, when an undistracted mind's way of holding has a well-balanced alertness, the undistractedness prevents the excitation and the alertness prevents the laxity. This is a very important point. Therefore it is said that one must become superbly masterful in how to maintain balance, by thinking "When the mind is tightened this much, excitation arises, so I will relax it a little" and by thinking "When it is loosened this much, it becomes lax, so I will tighten it a little." [441] One should master maintaining balance in this way: each time clarity arises, guard against excitation and cultivate stability; and each time stability arises, guard against laxity and develop clarity. The three expressions "arousing the mind," "tightening," and "firming" are synonymous, and the three expressions "deflating the mind," "loosening," and "slackening" are synonymous. Thus one must recognize laxity and excitation and know how to rely on their antidotes.

However, if one relies too much on meta-awareness when laxity and excitation are not actually arising, then there is the danger of causing the basic object of meditation to dissipate. When the faults of laxity and excitation are absent, the main method is to remain serene. This applies not only to stabilizing meditation but also to analytical meditation. Accordingly, one should practice while being free of both laxity and excitation.

The Nine Stages of Mental Stabilization

When calm abiding is cultivated by training in the stages for developing a single-pointed mind through relying properly on mindfulness and meta-awareness, the mind gradually has fewer distractions from laxity and excitation, and the mental quality of stability increases. As for how this process unfolds, texts such as *Ornament for the Mahāyāna Sūtras*, *Śrāvaka Levels*, and the *Compendium of Knowledge* present how the nine stages of mental stabilization arise in a gradual sequence.

Among these, the first is *mental placement*. This is concentration that completely withdraws the mind from external sensory objects and fixes it on a suitable meditation object. Initially the mind is simply placed on the meditation object. [442] It is like a bee alighting on a flower, settling momentarily and unable to stay put for long. At this point, the mind is simply placed on the meditation object momentarily, and being predominantly unstable, it cannot stay there for long. In this stage, it seems as if conceptualization is increasing. However, conceptualization has not increased; rather, this is the experience of noticing conceptualization. This first mental stabilization is accomplished through the first of the six powers, the *power of learning* the instructions from others.

The second is *continuous placement*. This is concentration that enables the mind directed during the first stage to remain a little longer on the meditation object without being distracted elsewhere. This mind is not like a bee alighting momentarily on a flower. Rather, this mind can remain somewhat continuously on the meditation object, and so it is called *continuous* placement. This stage is attained through the *power of reflection*.

The third is *patched placement*. This refers to concentration that, due to having become fairly familiar with the object, recognizes when forgetfulness distracts the mind, in the sense of causing it to be pulled away from the object, and places it again on that same meditation object. When the mind is pulled away from the object, it immediately understands that it is distracted in such a way and can bring itself back to the meditation object. It is like being able to rein in a horse when it wanders off the trail. Regarding distraction and stability in this context, distraction is less frequent, and stability is more frequent, so the mind mostly abides on the meditation object. [443] Each time the mind does not stay on the meditation object but wanders off, it is immediately placed again on the meditation object, and so it is called *patched* placement.

The fourth is *close placement*. Having generated strong mindfulness, the mind withdraws again and again from reaching out to broad or extensive objects and instead makes the object of meditation subtle; the mind is thereby greatly improved. Although a little laxity and excitation do occur, the meditation object is not lost owing to the generation of powerful mindfulness, and the mind can be contained within. From this point on, it is impossible to lose the meditation object, so this concentration is distinguished from the first three. Yet, while the basic object of meditation is not lost, since laxity and excitation arise strongly, one must rely on their antidotes. Both the third and the fourth mental stabilizations are accomplished through the *power of mindfulness*. From this point on, like a person who has grown up, it is said that mindfulness has fully matured, or that the power of mindfulness is complete.

The fifth is *taming*. In dependence on powerful meta-awareness, one comes to know by experience the good qualities of concentration, and the mind is uplifted with joy. Supported by this joy, meta-awareness monitors whether the occurrence of subtle laxity increases and accompanies this concentration. This involves recognizing the danger here of subtle laxity owing to the mind being overly drawn inward during the fourth stage. However, from this point on, although there is a danger of subtle laxity, there is no danger of the gross kind. Thus there is a difference between this stage and the earlier stage as to the occurrence of gross laxity and excitation. [444]

The sixth is *pacification*. This is concentration that, in reliance on powerful meta-awareness, experientially knows the faults of distraction and stops it. On the fifth stage one had to guard against subtle laxity, but here there is the danger of subtle excitation, and at this stage the danger of that is also recognized and avoided. Among the six powers, both the fifth and the sixth stages are accomplished through the *power of meta-awareness*. From this point on, the power of meta-awareness is said to be complete. The presence or absence of the danger of subtle laxity and subtle excitation is what distinguishes the fifth and the sixth stages.

The seventh is *complete pacification*. During the previous stage, pacification, one is guarding against mental scattering from the meditation object. What occurs now is attachment to sense desire, mental discomfort, dullness, sleepiness, and so on. To pacify those flaws, this level of concentration involves meditating on unattractiveness, love, and light, respectively. At this stage, this form of concentration is sustained for a long time through the power of diligence. During the fifth and sixth stages of mental stabilization, there is a fear of being impeded by laxity and excitation. However, during this seventh stage, although laxity and excitation may arise, they can be stopped by diligence; thus it is said that laxity and excitation cannot create any significant hindrance.

The eighth is *single-pointed attention*. This is concentration that continues for a long time and that, through deliberate application of mindfulness and meta-awareness, cannot be hindered by opposing factors such as laxity, excitation, and so on. Both this and the ninth mental stabilization are similar in that they cannot be hindered by conditions opposing concentration such as laxity and excitation. [445] However, the eighth mental stabilization has to continuously apply mindfulness and meta-awareness, whereas the ninth need not do so.

The progression of these stages of mental stabilization may be illustrated as follows: at first one's enemies are powerful, then after a while some of their power declines, and finally they lose all their power. In a similar way, the power of laxity and excitation gradually declines. Among the six powers, the seventh and eighth mental stabilizations are accomplished through the *power of diligence*.

The ninth is *balanced placement*. This is concentration that, relying on repeated familiarization with the meditative object that has been single-pointedly focused on before, remains placed in equipoise on the meditation object exactly as one wishes without any need for gross effort. This ninth stage is said to be accomplished through the *power of complete familiarity*. At this stage, not only is there freedom from laxity and excitation, but there is also an ability to sustain continuity without needing to make any effort using diligence. A kind of independence is attained in one's mind, and the mind can stay on the object of meditation for just as long, or as short, as it is placed there.

For more about these nine mental stabilizations, one should consult *Ornament for the Mahāyāna Sūtras*, starting with the passage that begins "After having directed your attention to the meditation object,"[534] and other such texts.

Four types of attention are present within these nine stages of mental stabilization. At the time of the first two stages of mental stabilization, strong effort is needed, so there is *tight focus*. [446] Then, on the next five stages of mental stabilization, concentration cannot be sustained for long owing to interruption by laxity and excitation, so there is *intermittent focus*. Then on the eighth stage of mental stabilization, where laxity and excitation cannot cause interruption, concentration can be sustained for a long time, so there is *uninterrupted focus*. Finally, on the ninth stage of mental stabilization, there is no interruption and no need to rely on continuous effort, so there is *effortless focus*.

Śrāvaka Levels explains how to identify the nine stages of mental stabilization and how they are accomplished by the six powers, the way in which the four types of attention are present, and the way in which genuine calm abiding is generated:

> What are the nine types of mental stabilization? Here the monk draws his mind within and places it there; he places it correctly, he places it by gathering it, he places it closely, he tames it, he pacifies it, he completely pacifies it, he makes it one-streamed, and he turns it into concentration.

And:

> He accomplishes these nine stages of mental stabilization through the six powers: the power of learning, the power of reflection, the power of mindfulness, the power of meta-awareness, the power of diligence, and the power of complete familiarity. Through the power of learning and the power of reflection, owing to what he has heard and what he has contemplated on that occasion, he accomplishes, respectively, the first stage of mental placement by drawing the mind within and placing it on the meditation object, and then the stage of correct placement by way of focusing on that very object. [447] Through the power of mindfulness based on focusing in that manner, [his mind] completely withdraws and does not waver, whereby he accomplishes the stages of gathered placement and close placement. Then, through the power of meta-awareness, mental scattering toward the signs of secondary afflictions and conceptualization are prevented, whereby he accomplishes the stages of taming and pacification. Through the power of diligence, the full arising of the two [laxity and excitation] is prevented, whereby he accomplishes the stages of [complete pacification and] one-streamed focus. Through the power of complete familiarity, he accomplishes concentration.
>
> One should know that within these nine stages of mental stabilization, there are four types of attention: tight focus, intermittent focus, uninterrupted focus, and effortless focus. Regarding these, in the stages of mental placement and correct placement there is the attention of tight focus; in the stages of gathered placement, close placement, taming, pacification, and complete pacification there is the attention of intermittent focus; in the stage of one-streamed focus there is the attention of uninterrupted focus; and in the stage of concentration there is the attention of effortless focus. Accordingly, given that these types of attention occur within the continuum of the nine stages of mental stabilization, they are in the class of calm abiding.[535] [448]

The Meaning of the Calm Abiding Illustration

[449] A meditator who has the collection of six prerequisites for calm abiding must understand the faults that inhibit concentration—such as laziness, forgetting the instructions, laxity and excitation, not applying, and applying—and accomplish calm abiding in reliance on the eight applications that are their antidotes. The antidote to the first fault is fourfold: faith, aspiration, pliancy, and effort. The antidote to the second fault is mindfulness. The antidote to the third fault is meta-awareness. The antidote to the fourth fault is application. The antidote to the fifth fault is equanimity. Moreover, when one understands the divisions into nine stages of mental stabilization, how these are accomplished by the six powers, and how they are included within the four types of attention, it is said that faultless concentration is easily accomplished. These processes are compared to the training of a wild elephant, as illustrated in the drawing on the facing page, and we now explain briefly the meaning of this illustration.

The six sections of the path illustrate the six powers. The noose symbolizes mindfulness, and the hook symbolizes meta-awareness. The presence or absence of flames as well as their size variations, from the first stage to the seventh stage of mental stabilization, symbolize the different amounts of effort present in one's application of mindfulness and meta-awareness. The elephant represents the mind, and the extent of its black color symbolizes the degree of laxity; the monkey represents mental scattering, and the extent of its black color symbolizes the degree of excitation. The objects of the five senses represent the objects of excitation. From the second section of the path, a white patch appears on the elephant's head and gradually increases in size, which illustrates an increase in mental clarity and stability. On the third section of the path there appears a rabbit, black in color, which symbolizes subtle laxity. Its appearance here indicates recognition of the distinction between gross and subtle laxity, and its looking back illustrates mental distraction. On the fourth section, the monkey trailing behind illustrates that the earlier strength of excitation has diminished. The monkey plucking a fruit from the wish-fulfilling tree indicates that although, in the context of cultivating calm abiding, the scattering of one's mind even toward a virtuous object must be avoided, this is not the case in other contexts. Thus it symbolizes reaping the fruit of the two aims of accomplishing one's own and others' welfare. On the fifth section, the black color of the elephant has faded away, and the monkey is absent. These represent concentration with uninterrupted focus, which, having relied a little on mindfulness and meta-awareness initially, cannot be interrupted by scattering, laxity, and excitation.

How Calm Abiding Is Attained through Meditation

[450] How genuine calm abiding comes to be attained through the nine stages of mental stabilization is as follows. Having well understood the aforementioned process for cultivating concentration, one then engages in that process, whereby the nine stages are generated sequentially. Then on the stage of the ninth mental stabilization, one is able to sustain a session for a long time free from subtle laxity and excitation. Yet even though one has attained a level of concentration with spontaneous focus that does not require the effort of continually relying on mindfulness and meta-awareness, if pliancy is not attained, then this is only an approximate calm abiding, not genuine calm abiding. The *Great Sūtra on Great Emptiness* says:

> O monks, also by the joy and bliss born of seclusion, the body is manifestly saturated, completely saturated, completely satisfied, and completely permeated. There is not even a small part of the entire body that is not pervaded or does not become pervaded by the joy and bliss born of seclusion. Ānanda, it is thus: here the monk draws his mind only within and places it there; he places it correctly, he places it by gathering it, he places it closely, he tames it, he pacifies it, he completely pacifies it, he makes it one-streamed, and he turns it into concentration.[536] [451]

Similarly, the *Unraveling the Intention Sūtra* says: "Bhagavan, until the bodhisattva achieves bodily pliancy and mental pliancy, what do we call the meditative attention that he uses when he turns his attention inward and focuses on his mind? Maitreya, it is not calm abiding, but we may say it occurs in conjunction with an aspiration conducive to calm abiding."[537] Also, *Ornament for the Mahāyāna Sūtras* says:

> Upon attaining great pliancy
> of one's body and mind,
> one is said to have attention.[538]

Stages of Meditation says: "When you engage in calm abiding meditation like this, where your body and mind have become pliant and your mind has the power to focus on the meditation object exactly as desired, then you should know that you have accomplished calm abiding."[539] Thus it is taught in many scriptural sources that both pliancy and mastery of abiding on the meditation object are required. [452]

So how is pliancy attained? The *Compendium of Knowledge* explains pliancy in general: "What is pliancy? It is a serviceability of body and mind that has the function of dispelling all hindrances, since it interrupts the continuum of bodily and mental dysfunction."[540] Here "bodily and mental dysfunction" refers to the unfitness of one's body and mind to be employed in accomplishing virtuous deeds at will. Bodily and mental pliancy, the antidote to this, makes the body and mind highly suitable to be employed in virtue by removing bodily and mental dysfunction.

There are two types of pliancy: bodily pliancy and mental pliancy. Bodily pliancy, gained through the power of concentration, is posited as a suitability of the body to be employed in virtuous activities as desired that, having removed the faults of bodily unserviceability, makes the body light like cotton. Mental pliancy is posited as a serviceability of the mind gained through the power of concentration that, having removed mental dysfunction, allows the mind to focus on the meditation object without impediment. Sthiramati's *Commentary on the Thirty Stanzas* says:

> The serviceability of the body is that which gives rise to lightness and flexibility in one's physical deeds. The serviceability of the mind, which occurs in a state of correct attention, is another element [i.e., it is different from bodily pliancy]; it is a mental factor that is the cause of delight and suppleness. [453] When the mind possesses that, it focuses without resistance on the meditation object. For this reason it is called "serviceability of the mind."[541]

Bodily pliancy is not a mental factor; it is a very pleasant tactile sensation within one's body.[542] *Commentary on the Thirty Stanzas* says: "One should know bodily pliancy to be a special tactile sensation accompanied by joy. For it says in a sūtra, 'When the mind is joyful, the body becomes pliant.'"[543] *Mental pliancy* refers to pliancy that is included within the ten virtuous mental factors. An early portent of the attainment of this special pliancy is that the person who is striving to attain concentration experiences a sense of heaviness in the brain. This is not an unpleasant heaviness. It is like the pleasant sensation of heaviness that occurs when someone's warm hand touches one's newly clean-shaven head. As soon as that experience arises, the meditator becomes free from mental dysfunction that prevents him or her from taking delight in abandoning the mental afflictions; its antidote, mental pliancy, arises first. *Śrāvaka Levels* says:

> An early indication of the imminent arising of a gross, easily discernible single-pointedness of mind, as well as mental and bodily pliancy, is a sensation of heaviness; but this is not characterized by anything that is harmful. [454] As soon as this arises, mental dysfunction within the afflictive class, which hinders delight in eliminating afflictions, is abandoned; and its remedy, mental serviceability and mental pliancy, arises.[544]

Then, through the force of the arising of pliancy that is mental serviceability, the body is pervaded by an energy wind that is the cause of bodily pliancy. When this occurs, it eliminates bodily dysfunction and gives rise to bodily pliancy; and having saturated the entire body, the meditator seems to be filled with the force of this serviceable energy wind. *Śrāvaka Levels* says: "Owing to that [pliancy] having arisen, energy winds—included among the great elements—that are conducive to the arising of bodily pliancy course through the body. When they flow, bodily dysfunction within the afflictive class, which hinders delight in eliminating afflictions, is abandoned, and its remedy, bodily pliancy, saturates the entire body, and the meditator seems to be filled with that energy."[545]

Accordingly, having achieved bodily pliancy, there arises a great experience of bliss in the body through the force of the energy wind; this is called the *bliss of bodily pliancy*. On the basis of having achieved this bliss of bodily pliancy, [455] there also arises an exceptional experience of bliss and joy in the mind; this is called the *bliss of mental pliancy*. Among these two types of pliancy, mental pliancy arises first, and among these two types of bliss, the bliss of bodily pliancy arises first. Thus one must distinguish between bodily and mental pliancy on the one hand and the bliss of bodily and mental pliancy on the other.

When bodily pliancy initially arises, a great experience of bliss arises in the body through the force of the energy wind. Then, on that basis, an exceptionally great experience of bliss and joy arises in the mind. After that, the power of the initial arising of pliancy gradually subsides, but pliancy is not extinguished. Instead, when that gross pliancy—which greatly agitates the mind—disappears, there arises a pliancy that is delicate like a shadow and conducive to unperturbed concentration. When excessive joy in the mind has subsided, the mind remains firmly on the meditation object, and one obtains calm abiding, which is free from the agitation caused by great joy.[546] As *Śrāvaka Levels* clearly says:

> When that [pliancy] initially arises, there appears the experience of mental joy, a great sense of mental bliss, attention to the presence of supreme mental joy, and fully manifest mental joy. After that, the power of the pliancy initially produced gradually becomes very subtle, and the body becomes endowed with shadow-like pliancy. [456] Mental joy of any type is removed, and the mind becomes stabilized with calm abiding; and with an aspect of exceptional peace, the mind focuses on the object of meditation. Thereafter this novice yogi has attention and is reckoned as "one having attention."[547]

Thus, by continually striving to develop concentration while abiding according to the prerequisites explained above, one will accomplish calm abiding; but if one drops the practice after doing it merely a few times, one will not accomplish it. Āryaśūra's *Compendium of the Perfections* says:

> Through yoga that is uninterrupted,
> strive to accomplish absorption.
> Just as one does not get fire
> from friction applied with repeated breaks,
> the same is true for yogic practice.
> So do not give up until the special state is attained.[548] [457]

Calm Abiding Focused on the Mind Itself

In general, there are countless possible meditation objects for establishing faultless concentration. To establish calm abiding by focusing on the mind, one should first fulfill all the prerequisites explained above and then practice as follows. Once the mind's nature appears in a coarse way, one should be content with just that much. If one meditates on that as the object, clarity will arise gradually. But if one emphasizes clarity from the outset, the meditation object will not be found.

Since the mind does not appear directly at first, one should focus on the image of the mind that appears as a representation in awareness. One's own mind is clear and cognizing, and since it does not exist in any physical manner, one cannot identify it by saying something like, "It has this color, this shape, and so on." When not examined, the mind remains unknown, and upon investigation, one finds nothing more than a clear emptiness. When examined further, one recognizes that it is clear and vivid, appearing as anything at all, and that it is the creator of all good and evil, all happiness and suffering. Recognizing it in this way, one gives rise to the forceful aspiration "I will hold my mind on this meditation object," and then one holds the mind single-pointedly on just that very meditation object. In brief, even if the meditation object thus recognized does not appear exactly with great clarity, if only a partial one or a mere general impression appears, one must be content with that image and hold the mind on it with the awareness "This is the meditation object." [458]

At that time, except for that meditation object itself being held as the object of awareness, one should not give the mind over to any other thoughts. Let go of all expectations focused on the doings and deeds of the three times, as in wishfully thinking "If only this desired thing would happen sooner or later," fearfully thinking "Suppose this bad thing were to happen sooner or later," and so forth. One should single-pointedly focus on the present moment of awareness that is uncontrived and characterized by simple clarity and knowing. Initially, one should not try to maintain this for long periods of time but only for short periods and only to the extent that one is able.

When stabilizing the mind on that object, the mind itself, one is not in a state of the mere cessation of all attention, a kind of blackout experience in which nothing at all appears, as when one faints or falls asleep. Instead, the mind should remain unperturbed, with a deep alertness but also somehow relaxed.

Focusing on the meditation object clearly, one sustains the experience through both a special mindfulness, which retains the object without forgetting, and meta-awareness, which observes whether any distraction toward something else arises. How mindfulness enables one not to forget the meditation object is not simply by recalling the object when one thinks about it or when someone asks about it. One's mindfulness persists unimpaired in such a way that the meditation object does not stop being an object of one's awareness; through that unimpaired mindfulness, one should remain undistracted.

Meta-awareness is used from time to time to monitor whether the mind is scattering. This does not mean that one takes one's mind off the meditation object and engages in monitoring with a new, separate cognition. [459] Rather, a subtle part of the mind engages in observation while the mind continues to hold the meditation object without losing it. For example, when walking along a path together with a friend, one keeps an eye on the path while looking at one's friend out of the corner of one's eye.

In brief, while single-pointedly focusing on the object of meditation, one examines one's experience. Upon considering that "When the mind is tightened or intensified to this extent, excitation occurs," one should relax the mind a little. Likewise, upon considering that "When the mind is relaxed to this extent, laxity occurs," one should tighten it a little. In this way one seeks mental stability by creating a balance between these two, so that the mind recedes from scattering and excitation. Every time mental stability occurs, be vigilant against the arising of laxity, and cultivate clarity endowed with a quality of vibrant awareness. By carefully maintaining both clarity and stability alternately like this, faultless concentration will arise.

Sustaining the practice in this way, when one tightens and intensely looks at the clear and cognizing nature of the mind, discursive thoughts may spontaneously arise. When that happens, one should recognize them immediately and forestall their continuity, forcefully interrupting them right then and there. Alternately, rather than eliminating whatever thoughts might occur, one could examine where they come from and where they disappear to. As one does this, owing to the force of having previously projected a strong intention to prevent the arising of discursive thoughts, those very thoughts are unable to continue by their own power, so they return to the main mind from which they arose. It is like when a skillful teacher sees his student distracted by playing, and instead of shouting "Stop that," he looks at the student sternly, and the student, becoming self-conscious, puts an end to the playing. [460]

Furthermore, there are six ways of settling the mind that help maintain our concentration. (1) Just as a cloudless sun remains supremely bright and clear, likewise one can rest the mind in the clear-light nature of the mind unobstructed by any conceptualization grasping at signs or by laxity and excitation. (2) Just as the powerful Garuḍa flies through the sky without needing to expend great effort by, for instance, beating his wings, likewise one can maintain the mind in a state that is neither too tight nor too loose, is endowed with mindfulness and meta-awareness, and has vibrant clarity at depth while being relaxed on the surface, and thus is sustained within the expanse of mindfulness. (3) Just as on the surface of a great ocean, when stirred by the wind, waves rise slightly but cannot disturb its depths, likewise when one rests one's mind on the meditation object, there may be slight movements of subtle thought, but one maintains concentration without being swayed in the least by coarse conceptualization. (4) Just as a small child regarding a temple does not scrutinize the details of the paintings but undistractedly takes in the more conspicuous parts, likewise when the mind focuses on the meditation object, one can remain single-pointedly on the meditation object without attachment or aversion or conceptualizing about any attractive or unattractive objects of the five sense consciousnesses that might appear. (5) Just as a bird does not leave any trace of its flight in the sky, likewise whatever feelings may arise—pleasant, unpleasant, or neutral—one's mind can be placed in equipoise without falling under the power of any of the three poisons—attachment, hatred, or delusion. [461] (6) Just as wool when combed through becomes soft and fluffy, likewise one's mind in meditative equipoise is free from the rough texture of laxity, excitation, and manifest mental afflictions.

Maintaining the practice in this way, one experiences the mind's nature, unobscured by anything, very clear and lucid, not existing as anything physical and thus empty like space. Whatever good or bad objects of the five senses arise, they arise clearly and vividly as appearances, like reflections in a clean mirror; they are experiences that cannot be identified as "this" or "that." Yet however stable such a concentration may be, if it is not sustained by the bliss of bodily and mental pliancy, then it is a single-pointed mind of the desire realm; but if it is sustained by such a bliss, then it is said to be calm abiding [which is a mental state of the form and formless realms].

25

Analysis and Insight

How Special Insight of Fine Investigation Is Cultivated

In general, the attainment of calm abiding must precede the attainment of special insight. *Engaging in the Bodhisattva's Deeds* says:

> Having understood that it is through special insight
> endowed with calm abiding that the mental afflictions are destroyed,
> calm abiding should first be sought.[549] [462]

The definition of special insight is: a type of wisdom that, from within calm abiding, finely investigates its object while sustained by the bliss of pliancy induced by having analyzed its meditation object. It is called *special insight* because sees its object in a special way—that is, in a way that is superior to calm abiding. It is superior to the nonanalytical state of calm abiding in that the object can appear more clearly when analyzed by special insight.

Special insight is attained through meditation in the following way. From within calm abiding, one meditates with finely investigating wisdom that analyzes its meditation object. When the power of that analysis induces the bliss of pliancy, this is posited as the attainment of genuine special insight. Before such pliancy arises, when one meditates with that finely analyzing wisdom, it is *approximate* special insight, but once that pliancy has arisen, it is *genuine* special insight. The definition of pliancy and how it arises are as explained above.

One may already have induced pliancy by calm abiding and that may remain undiminished, but here it is not a matter of just having pliancy in general. So what is it? When pliancy is induced by the power of analytical meditation, then the meditative awareness turns into special insight. The same point applies to both kinds of special insight: the special insight focusing on reality as manifold (conventional nature) and the special insight focusing on reality as it is (ultimate nature). The *Unraveling the Intention Sūtra* says: [463]

> Bhagavan, until the bodhisattva achieves pliancy of body and mind, what do we call the meditative attention that he uses to internally attend to the carefully contemplated phenomena as an image (*pratibimba*) that is the object of his concentration?
>
> Maitreya, it is not insight, but we may say it occurs in conjunction with an aspiration conducive to insight.[550]

Ratnākaraśānti's *Instructions for the Perfection of Wisdom* also says:

> Mental and bodily pliancy have been attained [through calm abiding], and one abides in them. Then one should attend to the carefully contemplated object as an internal image-object of concentration and finely investigate it. As long as mental and bodily pliancy have not arisen, this [analytical wisdom] is approximate insight; once they have arisen, it is genuine insight.[551]

Stages of Meditation employs an analogy to explain the need to meditate with both calm abiding and special insight:

> With special insight alone, lacking calm abiding, the yogi's mind will be distracted toward other objects; like an oil lamp in the wind, it will not be stable. For this reason, the light of wisdom will not be very bright. [464] In that case, one must rely equally on both. . . . With the power of calm abiding, the mind will not be moved by the wind of conceptions, like an oil lamp placed in a windless spot. Special insight eliminates the entire web of pernicious views; hence, one is not perturbed by others. The *King of Concentrations Sūtra* says: "With the power of

> calm abiding, one becomes undistracted; with special insight, one becomes like a mountain."[552]

In general, it is said that all forms of meditative concentration are encompassed by the categories of calm abiding and special insight. Just as a tree has many branches, leaves, and fruit yet all those come together as parts of the main tree, likewise all the various types of meditation are included within two classes. Specifically, as introduced above, either they are instances of *analytical meditation*, where one meditates by way of thinking about and analyzing the meditation object, using scriptural quotations, reasoning, and examples, or they are instances of *stabilizing meditation*, where one intently places the mind single-pointedly on the meditation object without using analysis. Analytical meditation encompasses the class of insight meditation, and stabilizing meditation encompasses the class of calm abiding meditation. *Stages of Meditation* says: "The Bhagavan taught bodhisattvas an unlimited variety of concentrations, such as the immeasurables and so on, [465] but all are encompassed by calm abiding and special insight. Therefore we will discuss just that path that is the union of calm abiding and special insight."[553] Vimalamitra's *Meaning of the Gradual Approach in Meditation* says: "Since all concentrations are included within those two, all yogis at all times definitely must rely on calm abiding and special insight."[554]

The Importance of Analytical Meditation

In general, one's degree of understanding that arises from learning about teachings corresponds to how many one has heard, and based on that, there is a corresponding degree of understanding that comes from critically reflecting on them. Similarly, the greater one's understanding, the greater one's practice of meditation, and based on this, one acquires correspondingly more ways to eliminate faults and actualize good qualities. Therefore learning and critical reflection are extremely important for meditation practice. The *Questions of Brahma Viśeṣacinti Sūtra* says: "They become learned so as to finely investigate phenomena in accordance with reality."[555] Also, *Stages of Meditation* says: [466] "With wisdom arisen from meditative cultivation, one should meditate solely on whatever one realized with wisdom arisen from learning and reflection, not on something else. It should be like racing a horse along the very course that it was introduced to. One must therefore engage in fine investigation of the ultimate."[556]

One might eschew the analytical meditation that uses the wisdom of fine investigation and instead consider it sufficient to stop thinking altogether. But to do so would be to abandon the root of perfect wisdom. *Stages of Meditation* explains: "By saying 'One should avoid thinking,' the wisdom defined as fine investigation of the ultimate is abandoned. Since the root of perfect wisdom is fine investigation of the ultimate, then if one abandons that, one severs its root and thereby abandons supramundane [perfect] wisdom."[557] The same text says: "If mindfulness and attention are not active, then [the practitioner] will be completely stupefied; as such, how could he be a yogi? By meditatively cultivating non-mindfulness and non-attention that lack fine investigation of the ultimate, one cultivates just stupidity. Hence the light of perfect wisdom is made to recede into the distance."[558] [467]

Thus one needs to engage in analytical meditation with the wisdom of fine investigation, over and over again, to counter the mental afflictions. One cannot eliminate the mental afflictions merely by stopping all mental activity; one must engage in analytical thinking, such as contemplating the faults of the afflictions. To defeat an enemy on the battlefield, one has to confront the enemy and disarm him; one cannot overcome an enemy by sitting down with one's eyes shut. The *Play of Mañjuśrī Sūtra* says: "Mañjuśrī asked, 'Daughter, how does a bodhisattva gain victory in battle?' The girl replied, 'O Mañjuśrī, when one analyzes all phenomena, they are not perceived.'"[559] *Stages of Meditation* says: "Likewise, a yogi opens his eye of wisdom and defeats his enemy—the mental afflictions—with the weapon of wisdom. He has no fear; he does not shut his eyes like a coward."[560]

While meditating, the choice to employ analytical or stabilizing meditation should be made on the basis of specific factors, such as the nature of the topic or focal object of one's meditation as well as the state of one's mind. For example, in meditations such as those for developing faith in one's guru or for cultivating compassion and tolerance, one mainly begins with analytical meditation. However, at the end of that analytical meditation, [468] having drawn forth an experience that transforms the mind, one must place the mind intently within that experience without discursive analysis. Likewise with the topic of death and impermanence, when thinking about it, one mainly does analytical meditation, simply analyzing the situation again and again using scriptural quotations and reasoning. Then when one finds conviction at the end of analysis, one mainly does stabilizing meditation by single-pointedly resting in the awareness that "this is the reality of death and impermanence." Thus even a single meditation topic like death and impermanence involves both analytical meditation and stabilizing meditation.

There are many scriptures that explain how to engage in analytical and stabilizing meditation, how all three—learning, reflection, and meditation—are essential, and how the mind needs to be trained by alternating analytical and stabilizing meditation. For example, *Ornament for the Mahāyāna Sūtras* says:

> At first, in dependence on learning, correct attention arises;
> from correct attention comes wisdom whose object is the
> ultimate.[561]

Likewise, *Stages of Meditation* says: "With wisdom arisen from meditative cultivation, one should meditate solely on whatever one realized with the wisdom arisen from learning and reflection, not on something else. It should be like racing a horse along the very course it was introduced to. One must therefore engage in fine investigation of the ultimate."[562] Vimalamitra's *Meaning of the Gradual Approach in Meditation* also says:

> When, through special insight meditation, wisdom greatly increases [469] and calm abiding decreases, the mind wavers like an oil lamp in the wind and thus cannot see reality clearly. Therefore one should do calm abiding meditation at that time. When calm abiding greatly increases and special insight decreases, then it is like being asleep so the mind cannot see reality clearly. Therefore one should do wisdom meditation at that time.[563]

Kamalaśīla's *Stages of Meditation* gives detailed explanations about the importance of both analytical meditation and stabilizing meditation, the difference between them, the unique influences of each, how to maintain both, and so on. Here is a short passage from *Stages of Meditation*:

The *Cloud of Jewels Sūtra* says:

> One who is skillful in dealing with faults engages in the yoga of meditating on emptiness in order to gain freedom from all proliferating tendencies. Meditating extensively on emptiness, he seeks whatever things scatter and attract the mind in terms of their essential nature, and seeking them in this way, he realizes that they are empty. Examining also the mind that seeks them, he realizes that it too is empty. And seeking also in terms of its essential nature the mind with which he examined the mind, he realizes that it is empty as well. Inquiring intensively in this way, he achieves [470] the yoga of signlessness.

This passage teaches that engaging in the yoga of signlessness is preceded by thorough investigation. It clearly demonstrates that it is impossible to achieve nonconceptuality by merely abandoning attention altogether and not analyzing the nature of things by means of wisdom.[564]

Stages of Meditation also says:

> Therefore, with one's mind held by the hand of concentration, one should use the instrument of extremely refined wisdom to extract the seed in the mind that gives rise to false conceptions regarding forms and so on. Accordingly, just as trees once uprooted do not grow again from the ground, false conceptions once uprooted will not grow again in the mind. For this very reason, the Bhagavan taught the path uniting calm abiding and special insight. Since those two [meditative states] are the cause of the nonconceptual, perfect wisdom, they are the key to abandoning the obscurations.
>
> Thus it was taught, "Abiding in ethical conduct, one attains concentration; having attained concentration, one engages in wisdom meditation; by means of wisdom, one attains immaculate exalted wisdom; one with immaculate exalted wisdom

has consummate ethical conduct." In other words, once calm abiding has stabilized the mind on the meditation object, wisdom thoroughly investigates it, [471] at which point the light of perfect, exalted wisdom arises. At that time, just as light clears away darkness, obscuration is dispelled. Like the eye and light, those two [calm abiding and special insight] are mutually compatible in the context of the arising of perfect, exalted, wisdom; they are not incompatible, like darkness and light. Concentration is not by nature darkness. Instead, it is defined by single-pointedness of mind. It is also taught that "In a state of concentration, one perfectly knows ultimate reality exactly as it is." Therefore concentration is fully compatible with wisdom and not incompatible.

One who is analyzing things with wisdom in a state of concentration may experience the nonperception of all things. That is the highest form of nonperception. For yogis, that kind of state characterized by calm abiding is effortless (*anābhoga*) because there is nothing else to be seen beyond it. This state is also called "calm" because in it all proliferating tendencies—characterized by conceptions of existence, nonexistence, and so on—are pacified. Thus, when a yogi analyzes with wisdom, he does not perceive any essential nature of an existent thing; he just does not have the conception of existence at that time. And he also does not have the conception of nonexistence.[565] [472]

In the sections above, we have briefly discussed a number of topics, including the instructions on how to cultivate flawless concentration—how to engage in analytical meditation when analytical meditation is required, how to engage in stabilizing meditation when stabilizing meditation is required, and how to combine analytical and stabilizing meditation when a union of the two is required—in accord with the practical instructions and interpretation of the sūtras presented in Maitreya's texts, in Asaṅga's texts on the *Levels of Yogic Deeds*, and in Kamalaśīla's *Stages of Meditation*. Our brief presentation also explored topics such as the definitions of calm abiding and special insight; the different types of meditation objects; the distinction between analytical meditation and stabilizing meditation; the way to establish calm abiding and special insight by means of meditation using the eight applications countering the faults and the nine stages of mental stabilization; the identification of laxity and excitation, both gross and subtle; the difference between laxity and dullness and the difference between scattering and excitation; the way to eliminate these faults by means of their respective antidotes; the identification of mindfulness and meta-awareness, the difference between them, and when to apply each; and how, when a novice first cultivates concentration, he or she should not be satisfied with mere stability alone but must develop vibrant clarity. As for a detailed presentation on meditation, with topics such as the characteristics of the person who meditates, the object meditated on, how to engage in meditation, and the way of developing particular levels of mind through meditation, this is found in texts such as Asaṅga's *Śrāvaka Levels*, Vasubandhu's *Treasury of Knowledge* and its *Autocommentary*, Śāntideva's *Compendium of Training*, and Kamalaśīla's *Stages of Meditation*, so one should understand these details from those sources. [473]

26

Mindfulness Meditation

HAVING EXPLAINED HOW to engage in analytical and placement meditations as a means for training the mind, we now illustrate the practice of meditation with a brief presentation of the four applications of mindfulness, which have been taught in numerous contexts in Buddhist texts as a way to counter the four distorted conceptions of the body and so on as pure, blissful, permanent, and a self. The *Sutta on the Application of Mindfulness* (*Satipaṭṭhāna Sutta*) appears in the Theravāda scriptures in a more extended version in the *Long Discourses* (sutta 22) and in another version in the *Middle-Length Discourses* (sutta 10). Presentations of the practice of the application of mindfulness can also be found in the *Connected Discourses* and the *Numerical Discourses*. Of these, the most well-known presentation is the one in the *Middle-Length Discourses*.

With the addition of the *Short Discourses* to these four sets of scripture, there are five major collections of suttas in the Pali canon. Although there are numerous parallels between the Pali and Tibetan canons, these collections were not translated into Tibetan as collections and so are not found in the Tibetan canon as such. According to the preambles of the *Sutta on the Application of Mindfulness* as found in the *Long* and *Middle-Length Discourses*, [474] these suttas were taught in the country of Kuru (in the west of the Ganges Plain). The *Sutta on the Application of Mindfulness* says:

> Monks, this is the one-way path for the purification of beings, for the surmounting of sorrow and lamentation, for the passing away of pain and dejection, for the attainment of the true way, for the realization of nibbāna—namely, the four applications of mindfulness. What are the four? Here, monks, in regard to the body a monk dwells contemplating the body, ardent,

> clearly comprehending, and mindful, having subdued longing and dejection in regard to the world. In regard to feelings he dwells contemplating feelings, ardent, clearly comprehending, and mindful, having subdued longing and dejection in regard to the world. In regard to the mind he dwells contemplating mind, ardent, clearly comprehending, and mindful, having subdued longing and dejection in regard to the world. In regard to phenomena he dwells contemplating phenomena, ardent, clearly comprehending, and mindful, having subdued longing and dejection in regard to the world.[566]

Opening with this brief summary, a more detailed presentation then explains the application of mindfulness to the body in the following sequence. First it shows how to cultivate mindfulness on the basis of one's breathing, whereby one focuses on the natural activity of breathing in and breathing out. [475] Then, by distinguishing outer from inner parts of the body and focusing on them separately, it explains how to become aware of the arising and disintegration of the body. This is followed by an explanation of how to develop full awareness of one's engagement in the four types of behavior—standing, walking, lying down, and sitting. Finally it explains how to become aware of the body as an impure substance, aware that the body is composed of the four elements, and how to meditate on impermanence by way of the nine signs of death, such as the decay and rotting of the body and so on. In the concluding section on mindfulness of the body, the same text says:

> In this way, in regard to the body he dwells contemplating the body internally, or he dwells contemplating the body externally, or he dwells contemplating the body both internally and externally. Or else he dwells contemplating in the body its nature of arising, or he dwells contemplating in the body its nature of vanishing, or he dwells contemplating in the body its nature of both arising and vanishing. Or else mindfulness that "there is a body" is simply established in him to the extent necessary for bare knowledge and repeated mindfulness. And he dwells independent, not clinging to anything in the world. That too is how a monk in regard to the body dwells contemplating the body.[567]

This sutta then teaches how to practice the application of mindfulness with feelings, regarding individual feelings of the three types—pleasant, painful, and neutral—and also worldly and transmundane pleasant feeling, worldly unpleasant feeling, and both worldly and transmundane neutral feeling. [476] Then it teaches how to practice the application of mindfulness with the mind, applying it to sixteen types of mind: minds with the three poisonous mental afflictions of attachment, anger, and delusion and minds that are free from them; and based on those six, minds that are contracted and distracted, exalted and unexalted, surpassed and unsurpassed, concentrated and unconcentrated, liberated and unliberated. Then, with regard to mindfulness of phenomena, it teaches how to meditate by relating to each of the following factors in a sequential order: the five hindrances (sense desire, hatred, both dullness and sleep, both excitation and regret, and doubt), the five aggregates, the six bases, the seven factors conducive to enlightenment, and the four noble truths. Finally, that sutta concludes with the benefits of meditating on the four applications of mindfulness. Here we have offered a short explanation as an introduction to this precious sutta.

Now we will look briefly at the presentations of the applications of mindfulness in the upper and lower Abhidharma texts. Asaṅga's *Compendium of Knowledge* states: "One should understand the applications of mindfulness in terms of their objects, their essential nature, their accompanying factors, their meditative cultivation, [477] and the fruits of meditation on them."[568] Accordingly, we will explain the applications of mindfulness in terms of these five points.

First, there are four *objects* of the applications of mindfulness: body, feelings, mind, and phenomena. Mindfulness practice is specified in terms of these four objects to help eliminate the four distorted apprehensions: (1) holding the body, which is unattractive, as attractive, (2) holding feelings, which are in the nature of suffering, as blissful, (3) holding the mind, which is impermanent, as permanent, and (4) holding phenomena, which are selfless, as possessing selfhood. The *Questions of Ratnacūḍa Sūtra* says:

> One who abides in the four applications of mindfulness completely abandons the four misconceptions. What are those four? In regard to the body, one who abides contemplating the body completely abandons distorted discernment of the unattractive

> as attractive. In regard to feelings, one who abides contemplating feelings completely abandons distorted discernment of suffering as happiness. In regard to the mind, one who abides contemplating the mind completely abandons distorted discernment of the impermanent as permanent. In regard to phenomena, one who abides contemplating phenomena completely abandons distorted discernment of the selfless as possessing selfhood. [478] Without distortion he establishes the perfect truth.[569]

Second, is the *essential nature* of the application of mindfulness. Mindfulness or wisdom takes one of the four—body, feelings, mind, or phenomena—as its focus and observes and analyzes it in terms of its specific characteristics, such as the body being impure and so on, or in terms of its general characteristics, such as being impermanent and so on. The *Compendium of Knowledge*, for example, says: "What is its essential nature? Wisdom and mindfulness."[570] The *Treasury of Knowledge* says:

> By thoroughly investigating both types of characteristics
> based on the body, feelings, mind, and phenomena,
> the wisdom arisen from learning and so on manifests.[571]

As for the etymology of the phrase "application of mindfulness" (Skt., *smṛtyupasthāna*), the Vaibhāṣikas consider it to mean that, just as when splitting a block of wood, one inserts a wedge in the crack to hold it open so that one hits the right place, wisdom engages its object through the power of mindfulness not forgetting the object or its aspects. Vasubandhu explains that initially wisdom finely investigates the meditation object, after which it is held by mindfulness. Thus "application of mindfulness" means that mindfulness is applied to the object of wisdom. [479] The *Treasury of Knowledge Autocommentary* says:

> Why is wisdom referred to as the "application of mindfulness"? The Vaibhāṣikas say, "Because mindfulness predominates. In other words, it is so called because [wisdom] engages [with its object] due to the functioning of the power of mindfulness. It is like a wedge that holds [open the crack] when splitting wood."

> However, the following makes sense: wisdom is the "application of mindfulness" in that this [wisdom] is the means by which mindfulness is applied [to the object].[572]

Third, the *accompanying factors* of the applications of mindfulness are the mind and mental factors that are concomitant with any of the four, the application of mindfulness to the body and so on. The *Compendium of Knowledge* says: "What are the accompanying factors? They are the concomitant mind and mental factors."[573]

Fourth, with respect to their *meditative cultivation*, Vasubandhu's *Treasury of Knowledge* says the following about how to meditate on the applications of mindfulness:

> By thoroughly investigating both types of characteristics
> based on the body, feelings, mind, and phenomena,
> the wisdom arisen from learning and so on manifests.[574]

As this states, there is the meditation both on the specific characteristics and on the general characteristics. [480]

Mindfulness of the Body

In regard to the body, the way to meditate on the specific characteristics is to meditate on the body as impure, as composed of the elements and their derivatives, and so forth. The way to meditate on general characteristics is to meditate on all conditioned things such as the body as impermanent and on all contaminated things as in the nature of suffering. *Sacred Vinaya Scripture* says: "In whatever place one stays, sitting cross-legged on the mat set down, mindfulness manifestly placed on one's upright body, one must abandon the eight tactile objects. One thinks, '[The body is like] a sickness, cancer, a weapon, poison, impermanent, suffering, merely empty, and selfless.'"[575] Indeed, since this body is in the nature of suffering, it is like a sickness; since its suffering is a strong sickness, it is like cancer; since this body produces suffering directly, it is like a weapon; since it produces suffering indirectly, it is like poison; since it disintegrates each moment, it is impermanent; since it is conjoined with the contamination of actions and mental afflictions, it is of the nature of suffering; since the body and so on is not established as belonging to an independent self, it is merely empty; since the body and so on is not established as being an independent self, it is selfless.

The antidote to a predominance of attachment is as follows. With resolute attention, one should meditate on the color, shape, touch, and honoring of the body as unattractive. And since the mind that meditates on that unattractiveness is of the nature of nonattachment, it serves as an antidote to attachment to the color of the body, the shape of the body, [481] the touch of the body, and the honoring of the body.

Among these four types of attachment to the body, as an antidote to attachment to the *color* of the body, one meditates discerning the body to be unattractive, focusing on the body, which is the object of attachment, appearing as blue, appearing as black, and appearing as red.[576] As an antidote to attachment to the *shape* of the body, one meditates discerning the body to be unattractive by focusing on the body, which is the object of attachment, appearing as wasting away and as broken up. As an antidote to attachment to the *touch*, or tactile feel, of the body, one meditates discerning the body to be unattractive by focusing on the body, which is the object of attachment, appearing as maggot-eaten and as a skeleton joined together with sinews. As an antidote to attachment to the *honoring* of the body, one meditates discerning the body to be unattractive by focusing on a dead, unmoving corpse. The *Perfection of Wisdom Sūtra in Eighteen Thousand Lines* says:

> O Subhūti, at such time as a bodhisattva, a great being, dwells in a charnel ground, upon seeing various bodies discarded in the charnel ground, which have been dead for a day, dead for two days, dead for three days, dead for four days, or dead for five days, with a decayed appearance, a blue appearance, a rotting appearance, [482] and a maggot-eaten appearance, he thinks "This body is also a phenomenon like that and has such a nature, so it is not free from that reality" and thus also meditates in that way with regard to this very body. O Subhūti, in this way, a bodhisattva, a great being, having eliminated covetousness for the world and mental discomfort, abides contemplating the body as a body with ardent meta-awareness and mindfulness. O Subhūti, furthermore, at such time as a bodhisattva, a great

being, seeing discarded human corpses that have been dead for six days, or dead for seven days, pecked and gnawed in various ways, whether by crows, eagles, golden-crested birds, vultures, vomit eaters, wolves, foxes, dogs, or types of worms, he thinks, "This body also is a phenomenon like that and has such a nature, so it is not free from that reality."[577]

The *Treasury of Knowledge* says:

> The beginner practices through [visualizing] bones as
> extended
> up to the ocean and then reducing it.
> It is said that an accomplished one practices by
> separating the foot and so on up to half the skull.
> One with perfected attention practices by holding the mind
> focused between the eyebrows.[578] [483]

Meditation on the unattractive with resolute attention visualizing the body as a skeleton is explained in three levels of practice: beginner-level yoga, the yoga of the accomplished one (Skt., *kṛtajaya*), and the yoga of perfected attention. In the first, one focuses the mind to the big toe of the foot, the forehead, or some other part of the body and visualizes that the flesh has rotted and decomposed and that it is just snow-white bones, like a skeleton. Then in stages after that, one visualizes the entire body as a skeleton, and then one visualizes that skeletons fill one's room until finally they pervade the earth reaching as far as the ocean. When one gathers the meditation object back, one withdraws it in stages from the outside, and having visualized one's own body alone as a skeleton, one focuses attention in this way again and again. This is how to practice beginner-level yoga.

In the second, one visualizes skeletons pervading everywhere. Then, as before, one gathers the meditation object back, and one visualizes the feet as bones. Releasing that visualization, one visualizes the remaining parts of the body other than the feet as bones and focuses the mind on those. Then gradually one visualizes that what remains is half the body as bones and focuses the mind on that. Finally, one continues up to the point where just half the skull is visualized as bones, and one focuses the mind. This is how to practice the yoga of the accomplished one.

In the third, one visualizes skeletons pervading everywhere. Then, as before, one gathers the meditation object back, and one gradually visualizes the remaining parts of the body as bones, and one focuses on that. Finally, releasing that visualization, one focuses the mind [484] on the place between the eyebrows as a remaining patch of bone merely the size of the big toe. This is how to practice the yoga of perfected attention.

Overall, there are nine ways to meditate on the body as unattractive. As an antidote to attachment to the color of the body, one meditates in three ways, cultivating the attitude that the body is unattractive by visualizing the body, which is the object of attachment, appearing as blue and so on. As an antidote to attachment to the shape of the body, one cultivates the attitude that the body is unattractive in two ways by focusing on the body, which is the object of attachment, appearing as wasted away and so on. As an antidote to attachment to the touch of the body, which is the object of attachment, one cultivates the attitude that it is unattractive in two ways, by visualizing the body appearing as maggot-eaten and so forth. As an antidote to attachment to the honoring of the body, one cultivates the attitude that it is unattractive in one way, by focusing on an unmoving corpse. And finally, as an antidote to attachment to all factors—the color and so forth of the body—one cultivates the attitude that the body is unattractive in one way, by visualizing the body as a skeleton. Since a skeleton doesn't have any of the four bases for attachment such as color, visualizing and meditating on the body as a skeleton is a suitable antidote to all four attachments.

The mind meditating on the body as unattractive is attending to what is visualized and superimposed onto the focus of one's attachment. Since it perceives, for example, just one facet of the form aggregate, such as a temporary color or shape, it is an antidote that works by merely suppressing a mental affliction; it is not an antidote that eliminates it completely. [485]

The antidote to a predominance of conceptualization is to meditate by counting the inflow and outflow of one's breath and placing one's mindfulness single-pointedly on that without distraction. The *Treasury of Knowledge* presents the meditation on the inflow and outflow of the breath in terms of six aspects, as follows.

With the body placed in the seven-point sitting posture, one focuses on the inflow and outflow of the breath while continuously maintaining single-pointed mindfulness. (1) With regard to *counting*, considering both the in-breath and the out-breath as one round, one mentally counts them from one up to ten, keeping the counting free from the three faults—confusion, excess, and omission. If one counts fewer than ten, the greater danger is that laxity might arise, and if one counts more than ten, the greater danger is excitation. If one happens to count two rounds as one, this is the fault of omission. If one counts one round as two, this is the fault of excess. If one takes the in-breath to be the out-breath, or the out-breath to be the in-breath, this is the fault of confusion.

(2) With regard to *monitoring*, when the breath flows in, it enters in the following order: the throat, the heart center, the navel, the groin, the thighs, the calves, the soles of the feet, and into the ground—as deep as one armspan if it is more powerful and as deep as one handspan if it is less powerful. When the breath flows out, it leaves from the soles of the feet and so on right up until it is exhaled from the nostrils—as far as one armspan if it is more powerful and as far as one handspan if it is less powerful. One's mind views this single-pointedly without distraction.

(3) Next, with regard to *fixing*, the breath is present from the tip of the nose to the soles of the feet, like a great garland. One's mind views this single-pointedly without distraction and cognizes whether the body is benefitted or harmed and whether it is made warmer or cooler. [486]

(4) Further, with regard to *investigating*, one thinks, "These in-breaths and out-breaths are not merely wind but are in the nature of the eight physical substances that are the four elements and their respective derivatives. Together with the mind and the mental factors based on them, I possess the five aggregates." So one investigates and views this single pointedly.

(5) Then, with regard to *transforming*, the mind that focuses on the breath transforms from within single-pointed mindfulness in that it trains in increasing virtue.

(6) Finally, with regard to *purifying*, the mind from within single-pointed mindfulness trains in increasing virtue more than before. The *Treasury of Knowledge Autocommentary* says:

> Mindfulness of the in-breath and the out-breath
> is said to have six aspects:
> counting, monitoring, fixing,
> investigating, transforming, and purifying.[579]

Mindfulness following the inflow and outflow of the breath focuses on one meditation object and experiences what is internal; and since the breath that is the basis of the meditation object has no color or shape, it can function as an antidote to conceptualization. Conversely, a mind meditating on unattractiveness focuses on such objects as colors and skeletons, experiencing what is external; this induces conceptualization, and so it cannot function as an antidote to conceptualization. [487] The *Treasury of Knowledge Autocommentary* says: "Those prone to conceptualization should engage in mindfulness of the inflow and outflow of the breath."[580]

Mindfulness of Feelings, Mind, and Phenomena

The way to meditate on the application of mindfulness regarding feelings is as follows. One meditates on feelings by examining their specific and general characteristics in order to see the faults of feelings, such as how pleasant feelings directly give rise to craving for the desirable and unpleasant feelings directly give rise to craving for cessation, and how they indirectly give rise to suffering and so on. The *Questions of Ratnacūḍa Sūtra* says:

> The bodhisattva does not develop attachment toward feelings that are pleasant because he has completely destroyed attachment. He does not develop anger toward feelings that are unpleasant because he has completely destroyed anger. He does not develop ignorance toward feelings that are neither pleasant nor unpleasant because he has completely destroyed ignorance. Due to that, whatever feeling he experiences, he knows and experiences them all as impermanent, knows and experiences them as suffering, knows and experiences them as selfless. He understands feelings that are pleasant to be impermanent. He understands feelings that are unpleasant to be painful. He understands feelings that are neither pleasant nor unpleasant to be peaceful. Thus whatever is pleasant is impermanent. Whatever is unpleasant [488] is suffering. Whatever is neither pleasant nor unpleasant is devoid of selfhood.[581]

The *Treasury of Knowledge* says:

The three bindings are due to the power of feeling.[582]

The way to meditate on the application of mindfulness regarding the mind entails examining the essential nature of mind by observing how it ceases moment by moment, how it cannot be shown, how it is unobstructive, how it is in the nature of clear light, how, from the point of view of its continuum, it stays constantly without any interruption, and so on. The *Kāśyapa Chapter Sūtra* says: "Kāśyapa, mind is like a continuous waterfall in that it does not remain still but dissolves upon arising. Kāśyapa, mind is like the wind, in that it travels far and cannot be held. Kāśyapa, mind is like the radiance of an oil lamp, in that it arises from causes and conditions."[583] In this way, based on learning, reflection, and so on, one analyzes and investigates both the specific and general characteristics of the mind's function, of the faults that occur under the power of the mental afflictions, of its effects, and so on. The *Kāśyapa Chapter Sūtra* says:

> Kāśyapa, since the mind has given rise to all suffering, it is like an enemy. [489] Kāśyapa, since the mind dissipates all roots of virtue, it is like a house of sand. Kāśyapa, since the mind discerns impermanence as permanence, it is like a dewdrop. Kāśyapa, since the mind discerns suffering as happiness, it is like a fishhook. Kāśyapa, since the mind discerns the selfless as self, it is like a dream. Kāśyapa, since the mind discerns the impure as pure, it is like a fly. Kāśyapa, since the mind causes many types of harm, it is like an adversary.[584]

The way to meditate on the application of mindfulness regarding phenomena, or *dharmas*, entails analyzing and investigating, based on learning and so on, what are to be abandoned (faults such as attachment and hatred) and what are to be adopted (good qualities such as love and compassion) from among all the mental factors and unassociated conditioning factors other than feelings. The *Questions of Ratnacūḍa Sūtra* says:

> In regard to phenomena, one abides in the application of mindfulness viewing phenomena, and one meditates on whatever factors are conducive to enlightenment while abandoning whatever factors are incompatible with enlightenment. Also,

> explicitly nurturing virtuous factors, [490] one does not fall into the view of permanence, and even abandoning nonvirtuous factors, one does not fall into the view of nihilism. For that which is without nihilism and without permanence is the middle way of the bodhisattvas, completely abandoning permanence and nihilism.[585]

In brief, one should meditate with mindfulness contemplating internal and external bodies to be impure as an antidote to apprehending the body as pure; and with mindfulness contemplating internal and external feelings to be suffering as an antidote to apprehending contaminated feelings as blissful; and with mindfulness contemplating internal and external minds to be momentary as an antidote to apprehending the mind as permanent; and with mindfulness contemplating all internal and external phenomena to be selfless as an antidote to apprehending phenomena as possessing selfhood. Also, one should meditate on love as an antidote to behavior with a predominance of hatred; on dependent arising as an antidote to being largely enveloped in obscuration; and on the extensive categories of the elements as an antidote to a predominance of pride.

When meditating on the impurities of the body and so on, it is essential to apply antidotes such as aspiration against faults like laziness. This point is explained as follows in the *Compendium of Knowledge*:

> Also meditation involves the following: aspiration, diligence, effort, enthusiasm, nondismay, nondiversion, mindfulness, meta-awareness, [491] and heedfulness. The cultivation of aspiration serves as an antidote to the secondary mental affliction of inattentiveness. The cultivation of diligence serves as an antidote to the secondary mental affliction of laziness. The cultivation of effort serves as an antidote to the secondary mental afflictions of laxity and excitation. The cultivation of enthusiasm serves as an antidote to the secondary mental affliction of discouragement. The cultivation of nondismay serves as an antidote to the secondary mental affliction of complete disenchantment due to harm. The cultivation of nondiversion serves as an antidote to the secondary mental affliction of being content with just a little progress. The cultivation of mindfulness serves

as an antidote to the secondary mental affliction of forgetting the instructions. The cultivation of meta-awareness serves as an antidote to the secondary mental affliction of mental depression due to downfall [from virtuous vows]. The cultivation of heedfulness serves as an antidote to the secondary mental affliction of discarding diligence.[586]

Fruits of Mindfulness Meditation

Fifth, the *fruits* of meditating on the impermanence of the body and so on are stated in the *Questions of Ratnacūḍa Sūtra*:

> What has the character of impermanence? Alas, this body is impermanent, it does not remain long, and it dies in the end. Knowing this truth, the bodhisattva does not engage in inappropriate livelihood for the sake of the body but extracts the essence [of this bodily existence]. He extracts three types of essence. What are those three? They are the essence of the body, the essence of resources, and the essence of livelihood. With the thought "This body is impermanent," [492] the bodhisattva pledges to be a servant and a student of all sentient beings. He strives to do whatever needs to be done. Thinking "This body is impermanent," he does not indulge in all the faulty conduct of the body, the guile, and the hypocrisy. Thinking "This body is impermanent," he becomes less concerned about his own life and he commits no evil, not even to save his life. Thinking "This body is impermanent," he does not indulge in craving and attachment to resources, for he gives away all his possessions.[587]

The *Explanation of the Compendium of Knowledge* says: "As for the fruits, in their respective order, the applications of mindfulness help one abandon the distorted conceptualizations [of the body and so on] as pure, blissful, permanent, and possessing selfhood."[588] Thus, as stated above, through meditative practice on these applications of mindfulness, all distorted minds thinking of the body as pure, of feelings as blissful, of the mind as permanent, and of phenomena as possessing autonomous selfhood will cease to arise. [493]

27

The Eight Worldly Concerns

AMONG THE VARIOUS METHODS taught in the Buddhist texts for training the mind, one important method concerns how to develop equanimity toward the eight worldly concerns. In general, most people tend to extol themselves and disparage others. They develop jealousy toward their superiors, competitiveness toward their equals, and pride toward their inferiors. They become vain when praised and angry when criticized. Therefore they experience various sufferings in the present and will do so in the future. Thus we need to prevent our own minds from falling under the sway of attachment, anger, and so forth, and we should not associate too much with those who are under the power of these afflictions but keep our distance from them. As it says in *Engaging in the Bodhisattva's Deeds*:

> They are jealous of superiors, competitive with equals,
> and proud toward inferiors; they develop conceit when praised
> and anger when something unpleasant is said to them;
> when will benefit ever be gained from childish beings?
>
> If one befriends the childish, then one will
> discuss with them the pleasures of worldly life
> while extolling oneself and criticizing others; [494]
> therefore nonvirtue will certainly ensue.[589]

So what are these eight worldly concerns? The *Questions of Ratnacūḍa Sūtra* says: "The eight worldly concerns are: gain and loss, fame and disrepute, pleasure and pain, praise and disparagement."[590] These are the eight concerns with which ordinary worldly beings are preoccupied and that hinder our minds from progressing toward better states: (1) concern about *gain*, when one likes getting things such as food, clothing, a dwelling, and so on, (2) concern about *loss*, when one dislikes not getting things such as food, clothing, and so on, (3) concern about *fame*, when one likes having a good reputation, (4) concern about *disrepute*, when one dislikes having a bad reputation, (5) concern about *pleasure*, when one likes pleasant feelings, (6) concern about *pain*, when one dislikes painful feelings, (7) concern about *praise*, when one likes to be praised, and (8) concern about *disparagement*, when one dislikes being disparaged. Nāgārjuna's *Friendly Letter* says:

> According to mundane understanding, gain and loss,
> pleasure and pain, fame and disrepute, and
> praise and disparagement are the eight worldly concerns; [495]
> equalize these so that they are not objects of one's mind.[591]

Here "equalizing" these eight worldly concerns means to maintain equanimity by preventing one's mind from falling under the sway of these concerns. This depends on understanding well the faults of the four likes and the four dislikes, as well as understanding their antidotes, which are explained in texts such as *Engaging in the Bodhisattva's Deeds*. Here we will select a few of these points and explain them briefly in their respective order.

How do we develop equanimity in the face of a great liking for material gain, such as for food and clothing? No matter who one may be, if one's mind has fallen under the sway of excessive delight in gaining things such as for food and clothes, then one should turn one's mind inward and contemplate as follows. "Suppose that one day, owing to this wicked [self-centered] mind, I strive with great effort—undergoing hardship and pain—to obtain the food, clothing, and so on that I crave. Even if I finally have a chance to enjoy these things, when eventually I die, I will be powerless to prevent both the material things acquired in the past as well as the pleasure of having enjoyed them from being anything but mere memories. They cannot be brought back again into my hands. It will be like awakening from a dream wherein I enjoyed the affluence of the gods." Furthermore, contemplate the fact that in pursuing these material gains, one must endure endless hosts of suffering both in trying to acquire them and in protecting them from others. Consider how in such a pursuit, to say nothing of the way it induces casual mistreatment of total strangers, [496] it is not inconceivable that someone becomes a monstrous individual who could even kill his or her own kind parents! *Engaging in the Bodhisattva's Deeds* advises that we contemplate along these lines:

> In a dream, one person experiences a hundred years
> of happiness and then wakes up.
> Another is happy for a moment
> and then awakes.
>
> Once they have awoken,
> will that happiness return for either?
> This illustrates what it is like at the time of death,
> whether one's life was short or long.[592]

And:

> Those distracted by their attachment to wealth
> have no chance to be free from the suffering of cyclic
> existence.[593]

And:

> One who for the sake of honor and profit
> may even kill his father and mother . . .[594]

So whoever one may be, whether the highest boss or the lowest servant, if one lacks self-awareness, one may fall under the sway of a great liking for material things. In such a case, not only does one gain no profit, but also, among the many faults that arise from that craving, one is led by the nose.[595] Therefore, as an antidote, one should practice the contentment of being "satisfied with enough" and not fall into either of the two extremes (whether greed or disgust) regarding food, clothing, and other material things.

Such contentment, as a virtue, [497] not only brings happiness to one's mind, it is also difficult to find even for those who are highly distinguished according to worldly standards. Also, since there is no pleasure in striving to acquire material things, and many unpleasant experiences are suffered for the sake of protecting them, those who are skillful in taking up what is good and rejecting what is evil abandon excessive craving for material resources. Its antidote is to generate contentment that is "satisfied with enough" by developing a balanced attitude toward material things. This attitude should definitely be integrated into one's mind, as expressed in *Engaging in the Bodhisattva's Deeds*:

> Even a king has difficulty finding
> such happiness and contentment as this.[596]

The *Questions of the Householder Vīradatta Sūtra* says:

> What happiness is there in seeking wealth?
> To guard it, one also experiences suffering.
> Therefore, having completely abandoned attachment,
> the wise are content with the basic necessities.[597]

How do we develop equanimity in the face of a great disliking for loss, as when one fails to get the desired food, clothing, and such? One might currently hold the following opinion. "Today's society enjoys unprecedented advances over the past, with developments in economics, science, technology, and so on. But not only is no respect accorded to those who have little in the way of material goods, [498] they are also looked down on. In contrast, society glorifies and accords great respect toward those who are well endowed with material possessions. So whoever one may be, it makes sense to dislike it when material goods such as [luxurious] food and clothing do not come into one's hands." As an antidote to this way of thinking, one should consider that there is in general no certainty about who in the world becomes an object of praise or of disparagement; our own experience confirms that both the poor and the rich can be disparaged. One should subdue one's dislike of material loss by thinking again and again about how difficult it is to experience joy in dependence on an ordinary, ill-natured individual who is very hard to get along with, no matter whether they are rich or poor. *Engaging in the Bodhisattva's Deeds* says:

> Belittling those who lack possessions
> and disparaging those who are wealthy,

how can those whose company is by nature painful
ever give rise to joy?[598]

How does one develop equanimity toward a great liking for fame? When one fails to take charge of one's mind and allows it to fall uncontrollably under the sway of excessive delight in fame, then one should turn one's mind inward and contemplate as follows. "With this distorted attitude, a time could come when, for the sake of fame, I might even squander all the wealth I have accumulated. For the sake of renown as a hero, without question, I could even one day provoke the loss of my own dear life." [499] Contemplate, "Empty words of fame, which I am so attached to, produce no happiness but only suffering. Also, at the time of my death, these words bring no happiness to anybody. Furthermore, great acclaim and fawning words, which I like so much, are of various kinds: they may or may not be beneficial to me, they may or may not be true about me, and they may or may not have been spoken deceptively." In these ways, one should carefully contemplate how unsuitable it is to develop a biased liking for something without even examining whether it is good or bad. *Engaging in the Bodhisattva's Deeds* says:

If for the sake of being famous
I lose my wealth or get myself killed,
what good are these words?
Once I have died, who will they delight?[599]

Mātṛceṭa's *Letter to King Kaniṣka* says:

Pleasant words may not be beneficial,
gentle words may not be true;
you should not keep in your heart
words that are not genuine.[600]

How do we develop equanimity in the face of a great disliking for disrepute? When one's mind has fallen under the sway of a great antipathy to infamy, then one should contemplate carefully as follows. "Because my conduct of body, speech, and mind [500] is disagreeable to the other person, he speaks badly of me. However, those empty unpleasant words never hurt even a small part of my body before and will not hurt it later, so there is no reason my mind should fall under the sway of rage and not remain in equanimity. Furthermore, some words I find unpleasant are motivated by beneficial intentions, and if I adapt my behavior according to their substance, not only would this bring peace temporarily but it would also lead to long-term benefit. Therefore it is not appropriate for me to dislike words solely because they are unpleasant without giving any thought to whether their substance has any merit." One should contemplate carefully in this way. *Engaging in the Bodhisattva's Deeds* says:

> Since unpleasant words
> do not harm my body,
> why, mind, do you get angry?[601]

Also, the *Letter to King Kaniṣka* says:

> Words intended to help one's friend
> may be unpleasant, but they bring benefit,
> gentleness, and bliss;
> so please take them to heart.[602] [501]

How do we develop equanimity when facing a great liking for pleasant feelings? When one's mind has fallen under the sway of excessive delight in pleasant feelings, one should contemplate carefully as follows. "I do not want pain, I only want pleasant feelings. But merely wanting that is not enough—I need a method to bring this about. Also, just any suitable method is not enough; to purchase beautiful ornaments, delicious foods, wines, and so on requires strenuous effort. This means that for the sake of pleasure, I must earn a living, cultivate land, build up a business, and so on. In brief, I cannot be sure that any work I do for the sake of pleasure will actually lead to happiness alone." One should contemplate carefully again and again in this way. *Engaging in the Bodhisattva's Deeds* says:

> One who wants just [transient] mental pleasure
> can devote himself to gambling, drinking, and so on.[603]

And:

> People engage in activities for the sake of happiness,
> but it is uncertain whether they will find it or not.[604]

How do we develop equanimity when facing a great disliking for painful feelings? When one's mind has fallen under the sway of a great antipathy to painful feelings, one should direct one's mind inward and contemplate as follows. "Regarding this suffering, the object for which I generate dislike, [502] if there is a way to change it, then I can do that, and there is no need to feel unhappy; and if, on the other hand, there is no way to change that painful thing, then there is no benefit whatsoever in feeling unhappy about it." *Engaging in the Bodhisattva's Deeds* says:

> If there is a remedy,
> what use is it to feel upset?
> If there is no remedy,
> what use is it to feel upset?[605]

How do we develop equanimity when facing a great liking for being praised? When one's mind has fallen under the sway of excessive delight in being praised, one should seize the reins of one's mind and contemplate carefully as follows. "Someone else's praise neither increases my merit nor extends my life, it does not make me stronger nor remove any sickness, and it does not bring about physical well-being. Even if it is not mere empty flattery, when I analyze carefully in what way it serves my purpose, I can see clearly that praise does not benefit me in any way at all. My mind is deceived by the small measure of reassurance praise gives me in this life and so on. Praise interests only those whose wisdom is befuddled, who engage in what is contrary to Dharma practice, who do not know even roughly what is to be accepted and rejected. [503] Once it is allowed to distract my mind, even my mindfulness becomes unstable under its sway." Contemplate carefully in this way again and again. *Engaging in the Bodhisattva's Deeds* says:

> The honor of praise and fame
> does not enhance my merit, my life,

my strength, my health,
nor my physical well-being.

When I know what is meaningful for me,
what purpose do these [praise and fame] have for me?[606]

Śrī Jaganmitrānanda's *Letter to King Candra* says:

This life's praise leads to happiness only
in the hearts of those seduced by the glorious light of fame.
These confused ones who strive for it
are distracted in mind and without mindfulness.[607]

How do we develop equanimity toward a great disliking for being disparaged? When one's mind has fallen under the sway of a great antipathy to being disparaged or reproached, then, without allowing one's mind [504] to succumb to attraction or aversion, one should carefully contemplate as follows. "There is not a single person in the world who is praised by everyone. My friends and companions praise me at the slightest opportunity, without investigating or even seeing any special quality in me. Indeed, I like it when I am not reproached by anyone, and I dislike it when I am reproached, even though it is unwarranted for me to dislike it when others reproach me simply because of opposing ways of thinking. Within the disparagement I dislike, there is disparagement by the wise and disparagement by the foolish. The wise are certainly justified in reproaching me, and so it is unsuitable for me to harbor resentment, thinking 'They disparage me' without entertaining even the slightest doubt about whether there is a good reason for it." One should contemplate carefully in this way. *Engaging in the Bodhisattva's Deeds* says:

If others revile me,
why am I joyful when lauded?
If others praise me,
why am I depressed when disparaged?[608]

Collection of Aphorisms says:

Whatever the foolish praise,
and whatever the wise reproach,
the wise one's reproach is worthy,
the foolish one's praise is not.[609] [505]

28

Increasing Good Qualities

As for how the mind comes to be transformed in dependence on the three levels of wisdom developed through learning, reflection, and meditative cultivation, the Buddhist texts explain that the growth and decline of good qualities rooted in the mind, such as wisdom and compassion, are not contingent on the growth or decline of the body but develop from habituation within the mind itself. *Exposition of Valid Cognition* says:

> Wisdom and so on
> will increase or decrease
> not through growth and decline of the body
> but through the functions of the mind itself.[610]

Competitive athletes can jump very high or run very fast with practice, so significant improvement is clearly possible on the basis of repeated training and practice. We observe that, in general, this kind of advance through training can be seen in the case of physical attributes. However, because attributes contingent on the body have limitations—even a highly trained athlete needs to exert effort every time he or she jumps or sprints—this kind of ability can never advance to the point of being limitless. Moreover, if we take the heating water as an example, its basis (the water) is unstable, so no matter how much the water is boiled, it ends only in evaporation; its hotness cannot cause it to burst into flame. [506]

In contrast, when qualities rooted in the mind, such as the antidotal forces of love and wisdom, become habitual, they arise effortlessly within one's mental continuum. And unlike physical abilities such as jumping or sprinting that require effort every single time, even a single initial effort in cultivating qualities of the mind can produce a continuum of numerous successive instances of similar type. It is like when fire burns firewood, it naturally keeps consuming the firewood until that becomes ash. Similarly, stains on a piece of gold are effortlessly purified by mercury. Furthermore, since good qualities of the mind are simply attributes that arise as the clear and cognizing nature of the mind, they do not dwell in an unstable basis, like heat in water. Owing to their basis, qualities such as these can develop without limit. Thus there is an enormous difference between qualities contingent on the body and those rooted in the mind in terms of how much they can increase. *Exposition of Valid Cognition* says:

> Love and so forth arise in the mind through habituation
> and naturally develop further,
> just as, owing to fire and so on, effects naturally develop
> further
> in wood, mercury, gold, and so on.
>
> The qualities that have arisen from that [habituation]
> arise as [the mind's] very nature.
> Therefore further and further efforts
> progressively enhance their excellence.
>
> And since the mental states of love and so on
> grow from earlier seeds of a similar type, [507]
> how could they be limited
> if they are being cultivated?
>
> A jump does not come in this way from an earlier jump.
> Since the strength and effort that cause it
> have a limited capacity,
> a jump is limited in its nature.[611]

Thus, when we speak of transforming our mind, it is not, for example, like upgrading the furnishings in our house, which involves changing things only in the physical environment. Any transformation in our mind has to come from changing our *outlook*—how we view our own self, others, and the external world—and from changing our everyday *attitudes*, changing our normal *temperament* in how we relate to our emotions, and changing our deeper *aspirations*. These are also the very reasons why the Buddhist texts present extensive methods for training the mind. There are techniques, such as those taught in the texts of mind training according to Mahāyāna teachings, on how to intentionally place all the blame ultimately on self-cherishing, how to view all beings as embodiments of kindness and cherish them, how to immediately take whatever one encounters into the path of mind training, how to take whatever adversities one meets as an aid for transforming the mind, and how to let go of adhering to objective true existence with respect to objects of both the outer and inner world. When people are able to conduct their lives each day on the basis of instructions that masters of the past have taught, such people have really begun a life of true happiness. All of this confirms that in the end, what is called happiness and well-being [508] must be defined mainly on the basis of one's own state of mind.

One might wonder, "Can nonvirtuous states of mind, such as the mental afflictions, be eliminated from their root? Or can they only be suppressed?" Buddhist texts state that faults of the mind such as attachment can not only be suppressed temporarily but can also be eliminated from their root—and they present numerous reasons to prove it. *Exposition of Valid Cognition*, for example, says:

> Anything that is susceptible to decrease and increase
> has counteragents (*vipakṣa*). Hence, by inculcating
> (*sātmībhāva*)
> [those counteragents] through habituation to them,
> the contaminants in some cases will be eliminated.[612]

When the wisdom realizing selflessness increases for some people, their faults of attachment and so on diminish. And since it is owing to the wisdom realizing selflessness increasing that their faults such as attachment diminish, it is the possession of wisdom itself that harms attachment and so on. Thus, by familiarizing oneself again and again with the wisdom realizing selflessness, it comes to be instilled in the mind itself. It is maintained that some people thereby come to extinguish the contaminants—the mental afflictions and so on—from their root.

Also, since the nature of the mind is clear light, the ultimate nature of the mind is that it is empty of true objective existence, not tainted by the stains [of the mental afflictions and their traces]. Not only this, but also the clear and cognizing nature of the mind itself is free from being tainted by the stains. [509] As discussed in the earlier section explaining the nature of mind, these stains of the mind are adventitious and can be removed. If mental afflictions such as hatred were integral to the mind, part of its essential nature, then they could not be separated from the mind's nature; in which case the mind would always have the character of hatred. That such is not the case is evident from everyday experience. Anger arises when its conditions are present, and anger does not arise when its conditions are not encountered. For this reason, we can infer that afflictions such as hatred are not part of the mind's essential nature and that, through the power of their antidotes, they can be removed. For example, when a pond is stirred it becomes murky, but the nature of the water itself is not murky, for when the mud settles down, the water returns to its clear state. Nāgārjuna's *Praise of Dharmadhātu* says:

> Leather armor can be cleansed by fire
> when contaminated by various stains,
> for when it is placed inside the fire,
> the stains are burned away but not the armor.
>
> Likewise with the mind, which is clear light,
> when contaminated by attachment and so on,
> the fire of wisdom incinerates the afflictions
> without incinerating its nature, the clear-light mind.[613]

When armor made from hardened leather and such like has become dirty, it is placed inside a fire, and the fire burns away the stains but does not burn the armor. [510] Likewise, when the stains of attachment, hatred, and so on within the mind, whose nature is clear light, are destroyed by an undistorted mind of wisdom, the stains are removed but not the mind of clear light.

Moreover, the root of all faults is said to be delusion grasping at true existence, and since this is a distorted cognition whose way of grasping is incorrect, there is no valid basis in reality for its way of grasping. In contrast, the forces opposing this grasping, the class of virtuous minds exemplified by the wisdom realizing the absence of objective true existence, possess valid grounding in reality. So if one cultivates continuous familiarity with these opposing forces, the stains of the mind can be removed. *Engaging in the Bodhisattva's Deeds* says:

> There is nothing whatsoever
> that does not become easier with habituation.[614]

Thus, as stated above, we need to identify the unwholesome states of mind on the basis of contemplating their faults; and once we recognize what the wholesome states of mind are, we need to cultivate familiarity with them. When the force of this practice is combined with contemplating how the wholesome states of mind have valid support, how they are qualities rooted in the mind, and how no conditioned phenomena ever remains static, the wholesome states of the mind will progressively increase. At the same time, the unwholesome states of mind will gradually diminish. This is analogous to how, with regard to a particular place or object, as heat increases, cold decreases. [511]

In the preceding sections we explained how Buddhist texts elucidate extensively the following points: how the false conceptualizations of inappropriate attention arise from ignorance in the form of unknowing as well as distorted conceiving; how afflictions such as attachment and anger arise from such conceptualizations; and how other afflictions such as pride, jealousy, and so on, which undermine the mind's composure, arise from attachment and anger. We also discussed how attachment gives rise to mental excitation, to distraction that diverts the mind uncontrollably outward, to laxity that causes the loss of ability to hold the object of meditation with firmness, and to mental dullness, which resembles a descent into darkness within one's mind and which makes the body and mind unserviceable. We explained the opposites of these factors, such as profound wisdom, which possess a capacity to distinguish the characteristics of objects without confusing them; and we also discussed love, empathy, tolerance, confidence, and so on. We also explained how qualities of the mind, such as concentration, come to arise through the cultivation and continued application of mindfulness and meta-awareness, enabling the mind to focus single-pointedly on a chosen object and further enhance such a capacity. We discussed how, since adventitious stains do not penetrate the mind's essential nature, the essential nature of mind is clear light;

and how, because there is no beginning to the continuity of the clear and cognizing mind, consciousness possesses a stable continuum; and how, once habituated, since qualities of the mind do not require renewed further effort, they can be increased without limit. We also discussed how, given that some of the mental factors naturally oppose one another, by increasing the force of one side, the strength of the other side [512] can be eliminated.

29

Concluding Topic: The Person or Self

In these first two volumes of *Science and Philosophy in the Indian Buddhist Classics*, we have organized the account of reality found in the texts of the great Indian Buddhist masters into five main topics. The first topic, the account of the objects of knowledge, was examined in volume 1. Here in volume 2 we examined the second topic, which is the account of the mind as the subject that knows them; the third topic, which is how the mind engages its objects; and the fourth topic, which concerns the methods that may be used by the mind to engage objects; in this last topic we included associated ancillary topics. Now in this short concluding section, we shall briefly discuss the fifth topic, the person—a cognizer of objects.

In the context of inquiry into how the mind engages its objects, as well as how it experiences pleasure, pain, and so on, the classical Indian philosophers examine whether there exists some entity that is the knower or the experiencer, such as the self or the person. Likewise, since everyday experience confirms that the sense of "I" emerges naturally, the question arises as to what exactly is the object of this thought "I am." [513] Questions such as these were addressed extensively by the early thinkers. We will discuss these matters in detail in volume 4, but we address them briefly here.

In general, one must posit the notion of a *person* who is the experiencer of pleasure and pain. One might ask, "Is such a person or self either the body or the mind, or is it something separate from them?" The way it appears to one's everyday cognitions, when one says, "I saw this flower," the thought occurs in one's mind that "I see it," even though it is visual cognition that actually does the seeing. Similarly, when the body is sick or when the mind is sad, one has the thought "I am sick" or "I am sad." Even so, one does not have the attitude that the body or the mind *is* one's self. Likewise, when one says "my body" or "my mind," what appears to the mind is something like a separate "I" that is the controller of the body and mind, or that is the owner of the body and the mind. Thus the non-Buddhist Indian schools of the past assert that the self is something entirely separate from the psychophysical constituents, or "aggregates." In contrast, the Buddhists maintain that what is called the *self* or the *person* is imputed in dependence on the collection of both the body and the mind, and therefore it is the same individual entity as the physical and mental aggregates [in that they cannot exist apart from each other but they are not identical]. [514]

What kind of thing, then, is this so-called self? Except for Buddhists, most religious traditions believe the essential nature of the self to be permanent, unitary, and independent. Buddhists assert that there is no such permanent, unitary, independent self, but they accept a mere self that exists in dependence on the aggregates. A sutta in the Theravāda scriptures says:

> Just as we refer to a chariot
> based on the group of parts,
> so, based on the aggregates,
> we have the convention "sentient being."[615]

Just as a "chariot" is imputed in dependence on the assembly of the chariot's parts, and to posit the entity of the chariot, there is no need for anything beyond the assembly of its parts, likewise, "self" or "person" is merely posited in dependence on its basis of imputation, the aggregates, and there is no need for anything beyond the basis of imputation, the aggregates.

Thus, according to Buddhist scriptures, the term *selflessness* refers to the absence of a self that is independent [515] and separate from the aggregates. However, in a more general sense, Buddhists do not deny that there is a self conventionally posited as an agent that comes and goes and so on and that is the experiencer of pleasure and pain. Now if there is no autonomous self separate from the aggregates of body and mind that owns or controls the body and mind, this raises many questions. How can all of a person's earlier and later experiences be posited as the experiences of one person? How can memory arise within a specific individual who, in relation to previous experiences, later thinks "I saw that" or "I did that"? Is the separateness of persons having different mindstreams defined exclusively in terms of the separateness of their physical bodies? These critical questions will be examined in volume 4.

At this point, one might wonder whether such a person or self has a beginning. Those who believe in a creator of the universe maintain that the self does have a beginning. For those who do not believe in such a creator—such as Buddhists, who assert that the external world and beings within arise purely in dependence on their own causes and conditions—the existence or nonexistence of a beginning to the self is a question of whether the aggregates, the designative basis of the self, have a beginning. This is because, for Buddhists, the self is imputed upon the aggregates. From among the five aggregates that are the basis of imputation of the person, if we take the consciousness aggregate as an example, what is referred to as "consciousness" or "mind" is something that is the nature of mere inner experience, which does not exist as material and is not characterized by color, shape, and so on. The substantial cause of a phenomenon that is merely in the nature of subjective experience must be of a similar kind as itself; [516] thus one cannot posit subtle physical particles to be the substantial cause of consciousness. If one were to trace the prior moments in the continuum of consciousness in general, and the subtle levels of mental consciousness in particular, there is no beginning. We already discussed why this is so in our brief presentation of past and future lives, an ancillary issue raised in the context of the conditions for consciousness. Since one cannot posit a beginning to consciousness, Buddhists maintain that one also cannot posit a beginning to a self that is imputed in dependence on the consciousness.

As for whether there is a final end to the self, the world's religious traditions have divergent views. Even among the Buddhist schools, some accept that the self has an end. However, Mahāyāna Buddhist schools maintain that since consciousness has neither beginning nor end, the imputed self also has neither beginning nor end.

The definition of *person* is: a living being that is imputed in dependence on any of the five aggregates. In general, the three terms *self*, *I*, and *person* are synonymous, in that they are alternative names for one and the same thing.

Although there is only one "I" for the continuum of a single person, such as Devadatta, many parts of "I" may be posited for that same person. In the case of the monk Devadatta, for example, the thought "I" can arise in relation to specific parts of him, such as his being a human being, a man, and an ordained monk. [517] And just as he has different aspects to his personal identity, so too for other people many parts of "I" can be posited. Also, the identity of "human being" is defined primarily on the basis of an assembly of aggregates that constitute a human body, and an animal is defined in dependence on an assembly of aggregates that constitute the body of such an animal. In the case of a human being too, based on different attributes of the aggregates, one can speak of being a male or a female, large or small, clever or foolish, and so on.

As for other terms for the *self* or the *person*, the *Teachings of Akṣayamati Sūtra* states:

> The sūtras that speak about self, about sentient being, living being, nourished being, creature, person, vital being, able being, agent, and experiencer, that speak by way of using such diverse terms and indicate there to be an owner even though there is no owner, these are provisional in their meaning.[616]

The meaning of these terms may be understood from the *Compendium of Enumerations*:

> *Self* is posited in terms of viewing the five aggregates of appropriation as the self or as belonging to the self. *Sentient being* is posited in terms of not knowing existent phenomena as they really are and being attached to them. *Vital being* is posited in terms of an aspect of mental cognition. *Able being* is posited in terms of becoming elevated or debased in relation to pride. [518] *Nourished being* is posited in terms of expanding due to the path of cyclic existence recurring. *Creature* is posited in terms of a living being's effort. *Person* is posited in terms of being dissatisfied and discontented with transmigrating again and again. *Living being* is posited in terms of being alive and staying alive. *One who is born* is posited in terms of being a phenomenon that arises [from causes] and so on.[617]

As for the Buddhist and non-Buddhist views on the self, as well as the proofs associated with them, these will be explained extensively in volume 4.

In the present work, we have assembled within a single collection many important subjects sourced from the word of the Buddha and especially from the texts of great ancient Indian Buddhist thinkers extant in Tibetan translations that pertain to accounts of reality and profound philosophical views. We have organized these presentations into a specific order and have provided accompanying commentarial explanations.

This completes the volumes presenting science from the series called *Science and Philosophy in the Indian Buddhist Classics.*

Notes

1. In its original Tibetan, the text of His Holiness's introduction covers both volumes on science and therefore appears only at the beginning of the first volume. In our translation, however, we put the sections of the introduction relevant to the physical world in volume 1 and the sections on the mind sciences in volume 2 so that each of the two volumes now carries an introduction from the Dalai Lama. The opening and closing of the introduction are the same in both volumes. All notes to the introduction are from its translator.
2. For an accessible and personal account of the Dalai Lama's encounter with science, and especially his reflections on the convergence between science and Buddhism with respect to their method, see Dalai Lama 2005, especially chapter 2.
3. The proceedings of many of these Mind and Life Dialogues are available in books. For details, visit https://www.mindandlife.org/books/.
4. The Dalai Lama outlines his views on what he understands by secular ethics in two important books, *Ethics for the New Millennium* and *Beyond Religion: Ethics for a Whole World.*
5. In Mahāyāna sources, buddhahood is characterized by two dimensions—the *truth body*, which is the enlightened being's ultimate reality, and the *form body*, which refers to the manifestation of that ultimate reality. When further elaborated, the truth body has two aspects—the wisdom aspect and its underlying natural expanse—while the form body is differentiated into an enjoyment body (a celestial or subtle form) and an emanation body (a form visible to ordinary humans).
6. These two are in fact part of three criteria of existence, with the third being that "it must not be invalidated by any analysis pertaining to the ultimate reality of things." Though implicit in the Indian Madhyamaka sources, they are made explicit and extensively discussed in Tsongkhapa's writings. For a brief contemporary explanation of the criteria of existence according to Tsongkhapa's Madhyamaka, see Jinpa 2002, 155–57.
7. See pages 215–30 for explanation of these eighty conceptions and the subtle states of mind during the four empty states.
8. This particular argument making the case for the potential for limitless enhancement of qualities of the mind, such as compassion, is most developed

in chapter 2 of Dharmakīrti's *Exposition of Valid Cognition* (*Pramāṇavārttika*). For an English translation of the relevant sections of Dharmakīrti's text based on its reading by the Tibetan thinker Gyaltsab Jé, see Jackson 1993, pp. 309–18.

9. This threefold classification of all forms of sound inference is most developed in Dharmakīrti's philosophy. For a critical analysis of Dharmakīrti's views on the correct forms of logical reason, especially the logical basis of inference, see Dunne 2004, chap. 3.
10. For a detailed presentation of the three classes of correct inferences, including especially the diverse subdivisions within the third category of nonperception inference, based on Tibetan interpretations, see Rogers 2009, 149–273.
11. For a lucid account on this consequence-revealing form of reasoning, see Perdue 2014, chap. 6.
12. Nāgārjuna's analytic corpus logically presenting the view that nothing possesses intrinsic existence includes his (1) *Fundamental Treatise on the Middle Way* (*Mūlamadhyamakakārikā*), (2) *Sixty Stanzas of Reasoning* (*Yuktiṣaṣṭikā*), (3) *Seventy Stanzas on Emptiness* (*Śūnyatāsaptati*), (4) *Treatise on Pulverizing [the Views of Opponents]* (*Vaidalyaprakaraṇa*), and (5) *Averting Objections* (*Vigrahavyāvartanī*). Sometimes his *Precious Garland* (*Ratnāvalī*) is also added, making the corpus a collection of six works.
13. See the essay that introduces part 1 for a brief explanation of the four schools of classical Buddhist philosophy.
14. Dharmakīrti develops these two presentations of the twelve links of dependent origination, in sequential and reversal orders, in chapter 2 of his *Exposition of Valid Cognition* (*Pramāṇavārttika*).
15. "Trailblazers" translates the Tibetan phrase *shing rta srol 'byed*, literally "chariotway makers," and conveys the idea that Nāgārjuna and Asaṅga paved the great pathways for the flourishing of Mahāyāna Buddhism. In Indian Mahāyāna sources, both these masters came to be deeply revered and mythologized with widespread recognition of both having been in fact prophesized by the Buddha himself.
16. Śāntideva himself mentions (in *Engaging in the Bodhisattva's Deeds*, 5.106) his *Compendium of Sūtras*, but this work seems to be lost.
17. Dignāga wrote several texts on logic and epistemology carrying the suffix "analysis" (*parīkṣā*) in their titles. These include the *Sāmānyaparīkṣā*, *Nyāyaparīkṣā*, *Vaiśeṣikaparīkṣā*, and *Sāṃkhyaparīkṣā*, all of which are no longer extant.
18. *Dhammapada*, 14.5. This verse also appears in the Tibetan canon in the *Sūtra of Individual Liberation*, *Prātimokṣa-sūtra*, Toh 2, 20b, Pd 5:49. The *Dhammapada* is a much-loved text that can be found in the Khuddaka Nikāya, a division of the Pali canon. It contains 423 verses in 26 chapters, arranged according to topics. There are a number of other recensions of this text, or texts closely related to it that may share a common ancestor, such as the Gāndhāri *Dhar-*

mapada, the Patna *Dharmapada*, and the *Udānavarga*. The last of these has been translated into Tibetan.

19. *Udānavarga*, 31.1. Toh 326, 243a, Pd 72:681.
20. Śāntideva, *Bodhisattvacaryāvatāra*, 5.6. Toh 3871, 19a, Pd 61:970.
21. The Tibetan term translated as "clear" is *gsal ba* (Sanskrit, *prakāśa*). Although the Sanskrit term means "luminous," the Tibetan term also conveys the sense of being "clear" or pellucid.
22. *Dharmadhātuprakṛtyasambhedanirdeśa*. Toh 52, 143a, Pd 40:398.
23. *Trāyastriṃśatparivarta-sūtra*, chap. 1. Toh 223, 129b, Pd 63:349.
24. *Aṅguttara Nikāya* 1.51.
25. *Aṣṭasāhasrikāprajñāpāramitā-sūtra*, chap. 1. Toh 12, 3a, Pd 33:6.
26. *Laṅkāvatāra-sūtra*, chap. 9. Toh 107, 186b, Pd 49:457.
27. *Tathāgatagarbha-sūtra*. Toh 258, 251a, Pd 66:690.
28. *Sarvabuddhaviṣayāvatārajñānālokālaṃkāra-sūtra*, chap. 2. Toh 100, 290b, Pd 47:762.
29. Maitreya, *Uttaratantra*, 1.30. Toh 4025, 56a, Pd 70:938.
30. Maitreya, *Uttaratantra*, 1.65. Toh 4025, 57b, Pd 70:941.
31. Dharmakīrti, *Pramāṇavārttika*, 2.210. Toh 4210, 115b, Pd 97:519.
32. Dharmakīrti, *Pramāṇavārttika*, 3.481. Toh 4210, 136b, Pd 97:568.
33. When characterizing the process of cognition in general, several types of objects are presented: the appearing object (*snang yul*), the observed object (*gzung yul*), the engaged object (*'jug yul*), and the conceived object (*zhen yul*).
34. Dharmakīrti, *Pramāṇavārttika*, 2.208. Toh 4210, 115b, Pd 97:518.
35. Śāntarakṣita, *Madhyamakālaṃkāra*, chap. 16. Toh 3884, 53b, Pd 62:896.
36. Śāntarakṣita, *Madhyamakālaṃkāravṛtti*. Toh 3885, 60b, Pd 62:916.
37. Kamalaśīla, *Madhyamakālaṃkārapañjikā*, chap. 1. Toh 3886, 94b, Pd 62:1007.
38. There is, however, an important Buddhist school, Cittamātra (Mind Only), that espouses an ontology in which consciousness is the one and only fundamental feature of reality.
39. Highest yoga tantra refers to the highest of the four classes of Buddhist tantra. The texts and associated teachings of these four classes belong to what is known as the Vajrayāna, which the Indo-Tibetan tradition views as representing the most advanced teachings of the Buddha.
40. See the last chapter of volume 1.
41. In Buddhist philosophy, the subtle sense organs are invisible and are located within their visible counterparts, such as the eyeball. However, the mental sense faculty is not an organ, it is a previous moment of mind.
42. For further details about the different types of winds and channels, see chapter 30 of volume 1 and part 3 of the present volume.
43. This summary is based on Chapa's sevenfold typology. For a presentation of Chapa's account of the order in which these minds engage an object, see part 4.
44. Dignāga, *Pramāṇasamuccaya*, 1.4–6. Toh 4203, 1a, Pd 97:3.
45. Dignāga, *Pramāṇasamuccayavṛtti*, chap. 1. Toh 4204, 15b, Pd 97:60.

46. Dharmakīrti, *Pramāṇaviniścaya*, chap. 1. Toh 4211, 154b, Pd 97:619.
47. *Saṃdhinirmocana-sūtra*, chap. 5. Toh 106, 12b, Pd 49:29. "Appropriating consciousness" (*len pa'i rnam par shes pa*) is another term for store consciousness (*kun gzhi rnam par shes pa*; Skt., *ālayavijñāna*). It is called "appropriating" because it is what appropriates the new life at the time of rebirth.
48. Although this scriptural quotation is well known, the sūtra that it comes from is not generally identified in epistemological literature. However, Jñānavajra's *Heart Ornament of the Tathāgata: Commentary on Descent into Laṅkā Sūtra*, chap. 2, says with regard to this unidentified quotation: "When Buddha said, 'Monks, consciousness knowing form is of two types: based on name (*ming*) [i.e., eye (*mig*) sense power] and on mental [sense power] that arises from it,' and so on, this is categorizing perception." There is much debate in Buddhist epistemological texts about the presence of mental perception in an ordinary person. Toh 4019, 65a, Pd 70:155.
49. Dharmakīrti, *Pramāṇaviniścaya*, chap. 1. Toh 4211, 158a, Pd 97:628.
50. Dharmakīrti, *Pramāṇavārttika*, 3.243. Toh 4210, 127b, Pd 97:547.
51. Vasubandhu, *Abhidharmakośabhāṣya*, chap. 2. Toh 4090, 99a, Pd 79:245. Commenting on *Treasury of Knowledge*, 2.61c.
52. Dharmakīrti, *Hetubindu*. Toh 4213, 253a, Pd 97:844.
53. Dharmakīrti, *Pramāṇavārttika*, 3.461. Toh 4210, 136, Pd 97:567.
54. See pages 256–62 below.
55. Dignāga, *Ālambanaparīkṣā*, v. 6. Toh 4205, 86a, Pd 97:431. Some earlier Tibetan commentaries on the meaning of *Investigation of the Object* give the following explanation. Blue, for instance, appearing as external to the sense consciousness, is the object (*don*) of that sense consciousness, and since it is the focal object and the condition for that consciousness, it is the *objective condition*. Although in general the term *object* (*don*) refers to many kinds of objects, in this context it is said to refer to the focal object condition. Suppose someone asks, "Why is it the focal object?" This text says: "Because it is the nature of consciousness." It follows that it is its focal object because it is the nature of the object of that consciousness. Suppose someone asks, "Why is it the condition?" This text says: "Because unmistakably part of that alone, it is the condition." Because the blue appearing to a sense consciousness is a part of the nature of that consciousness alone, it is correctly considered the condition, even though it occurs at the same time as the consciousness. "Since it arises when there is the object / but does not arise when there is not, / it has unerring forward and reverse [pervasions]."
56. Dignāga, *Ālambanaparīkṣā*, v. 7. Toh 4205, 86a, Pd 97:431.
57. Āryadeva, *Catuḥśataka*, 7.10. Toh 3846, 8b, Pd 57:798.
58. Dharmakīrti, *Pramāṇavārttika*, 2.166. Toh 4210, 113b, Pd 97:515.
59. Dharmottara, *Paralokasiddhi*. Toh 4251, 259a, Pd 106:676.
60. The diversity of senses of *conceptuality* derives in part from the Tibetan use of a single term, *rtog pa*, to render two distinct Sanskrit terms, *vikalpa* and *vitarka*.

The former term, along with other derivatives of the verbal root (*kḷp*), is the one that is typically rendered as "conceptuality" in English translations of Sanskrit philosophical texts. In contrast, *vitarka*, according to the Abhidharma system of classifying mental factors, refers to an initial engagement of or inquiry into an object that is not sufficient for full-blown *vikalpa* or conceptuality.

61. Vasubandhu, *Abhidharmakośa*, 2.33. Toh 4089, 5a, Pd 79:11.
62. Maitreya, *Madhyāntavibhāga*, 1.9. Toh 4021, 40a, Pd 70:903. The Sanskrit term used here is *parikalpa*, which is unrelated to *vitarka* cited just above, but is formed from the same verbal root *(kḷp)* as *vikalpa*.
63. Dignāga, *Pramāṇasamuccaya*, 1.3. Toh 4203, 1a, Pd 97:3. Here the Sanskrit term is *kalpanā*, from the same verbal root (*kḷp*) as *vikalpa*.
64. *Laṅkāvatāra-sūtra*, chap. 6. Toh 107, 145a, Pd 49:358.
65. Dharmakīrti, *Pramāṇavārttika*, 3.283. Toh 4210, 129a, Pd 97:551.
66. Dharmakīrti, *Pramāṇavārttika*, 3.299. Toh 4210, 130a, Pd 97:552.
67. An *object universal* (*don spyi*) is a universal referring to objects, whereas a *word universal* (*sgra spyi*) is the universal aspect of a word that is expressed through the specific utterances that are its tokens.
68. Dharmakīrti, *Pramāṇaviniścaya*, chap. 1. Toh 4211, 154b, Pd 97:619. This same passage was quoted in the previous chapter on page 51.
69. Dharmakīrti, *Pramāṇaviniścaya*, chap. 1. Toh 4211, 155a, Pd 97:620.
70. Dharmakīrti, *Pramāṇaviniścaya*, chap. 1. Toh 4211, 155a, Pd 97:620.
71. Dharmakīrti, *Pramāṇaviniścaya*, chap. 1. Toh 4211, 155b, Pd 97:621.
72. See page 65, which references Dignāga, *Pramāṇasamuccaya*, 1.3. Toh 4203, 1a, Pd 97:3.
73. Dharmakīrti, *Pramāṇavārttika*, 3.287. Toh 4210, 129b, Pd 97:551.
74. Dharmakīrti, *Pramāṇaviniścaya*, chap. 1. Toh 4211, 154b, Pd 97:619.
75. The term *sgra don* (*śabdārtha*) is ambiguous according to whether the Sanskrit term *śabdārtha* or its Tibetan equivalent is interpreted as a conjuctive (*dvanda*) compound or a genitive (*tatpuruṣa*) compound. Thus we have, respectively, either "word and referent" or "the referent of the word."
76. Dignāga, *Pramāṇasamuccayavṛtti*, chap. 1. Toh 4204, 15a, Pd 97:59.
77. The classical Buddhist texts usually mention four kinds of inappropriate attention: holding impermanent things as permanent, holding things in the nature of suffering as pleasurable, holding impure things as pure, and holding selfless things as self-existent.
78. Dharmakīrti, *Pramāṇavārttika*, 3.290–91. Toh 4210, 129b, Pd 97:552.
79. Dharmakīrti, *Nyāyabindu*, chap. 1. Toh 4212, 231a, Pd 97:812.
80. Dharmakīrti, *Pramāṇavārttika*, 2.5. Toh 4210, 107b, Pd 97:500.
81. Dharmakīrti, *Pramāṇavārttika*, 2.1. Toh 4210, 107b, Pd 97:500
82. Dharmakīrti, *Pramāṇavārttika*, 2.1. Toh 4210, 107b, Pd 97:500.
83. Dharmakīrti, *Pramāṇavārttika*, 2.5. Toh 4210, 107b, Pd 97:500.
84. Dharmakīrti, *Pramāṇaviniścaya*, chap. 1. Toh 4211, 152b, Pd 97:619.
85. A *subsequent cognition* is a cognition that follows on a nonconceptual

perception, and it often takes the form of a conceptual judgment that interprets the content of the nonconceptual perception. According to the approach discussed here, the nonconceptual perception (for example, of a sound) would count as a valid cognition because it is both nondeceptive and newly realizing. However, a subsequent cognition (for example, the judgment that "sound is impermanent") is not newly realizing because its object (namely, the sound) has already been apprehended by the perception.

86. A conceptual cognition apprehending a pot mistakes the object universal of a pot for the pot itself.
87. *'Dzin stangs kyi yul.* See chapter 17 for further explanation regarding the object as cognized versus the appearing object.
88. Dharmakīrti, *Nyāyabindu*, chap. 1. Toh 4212, 231a, Pd 97:812.
89. Asaṅga, *Yogācārabhūmi*, chap. 17. Toh 4035, 191a, Pd 97:1124.
90. Śāntarakṣita, *Madhyamakālaṃkāra*, v. 44. Toh 3884, 54b, Pd 62:899.
91. *Laṅkāvatāra-sūtra*, chap. 2. Toh 107, 91a, Pd 59:226. "Delusion" (*gti mug*; Skt., *moha*) here means *ignorance* (*ma rig pa*; Skt., *avidyā*), which is the root of cyclic existence, *saṃsāra*.
92. Dignāga, *Pramāṇasamuccaya*, 1.7. Toh 4203, 2a, Pd 97:4.
93. The first three of these were discussed in chapter 3 above.
94. Asaṅga, *Abhidharmasamuccaya*, chap. 1. Toh 4049, 53a, Pd 76:137.
95. *Laṅkāvatāra-sūtra*, chap. 8. Toh 107, 162b, Pd 49:400.
96. Asaṅga, *Mahāyānasaṃgraha*, 1. Toh 4048, 5a, Pd 49:400.
97. Maitreya, *Madhyāntavibhāga*, 1.9. Toh 4021, 40a, Pd 70:103.
98. *Main mind* (*gtso sems*) is a technical term used in this context to distinguish between consciousness per se and its various functions or attributes.
99. Yaśomitra, *Abhidharmakośavyākhyā-sphuṭārthā*, chap. 2. Toh 4092, 129b, Pd 80:309. This is commenting on *Treasury of Knowledge* 2.34d.
100. Asaṅga, *Yogācārabhūmi*. Toh 4035, 30a, Pd 72:740.
101. Although this sūtra quotation is very well known, we do not find its source identified in any commentary. In some contexts it reads, "Sentient beings each have a single stream of consciousness," whereas in other contexts one finds, "There is no time and no place where two primary consciousnesses occur together at the same time."
102. Dharmakīrti, *Pramāṇavārttika*, 3.178. Toh 4210, 125a, Pd 97:542.
103. Vasubandhu, *Madhyāntavibhāgabhāṣya*. Toh 4027, 3a, Pd 71:7.
104. Sthiramati, *Madhyāntavibhāgaṭīkā*, chap. 1. Toh 4032, 204b, Pd 71:531.
105. Vasubandhu, *Abhidharmakośa*, 2.34. Toh 4089, 5a, Pd 79:11.
106. Pūrṇavardhana, *Abhidharmakośaṭīkā-lakṣaṇānusāriṇī*, chap. 2. Toh 4093, cu:149b, Pd 81:377. This is commenting on *Treasury of Knowledge* 2.34d.
107. Asaṅga, *Abhidharmasamuccaya*, 1. Toh 4049, 72a, Pd 76:183.
108. Jinaputra, *Abhidharmasamuccayabhāṣya*, 1. Toh 4053, 176b, Pd 76:1410.
109. Asaṅga, *Viniścayasaṃgraha*, chap. 7. Toh 4038, zhi:60a, Pd 74:141.
110. Sthiramati, *Madhyāntavibhāgaṭīkā*, chap. 1. Toh 4032, 204, Pd 71:532.

111. Literally, "different isolates" (*ldog pa tha dad*), which means that although the mind and its concomitant factors constitute a single entity, they have different conceptual identities.
112. Vasubandhu, *Abhidharmakośa*, 2.23. Toh 4089, 4b, Pd 79:10.
113. The *Great Explanation* (*Mahāvibhāṣā*) is a text of the third period of Sarvāstivāda Abhidharma literature that both expounds and expands upon earlier canonical texts. It represents the position of the Kashmiri Sarvāstivādins. The seven canonical scriptures of the Abhidharma preserved within the Sarvāstivāda tradition are: (1) *Attainment of Knowledge* (*Jñānaprasthāna*), attributed to Kātyāyanīputra, (2) *Topic Divisions* (*Prakaraṇapāda*), attributed to Vasumitra, (3) *Compendium of Consciousness* (*Vijñānakāya*), attributed to Devaśarman, (4) *Aggregate of Dharma* (*Dharmaskandha*), attributed to Śāriputra (Sanskrit recension) or to Mahāmaudgalyāyana (Chinese recension), (5) *Treatise on Designation* (*Prajñaptiśāstra*), attributed to Maudgalyāyana (Sanskrit recension) or Mahākātyāyana (Chinese recension), (6) *Compendium of Elements* (*Dhātukāya*), attributed to Pūrṇa (Sanskrit recension) or Vasumitra (Chinese recension), and (7) *Statements on the Types of Being* (*Saṅgītiparyāya*), attributed to Mahākauṣṭhila (Sanskrit recension) or Śāriputra (Chinese recension). None of these seven works or the *Great Treatise* found their way into the Tibetan canon. A set of seven canonical Abhidhamma scriptures is similarly preserved within the Theravāda tradition.
114. Here the verses cited as *Verse Summary*, identifying the fifty-one mental factors, are taken from the *Verse Summary of Mind and Mental Factors* by eighteenth-century Geluk scholar Yongzin Yeshe Gyaltsen.
115. Asaṅga, *Abhidharmasamuccaya*, chap. 1. Toh 4049, 45b, Pd 76:118.
116. Āryadeva, *Catuḥśataka*, 6.10. Toh 3846, 7b, Pd 57:796.
117. Asaṅga, *Abhidharmasamuccaya*, chap. 1. Toh 4049, 45b, Pd 76:119.
118. Asaṅga, *Abhidharmasamuccaya*, chap. 1. Toh 4049, 48a, Pd 76:125.
119. Asaṅga, *Viniścayasaṃgraha*, chap. 7. Toh 4038, zhi:59a, Pd 74:140.
120. Vasubandhu, *Abhidharmakośa*, 4.1. Toh 4089, 10b, Pd 79:24.
121. Asaṅga, *Abhidharmasamuccaya*, chap. 1. Toh 4049, 48b, Pd 76:125.
122. Asaṅga, *Abhidharmasamuccaya*, chap. 1. Toh 4049, 48b, Pd 76:125.
123. Jinamitra, *Yogācārabhūmivyākhyā*. Toh 4043, 77a, Pd 75:210.
124. Asaṅga, *Viniścayasaṃgraha*, chap. 7. Toh 4038, zhi:58b, Pd 74:138.
125. Asaṅga, *Viniścayasaṃgraha*, chap. 7. Toh 4038, zhi:59b, Pd 74:140.
126. Asaṅga, *Abhidharmasamuccaya*, chap. 1. Toh 4049, 48b, Pd 76:126.
127. Asaṅga, *Viniścayasaṃgraha*, chap. 7. Toh 4038, zhi:59a, Pd 74:140.
128. Asaṅga, *Abhidharmasamuccaya*, chap. 1. Toh 4049, 48b, Pd 76:126.
129. Asaṅga, *Viniścayasaṃgraha*, chap. 7. Toh 4038, zhi:59a, Pd 74:140.
130. Śāntideva, *Śikṣāsamuccayakārikā*, v. 8. Toh 3939, 2a, Pd 76:1006. The Tibetan term *gus pa* is usually translated as "respect," but here it has the connotation of taking something seriously rather than having an attitude of reverence, so "appreciation" is used instead.

131. Śāntideva, *Bodhisattvacaryāvatāra*, 5.30. Toh 3871, 11a, Pd 61:972.
132. Asaṅga, *Abhidharmasamuccaya*, chap. 1. Toh 4049, 48b, Pd 76:126.
133. This is not so according to Theravāda commentaries on Abhidhamma, which posit mindfulness to be a virtuous mental factor only (see page 186 below).
134. Asaṅga, *Viniścayasaṃgraha*, chap. 7. Toh 4038, zhi:59b, Pd 74:140.
135. Śāntideva, *Bodhisattvacaryāvatāra*, 5.23. Toh 3871, 11a, Pd 61:972.
136. *Gaganagañjaparipṛcchā-sūtra*, chap. 7. Toh 148, 313b, Pd 57:104.
137. Maitreya, *Mahāyānasūtrālaṃkāra*, 17.25. Toh 4020, 22a, Pd 70:152.
138. Asaṅga, *Abhidharmasamuccaya*, chap. 1. Toh 4049, 48b, Pd 76:126.
139. Asaṅga, *Viniścayasaṃgraha*, chap. 7. Toh 4038, zhi:59b, Pd 74:140.
140. *Mahāparinirvāṇa-sūtra*, chap. 11. Toh 119, 151a, Pd 53:353.
141. The term for learning literally means "hearing" or "listening," since the paradigmatic case of learning involves instruction through oral discourse.
142. Maitreya, *Uttaratantra*, 5.6. Toh 4025, 72a, Pd 70:976.
143. Asaṅga, *Abhidharmasamuccaya*, chap. 1. Toh 4049, 48b, Pd 76:126.
144. Vasubandhu, *Abhidharmakośabhāṣya*, chap. 1. Toh 4090, 27, Pd 79:67. The citation is from the commentary on *Abhidharmakośa* 1.2b.
145. *Sūtrālaṃkāravyākhyā*. Toh 4026, 165b, Pd 70:1220.
146. Asaṅga, *Viniścayasaṃgraha*, chap. 7. Toh 4038, zhi:59a, Pd 74:139.
147. One might find it easier to use the term *intelligence* here rather than *wisdom*. However, *wisdom* is preferable in most contexts.
148. *Bodhisattvagocara-sūtra*, i.e., *Sūtra Teaching the Miracles in the Range of a Bodhisattva's Skillful Means* (*Bodhisattvagocaropāyaviṣayavikurvāṇanirdeśa-sūtra*), chap. 2. Toh 146, 92b, Pd 57:244.
149. Asaṅga, *Abhidharmasamuccaya*, chap. 1. Toh 4049, 48b, Pd 76:126.
150. Asaṅga, *Abhidharmasamuccaya*, chap. 1. Toh 4049, 48b, Pd 76:126.
151. Asaṅga, *Abhidharmasamuccaya*, chap. 1. Toh 4049, 48b, Pd 76:126.
152. Asaṅga, *Abhidharmasamuccaya*, chap. 1. Toh 4049, 48b, Pd 76:126.
153. Asaṅga, *Abhidharmasamuccaya*, chap. 1. Toh 4049, 48b, Pd 76:126.
154. Asaṅga, *Abhidharmasamuccaya*, chap. 1. Toh 4049, 48b, Pd 76:126.
155. Asaṅga, *Abhidharmasamuccaya*, chap. 1. Toh 4049, 49a, Pd 76:127.
156. Asaṅga, *Abhidharmasamuccaya*, chap. 1. Toh 4049, 49a, Pd 76:127.
157. Asaṅga, *Abhidharmasamuccaya*, chap. 1. Toh 4049, 49a, Pd 76:127.
158. Asaṅga, *Abhidharmasamuccaya*, chap. 1. Toh 4049, 49a, Pd 76:127.
159. Asaṅga, *Abhidharmasamuccaya*, chap. 1. Toh 4049, 49a, Pd 76:127.
160. Asaṅga, *Śrāvakabhūmi*. Toh 4036, 144b, Pd 73:356.
161. Asaṅga, *Abhidharmasamuccaya*, chap. 1. Toh 4049, 49a, Pd 76:127.
162. Maitreya, *Abhisamayālaṃkāra*, chap. 1. Toh 3786, 3b, Pd 49:8.
163. Haribhadra, *Abhisamayālaṃkāravṛtti*, chap. 1. Toh 3793, 91b, Pd 52:237.
164. Śāntideva, *Bodhisattvacaryāvatāra*, 8.95–96. Toh 3871, 27a, Pd 61:1008.
165. Dharmakīrti, *Pramāṇavārttika*, 2.221. Toh 4210, 116a, Pd 97:520.
166. Śāntideva, *Bodhisattvacaryāvatāra*, 8.129. Toh 3871, 28b, Pd 61:1011.
167. *Sarvapuṇyasamuccayasamādhi-sūtra*, chap. 2. Toh 134, 119a, Pd 56:311.

168. *Akṣayamatinirdeśa-sūtra*, chap. 4. Toh 175, 131a, Pd 60:329.
169. Pṛthivībandhu, *Pañcaskandhabhāṣya*. Toh 4068, 56a, Pd 77:809.
170. *Saddharmasmṛtyupasthāna-sūtra*, chap. 22. Toh 287, sha:73a, Pd 71:168.
171. *Mañjuśrīvikrīḍita-sūtra*, chap. 2. Toh 96, 231b, Pd 46:628.
172. Candrakīrti, *Madhyamakāvatāra*, 6.205. Toh 3861, 214a, Pd 60:542.
173. Prajñākaramati, *Bodhicaryāvatārapañjikā*, chap. 6. Toh 3872, 112a, Pd 61:1218.
174. Śāntideva, *Bodhisattvacaryāvatāra*, 6.10. Toh 3871, 15a, Pd 61:981.
175. Or, rather than a quality needed to penetrate the nature of reality that is very difficult to face, this type of forbearance can be interpreted as the result of having understood how things are in reality, thus enabling one to develop a more stoic approach to life in general.
176. *Tathāgatamahākaruṇanirdeśa-sūtra*, chap. 2. Toh 147, 159b, Pd 57:418.
177. *Saddharmasmṛtyupasthāna-sūtra*, chap. 22. Toh 287, sha:45a, Pd 71:104.
178. Śāntideva, *Bodhisattvacaryāvatāra*, 6.2. Toh 3871, 14b, Pd 61:980.
179. Candrakīrti, *Madhyamakāvatāra*, 3.8. Toh 3861, 203a, Pd 60:517.
180. Śāntideva, *Bodhisattvacaryāvatāra*, 5.12–13. Toh 3871, 10b, Pd 61:971.
181. Candrakīrti, *Madhyamakāvatāra*, 3.4. Toh 3861, 203a, Pd 60:517.
182. Āryadeva, *Catuḥśataka*, 5.9. Toh 3846, 6a, Pd 57:793.
183. Candrakīrti, *Catuḥśatakaṭīkā*, chap. 5. Toh 3865, 96b, Pd 60:1163.
184. Śāntideva, *Bodhisattvacaryāvatāra*, 6.41. Toh 3871, 16a, Pd 61:983.
185. Śāntideva, *Bodhisattvacaryāvatāra*, 6.39–40. Toh 3871, 16a, Pd 61:983.
186. *Sarvapuṇyasamuccayasamādhi-sūtra*, chap. 2. Toh 134, 119a, Pd 56:311.
187. See Sthiramati, *Commentary on the Thirty Stanzas* (*Triṃśikābhāṣya*) on *Triṃśikā* verse 11c: *kaṃ ruṇaddhīti karuṇā / kam iti sukhasyākhyā sukhaṃ ruṇaddhīty arthaḥ /*.
188. Yaśomitra, *Abhidharmasamuccayabhāṣya*. Toh 4053, 126a, Pd 76:1286.
189. Vasubandhu, *Abhidharmakośabhāṣya*, chap. 8. Toh 4090, 77a, Pd 79:866. This is commenting on *Treasury of Knowledge* 8.29c.
190. *Pañcaviṃśatisāhasrikāprajñāpāramitā-sūtra*, chap. 38. Toh 9, 358a, Pd 27:793.
191. *Saddharmasmṛtyupasthāna-sūtra*, chap. 22. Toh 287, sha:50b, Pd 71:116.
192. *Samādhirāja-sūtra*, chap. 38. Toh 127, 150a, Pd 55:361.
193. Asaṅga, *Abhidharmasamuccaya*, chap. 2. Toh 4049, 78b, Pd 76:199.
194. Yaśomitra, *Abhidharmasamuccayabhāṣya*, chap. 2. Toh 4053, 187b, Pd 76:1438.
195. Asaṅga, *Abhidharmasamuccaya*, chap. 1. Toh 4049, 59a, Pd 76:127.
196. In the present volume, the term *anger* (*khong khro*; Skt., *krodha*; Pali, *kodha*) refers only to afflictive anger, which is very much like hatred in that it necessarily includes harmful intent.
197. Asaṅga, *Abhidharmasamuccaya*, chap. 1. Toh 4049, 59a, Pd 76:127.
198. Nāgārjuna, *Ratnāvalī*, 5.31. Toh 4158, 123a, Pd 96:327.
199. *Saddharmasmṛtyupasthāna-sūtra*, chap. 22. Toh 287, sha:44b, Pd 71:103.
200. Śāntideva, *Bodhisattvacaryāvatāra*, 6.3–6. Toh 3871, 14b, Pd 61:980.
201. Asaṅga, *Abhidharmasamuccaya*, chap. 1. Toh 4049, 49a, Pd 76:128.
202. Nāgārjuna, *Ratnāvalī*, 5.12. Toh 4158, 122b, Pd 96:326.

203. Asaṅga, *Abhidharmasamuccaya*, chap. 1. Toh 4049, 49a, Pd 76:128.
204. Asaṅga, *Viniścayasaṃgraha*, chap. 9. Toh 4038, zhi:84a, Pd 74:201.
205. Dharmakīrti, *Pramāṇavārttika*, 2.215. Toh 4210, 115b, Pd 97:519. The Tibetan version deviates slightly from the Sanskrit version, but the latter has been followed here. See Eltschinger 2009, 41. A translation following the Tibetan would read: "Because [ignorance] opposes wisdom, because it is cognitive due to being a mental factor, and because [the Buddha] said that ignorance is distorted cognition, any other explanation is incorrect."
206. Dharmakīrti, *Pramāṇavārttika*, 1.224. Toh 4210, 103a, Pd 97:489.
207. Candrakīrti, *Madhyamakāvatāra*, 6.120. Toh 3861, 210a, Pd 60:532.
208. Asaṅga, *Abhidharmasamuccaya*, chap. 1. Toh 4049, 49b, Pd 76:128.
209. The view of the perishable collection may be interpreted in at least two ways: either in accordance with the Prāsaṅgika-Madhyamaka system, which posits the *self* and *belonging to self* to be the direct object of that view, or in accordance with other systems such as the Yogācāra, which posits the aggregates to be the direct object of that view.
210. Asaṅga, *Abhidharmasamuccaya*, chap. 1. Toh 4049, 49b, Pd 76:128.
211. Asaṅga, *Abhidharmasamuccaya*, chap. 1. Toh 4049, 49b, Pd 76:128.
212. Asaṅga, *Abhidharmasamuccaya*, chap. 1. Toh 4049, 49b, Pd 76:129.
213. Asaṅga, *Abhidharmasamuccaya*, chap. 1. Toh 4049, 49b, Pd 76:128.
214. Asaṅga, *Abhidharmasamuccaya*, chap. 1. Toh 4049, 49b, Pd 76:128.
215. Asaṅga, *Abhidharmasamuccaya*, chap. 1. Toh 4049, 50b, Pd 76:130.
216. Asaṅga, *Abhidharmasamuccaya*, chap. 1. Toh 4049, 50b, Pd 76:130.
217. Asaṅga, *Abhidharmasamuccaya*, chap. 1. Toh 4049, 50b, Pd 76:130.
218. Asaṅga, *Abhidharmasamuccaya*, chap. 1. Toh 4049, 50b, Pd 76:131.
219. Asaṅga, *Abhidharmasamuccaya*, chap. 1. Toh 4049, 50b, Pd 76:131.
220. Asaṅga, *Abhidharmasamuccaya*, chap. 1. Toh 4049, 50b, Pd 76:131.
221. Asaṅga, *Abhidharmasamuccaya*, chap. 1. Toh 4049, 50b, Pd 76:131.
222. Asaṅga, *Abhidharmasamuccaya*, chap. 1. Toh 4049, 50b, Pd 76:131.
223. Asaṅga, *Abhidharmasamuccaya*, chap. 1. Toh 4049, 51a, Pd 76:131.
224. *Adhyāśayasañcodana-sūtra*, chap. 25. Toh 69, 144b, Pd 43:399.
225. Asaṅga, *Abhidharmasamuccaya*, chap. 1. Toh 4049, 51a, Pd 76:132.
226. Asaṅga, *Abhidharmasamuccaya*, chap. 1. Toh 4049, 51a, Pd 76:132.
227. Asaṅga, *Abhidharmasamuccaya*, chap. 1. Toh 4049, 51a, Pd 76:132.
228. Asaṅga, *Abhidharmasamuccaya*, chap. 1. Toh 4049, 51a, Pd 76:132.
229. Asaṅga, *Abhidharmasamuccaya*, chap. 1. Toh 4049, 51a, Pd 76:132.
230. Asaṅga, *Abhidharmasamuccaya*, chap. 1. Toh 4049, 51a, Pd 76:132.
231. *Saddharmasmṛtyupasthāna-sūtra*, chap. 4. Toh 287, ya:128a, Pd 68:340.
232. Asaṅga, *Abhidharmasamuccaya*, chap. 1. Toh 4049, 51a, Pd 76:132.
233. Śāntideva, *Bodhisattvacaryāvatāra*, 7.2. Toh 3871, 20a, Pd 61:992.
234. Asaṅga, *Abhidharmasamuccaya*, chap. 1. Toh 4049, 51a, Pd 76:132.
235. Asaṅga, *Abhidharmasamuccaya*, chap. 1. Toh 4049, 51a, Pd 76:133.
236. Asaṅga, *Abhidharmasamuccaya*, chap. 1. Toh 4049, 51b, Pd 76:133.

237. Asaṅga, *Abhidharmasamuccaya*, chap. 1. Toh 4049, 51b, Pd 76:133.
238. Asaṅga, *Abhidharmasamuccaya*, chap. 2. Toh 4049, 79a, Pd 76:199.
239. Asaṅga, *Abhidharmasamuccaya*, chap. 1. Toh 4049, 51b, Pd 76:134.
240. Yaśomitra, *Abhidharmasamuccayabhāṣya*, chap. 1. Toh 4053, 7b, Pd 76:971.
241. Asaṅga, *Abhidharmasamuccaya*, chap. 1. Toh 4049, 52a, Pd 76:134.
242. Yaśomitra, *Abhidharmasamuccayabhāṣya*, chap. 1. Toh 4053, 8a, Pd 76:972.
243. Asaṅga, *Abhidharmasamuccaya*, chap. 1. Toh 4049, 52a, Pd 76:134.
244. Asaṅga, *Yogācārabhūmi*. Toh 4035, 5b, Pd 72:681.
245. Vasubandhu, *Pañcaskandhaprakaraṇa*. Toh 4059, 12b, Pd 77:38.
246. Vasubandhu, *Abhidharmakośa*, 2.23. Toh 4089, 4b, Pd 79:10.
247. Vasubandhu, *Abhidharmakośa*, 2.24. Toh 4089, 4b, Pd 79:10. There are some differences between traditions. The *Treasury of Knowledge* presents ten omnipresent mental factors, which is the same as in the *Sūtra on the Application of Mindfulness*. As outlined above, the *Compendium of Knowledge* lists five omnipresent mental factors. The Theravāda text *Compendium of Abhidhamma* by Anuruddha adds *life force* and *single-pointedness* to those five, making a total of seven omnipresent mental factors.
248. Vasubandhu, *Abhidharmakośabhāṣya*, chap. 2. Toh 4090, 64b, Pd 79:161.
249. Vasubandhu, *Abhidharmakośa*, 2.25. Toh 4089, 5a, Pd 79:10.
250. Vasubandhu, *Abhidharmakośa*, 2.26. Toh 4089, 5a, Pd 79:10.
251. Vasubandhu, *Abhidharmakośa*, 2.26. Toh 4089, 5a, Pd 79:10.
252. Vasubandhu, *Abhidharmakośa*, 2.32. Toh 4089, 5a, Pd 79:11.
253. Vasubandhu, *Abhidharmakośa*, 2.27. Toh 4089, 5a, Pd 79:10.
254. Pūrṇavardhana, *Abhidharmakośaṭīkā-lakṣaṇānusāriṇī*, chap. 2. Toh 4093, cu:141b, Pd 81:355.
255. Vasubandhu, *Abhidharmakośa*, 2.33. Toh 4089, 5a, Pd 79:11.
256. For further discussion about the different types of meditative absorption, see chapter 24.
257. Yaśomitra, *Abhidharmakośavyākhyā-sphuṭārthā*, 5. Toh 4092, 135a, Pd 80:1160.
258. "Fifty-eight mental factors are listed in the *Mahāvibhāṣa* in seven major groups, a grouping parallel to that found in the *Dhātukāya*." Potter 1996, 115.
259. These groupings of mental factors based on the *Dhātukāya*, one of the seven early treatises on Abhidharma (see note 113 above), are based on the summary of the text found in Potter 1996, 347–50.
260. Although many of the mental factors listed here appear to be descriptions of verbal or physical behavior, they should be understood to indicate the mental attitudes underlying or motivating such behavior. Some of the behaviors listed here appear in sections discussing conduct, which can be found in commentaries on early Buddhist sources such as the *Path of Purification* (*Visuddhimagga*) of Buddhaghosa (1991, 19): "Abstinence from such wrong livelihood as entails transgression of the six training precepts announced with respect to livelihood and entails the evil states beginning with 'scheming, talking, hinting,

belittling, pursuing gain with gain' (Majjhima Nikāya ii, 75) is *virtue of livelihood purification.*"

261. *Mnyam par bzhag pa'i sa*; Skt., *Samāhitābhūmi*. This is the sixth chapter of Asaṅga's *Levels of Yogic Deeds* (*Yogācārabhūmi*). The five hindrances (Skt., *nivaraṇa*) indicated here are (1) sensual desire, (2) hatred, (3) sloth and torpor, (4) restlessness and remorse, and (5) doubt.
262. Asaṅga, *Vastusaṃgraha*, part 2. Toh 4039, 193b, Pd 74:1224.
263. Asaṅga, *Vastusaṃgraha*, part 1. Toh 4039, 175a, Pd 74:1179.
264. Asaṅga, *Viniścayasaṃgraha*, chap. 22. Toh 4038, zi:199a, Pd 74:482.
265. Pṛthivībandhu, *Pañcaskandhabhāṣya*. Toh 4068, 69b, Pd 77:842.
266. Asaṅga, *Viniścayasaṃgraha*, chap. 7. Toh 4038, zhi:60b, Pd 74:143.
267. Asaṅga, *Viniścayasaṃgraha*, chap. 7. Toh 4038, zhi:62b, Pd 74:147.
268. Asaṅga, *Viniścayasaṃgraha*, chap. 7. Toh 4038, zhi:65a, Pd 74:154.
269. Vasubandhu, *Abhidharmakośa*, 8.29. Toh 4089, 24b, Pd 79:57.
270. This analysis of how to posit the substantial and imputed mental factors is in accordance with Chim Jampaiyang's *Ornament of Abhidharma Commentary* (2009), 116.
271. This is the same category as the ten extensive mental factors according to the *Treasury of Knowledge*; only the name of the category is different.
272. *Saddharmasmṛtyupasthāna-sūtra*, chap. 19. Toh 287, ya:302b, Pd 68:736.
273. *Saddharmasmṛtyupasthāna-sūtra*, chap. 4. Toh 287, ya:129b, Pd 68:343.
274. *Saddharmasmṛtyupasthāna-sūtra*, chap. 41. Toh 287, ra:244b, Pd 69:575.
275. *Saddharmasmṛtyupasthāna-sūtra*, chap. 41. Toh 287, ra:245a, Pd 69:576. Here in *Sūtra on the Application of Mindfulness*, these mental factors are called "extensive afflictions" (*nyon mongs pa'i sa chen po*); we should investigate whether they are called "extensive" (*chen po*) on account of accompanying all mental afflictions, as posited in the *Treasury of Knowledge*.
276. *Saddharmasmṛtyupasthāna-sūtra*, chap. 41. Toh 287, ra:245a, Pd 69:576.
277. The Tibetan text has *ston pa*, literally meaning "indication" or "revelation," which is most probably a scribal error. It should be *mi ston pa* ("nonrevelation"), with a negative particle *mi* as a prefix, and represents a variant translation of the same Sanskrit term (*mrakṣa*) translated elsewhere as "concealment."
278. *Saddharmasmṛtyupasthāna-sūtra*, chap. 41. Toh 287, ra:245b, Pd 69:577.
279. *Saddharmasmṛtyupasthāna-sūtra*, chap. 41. Toh 287, ra:245b, Pd 69:577.
280. *Saddharmasmṛtyupasthāna-sūtra*, chap. 41. Toh 287, ra:247b, Pd 69:582.
281. *Saddharmasmṛtyupasthāna-sūtra*, chap. 41. Toh 287, ra:247b, Pd 69:582.
282. *Saddharmasmṛtyupasthāna-sūtra*, chap. 41. Toh 287, ra:245a, Pd 69:576.
283. *Saddharmasmṛtyupasthāna-sūtra*, chap. 41. Toh 287, ra:244a, Pd 69:573.
284. The English translation here is based mostly on the reading of the Tibetan translation of this Pali text by Sempa Dorje (1988) but follows as closely as possible the translation from Pali by Wijeratne and Gethin (2007). Where the terminology differs, it can be correlated quite easily with Gethin's translation.

It may be helpful to note that what we translate here as "mental factors" is translated by Gethin as "mentalities."

285. Sempa Dorje 1988, 178. There is no numbering in the source text; numbering is provided here for the reader to easily identify the implied numbers.
286. Sempa Dorje 1988, 116.
287. Sempa Dorje 1988, 117.
288. Sempa Dorje 1988, 137.
289. The Tibetan term here is *chags pa dang bral*, which means "free from attachment." However, this Tibetan term seems to be based on a scribal error. It translates the Pali term *viratta*, which means "free from passion," but the Pali text reads *virati*, which means "abstinence." This could well have been an error present in the particular Pali text from which the Tibetan translation was made.
290. Sempa Dorje 1988, 153.
291. Sempa Dorje 1988, 171.
292. Sempa Dorje 1988, 174.
293. Nāgārjuna, *Ratnāvalī*, 5.2. Toh 4158, 122a, Pd 96:325.
294. The term *hypocrisy* (Tib., *tshul 'chos pa*) is used for the sixteenth and the thirty-third, though the descriptions differ.
295. The *Dharmasaṃgraha* is not included in the Tibetan canonical collection of treatises known as the Tengyur. It was later translated into Tibetan and into English.
296. Candrakīrti, *Pañcaskandhaprakaraṇa*. Toh 3866, 245a, Pd 60:1549.
297. Candrakīrti, *Pañcaskandhaprakaraṇa*. Toh 3866, 255b, Pd 60:1573.
298. Candrakīrti, *Pañcaskandhaprakaraṇa*. Toh 3866, 255b, Pd 60:1573.
299. Candrakīrti, *Pañcaskandhaprakaraṇa*. Toh 3866, 255b, Pd 60:1574.
300. Candrakīrti, *Pañcaskandhaprakaraṇa*. Toh 3866, 256a, Pd 60:1575.
301. Candrakīrti, *Pañcaskandhaprakaraṇa*. Toh 3866, 256a, Pd 60:1575.
302. Candrakīrti, *Pañcaskandhaprakaraṇa*. Toh 3866, 256a, Pd 60:1575.
303. Candrakīrti, *Pañcaskandhaprakaraṇa*. Toh 3866, 256a, Pd 60:1576.
304. Candrakīrti, *Pañcaskandhaprakaraṇa*. Toh 3866, 256b, Pd 60:1576.
305. The Buddha is commonly accepted to have taught the four noble truths: the truth of suffering, that of its cause, that of its cessation, and that of the path leading to its cessation. The first noble truth has three levels: the suffering of suffering, the suffering of change, and pervasive suffering.
306. Candrakīrti, *Pañcaskandhaprakaraṇa*. Toh 3866, 259a, Pd 60:1582.
307. Candrakīrti, *Pañcaskandhaprakaraṇa*. Toh 3866, 259a, Pd 60:1582.
308. Candrakīrti, *Pañcaskandhaprakaraṇa*. Toh 3866, 261b, Pd 60:1588.
309. Candrakīrti, *Pañcaskandhaprakaraṇa*. Toh 3866, 262a, Pd 60:1590.
310. Candrakīrti, *Pañcaskandhaprakaraṇa*. Toh 3866, 263a, Pd 60:1591.
311. Candrakīrti, *Pañcaskandhaprakaraṇa*. Toh 3866, 263a, Pd 60:1592.
312. Candrakīrti, *Pañcaskandhaprakaraṇa*. Toh 3866, 263a, Pd 60:1592.
313. Candrakīrti, *Pañcaskandhaprakaraṇa*. Toh 3866, 263a, Pd 60:1592.

314. Candrakīrti, *Pañcaskandhaprakaraṇa*. Toh 3866, 263b, Pd 60:1593.
315. Candrakīrti, *Pañcaskandhaprakaraṇa*. Toh 3866, 264a, Pd 60:1594.
316. Candrakīrti, *Pañcaskandhaprakaraṇa*. Toh 3866, 264a, Pd 60:1594.
317. Candrakīrti, *Pañcaskandhaprakaraṇa*. Toh 3866, 265a, Pd 60:1597.
318. As mentioned in the section discussing the *Sūtra on the Application of Mindfulness*, the Tibetan term *ston pa* seems to be a scribal error in this context—it appears to be missing a negative prefix, so we translate it as "nonrevelation" rather than as "revelation."
319. Literally, the Tibetan (*rlung*) and Sanskrit (*vāyu*) means "wind" or "air." In Buddhist tantric systems, this term refers to a type of flowing energy that constitutes and sustains the function of mind and body. See the discussion in chapter 16 below.
320. Pūrṇavardhana, *Abhidharmakośaṭīkā-lakṣaṇānusāriṇī*, chap. 1. Toh 4093, cu:51b, Pd 81:126. The texts within quotes are citations from Vasubandhu's *Treasury of Knowledge Autocommentary*, where it is commenting on the root text, 1.22.
321. Pūrṇavardhana, *Abhidharmakośaṭīkā-lakṣaṇānusāriṇī*, chap. 3. Toh 4093, cu:327a, Pd 81:811. The first text in quotes is a citation from *Treasury of Knowledge* 3.42d, and the second is a citation from the autocommentary on that verse.
322. Yaśomitra, *Abhidharmakośavyākhyā-sphuṭārthā*, chap. 1. Toh 4092, 39a, Pd 80:93. This is commenting on *Treasury of Knowledge* 1.20ab.
323. Asaṅga, *Viniścayasaṃgraha*, chap. 33. Toh 4038, zi:6a, Pd 74:758.
324. Vasubandhu, *Abhidharmakośa*, 3.42. Toh 4089, 8b, Pd 79:18.
325. Vasubandhu, *Abhidharmakośabhāṣya*, chap. 2. Toh 4090, 60b, Pd 79:150. The text in quotes is from Vasubandhu, *Treasury of Knowledge*, 2.16ab.
326. Asaṅga, *Yogācārabhūmi*. Toh 4035, 8a, Pd 72:687.
327. Vasubandhu, *Abhidharmakośabhāṣya*, chap. 8. Toh 4090, khu:79b, Pd 79:871. The text in quotes is from *Treasury of Knowledge* 8.33b.
328. Nāropa, *Pañcakramasaṅgrahaprakāśa*, v. 2. Toh 2333, 276a, Pd 26:1747. This verse was cited in volume 1, page 382, in the chapter on the subtle body.
329. Nāgārjuna, *Pañcakrama*, 1.3. Toh 1802, 45a, Pd 18:129.
330. Āryadeva, *Caryāmelāpakapradīpa*, chap. 4. Toh 1803, 79a, Pd 18:215.
331. Āryadeva, *Caryāmelāpakapradīpa*, chap. 4. Toh 1803, 79a, Pd 18:216.
332. *Vajramālā-tantra*. Toh 445, 276b, Pd 81:915.
333. Āryadeva, *Caryāmelāpakapradīpa*, chap. 3. Toh 1803, 70b, Pd 18:195.
334. Nāgabodhi, *Pañcakramārthabhāskaraṇa*, 1. Toh 1833, 211a, Pd 18:1333.
335. Nāgārjuna, *Pañcakrama*, 2.7–11. Toh 1802, 48a, Pd 18:137.
336. Nāgārjuna, *Pañcakrama*, 2.16–20. Toh 1802, 48a, Pd 18:137.
337. Nāgārjuna, *Pañcakrama*, 2.24–25. Toh 1802, 49b, Pd 18:137.
338. Nāgārjuna, *Pañcakrama*, 2.31. Toh 1802, 49a, Pd 18:139.
339. Āryadeva, *Caryāmelāpakapradīpa*, chap. 4. Toh 1803, 77b, Pd 18:212.
340. Āryadeva, *Caryāmelāpakapradīpa*, chap. 4. Toh 1803, 77b, Pd 18:212.

341. Āryadeva, *Caryāmelāpakapradīpa*, chap. 4. Toh 1803, 78a, Pd 18:213.
342. Āryadeva, *Svādhiṣṭhānakramaprabheda*, vv. 20–21. Toh 1805, 112b, Pd 18:308.
343. *Caturdevipariprcchā*, 1.18. Toh 446, 278a, Pd 81:949.
344. See Wallace 2010, on which this translation is partially based.
345. Puṇḍarīka, *Vimalaprabhā-kālacakratantraṭīkā*, 1. Toh 1347, 32b, Pd 6:757.
346. *Guhyasamāja-uttaratantra*. Toh 443, 154b, Pd 81:599.
347. Nāgabodhi, *Samājasādhanavyavasthāli*, chap. 4. Toh 1809, 130a, Pd 18:358.
348. Nāgabodhi, *Samājasādhanavyavasthāli*, chap. 4. Toh 1809, 130, Pd 18:359.
349. Nāgabodhi, *Samājasādhanavyavasthāli*, chap. 4. Toh 1809, 130b, Pd 18:359.
350. Nāgabodhi, *Samājasādhanavyavasthāli*, chap. 4. Toh 1809, 130b, Pd 18:359.
351. *Guhyasamāja-vajrajñānasammucaya*. Toh 447, 282b, Pd 81:963.
352. Buddhaśrījñāna, *Dvikramatattvabhāvanā*. Toh 1853, 15a, Pd 21:861.
353. This genre is also called "types of cognition" (Tib., *blo rigs*), where the second syllable exhibits an alternative spelling in Tibetan.
354. Contemporary researchers on Buddhist epistemology attribute this distinction between a given cognition's *observed object* versus its *engaged object* to Dharmottara.
355. Among Tibetan epistemologists, Sakya Paṇḍita and his followers did not construct a theory of implicit objects of cognition. The Kadampas Chapa Chökyi Sengé and Chomden Rikral, along with Tsongkhapa and his Gelukpa heirs, assert that cognitions can realize implicitly, based on their interpretation of Indian epistemologists such as Dharmottara, Śāntarakṣita, and Kamalaśīla.
356. Vasubandhu, *Abhidharmakośa*, 1.42. Toh 4089, 3b, Pd 79:7.
357. Yaśomitra, *Abhidharmakośavyākhyā-sphuṭārthā*, 1. Toh 4092, 74b, Pd 80:177.
358. Jitāri, *Sugatamatavibhaṅga*, v. 2. Toh 3899, 7b, Pd 63:884.
359. This statement appears slightly differently in the root-text verse cited above.
360. Jitāri, *Sugatamatavibhaṅgabhāṣya*. Toh 3900, 38b, Pd 63:961.
361. Bodhibhadra, *Jñānasārasamuccayanibandhana*. Toh 3852, 42a, Pd 57:891.
362. Bodhibhadra, *Jñānasārasamuccayanibandhana*. Toh 3852, 42b, Pd 57:892.
363. Śāntarakṣita, *Madhyamakālaṃkāra*, v. 19. Toh 3884, 53b, Pd 62:897.
364. Śāntarakṣita, *Madhyamakālaṃkāra*, v. 20. Toh 3884, 53b, Pd 62:897.
365. Dharmakīrti, *Pramāṇavārttikasvopajñavṛtti*, chap. 1. Toh 4216, 275b, Pd 97:934. Commenting on *Exposition of Valid Cognition* 1.46–47.
366. Dharmakīrti, *Pramāṇavārttikasvopajñavṛtti*, chap. 1. Toh 4216, 275a, Pd 97:932.
367. Dharmakīrti, *Pramāṇavārttika*, 1.43. Toh 4210, 96a, Pd 97:473.
368. Candrakīrti, *Prasannapadā*, chap. 1. Toh 3860, 25b, Pd 60:59. Candrakīrti is commenting here on the well-known sūtra citation "Someone with eye consciousness cognizes blue, but does not think 'This is blue,'" and suggests that the sūtra passage characterizes sensory cognitions as dumber than mental conceptual cognitions.
369. Dharmakīrti, *Pramāṇavārttika*, 1.50. Toh 4210, 96b, Pd 97:474.
370. Dharmakīrti, *Pramāṇavārttika*, 1.45. Toh 4210, 96b, Pd 97:473.

371. Dharmakīrti, *Pramāṇavārttika*, 1.48. Toh 4210, 96b, Pd 97:474.
372. Dharmakīrti, *Pramāṇavārttika*, 1.65ab. Toh 4210, 96b, Pd 97:475.
373. Chapa Chökyi Sengé 2006, 7.
374. The "three modes" refer to the threefold criteria of a correct inference. This topic is presented in part 5 below.
375. His Holiness the Dalai Lama has said that explaining the seven types of mind in this way provides an easy introduction.
376. Dharmakīrti, *Pramāṇavārttika*, 4.99cd. Toh 4210, 143a, Pd 97:583.
377. Jñānagarbha, *Satyadvayavibhaṅga*, v. 28. Toh 3881, 2b, Pd 62:757.
378. *Laṅkāvatāra-sūtra*, chap. 6. Toh 107, 254a, Pd 49:654.
379. Asaṅga, *Viniścayasaṃgraha*, chap. 11. Toh 4038, zhi:110a, Pd 74:264.
380. Śāntarakṣita, *Tattvasaṃgraha*, 23.3–4. Toh 4266, 54b, Pd 102:135. This has been translated from the Tibetan, but the Sanskrit differs slightly. Although Chapa's presentation of a mind of correct assumption is refuted by Sakya Paṇḍita in his *Treasure of the Science of Valid Cognition* (*Tshad ma rigs pa'i gter*), the logicians of Ngok Loden Sherab's Kadampa tradition, as well as Tsongkhapa's spiritual heirs, explicitly accept the notion of correct assumption.
381. The Tibetan editors of *Science and Philosophy in the Indian Buddhist Classics* described the two stanzas cited from Śāntarakṣita's *Tattvasaṃgraha* as presenting a Sāṃkhya assertion about valid cognition. But as Kamalaśīla makes clear in his commentary and as is indicated by the reference to "eternal unproduced speech," the stanzas refer to Mīmāṃsā assertions about the nature of speech-derived valid cognition. The English translation adopts the latter position here.
382. The Tibetan editors emphasize this point about a possible Indian source for the notion of correct assumption with the awareness that some Tibetan thinkers, especially the influential Sakya Paṇḍita, do not accept the notion to be coherent.
383. Dignāga, *Pramāṇasamuccaya*, 2.1. Toh 4203, 4a, Pd 97:8.
384. Dharmakīrti, *Pramāṇavārttika*, 4.48ab. Toh 4210, 141a, Pd 97:578.
385. Dharmakīrti, *Pramāṇavārttika*, 4.118cd. Toh 4210, 144a, Pd 97:585.
386. Dharmakīrti, *Pramāṇavārttika*, 1.214ab. Toh 4210, 102b, Pd 97:488.
387. Dharmakīrti, *Pramāṇavārttika*, 2.3ab. Toh 4210, 107b, Pd 97:500.
388. Dharmottara, *Pramāṇaviniścayaṭīkā*, chap. 1. Toh 4229, 9b, Pd 104:779. Among the Tibetan logicians, Sakya Paṇḍita maintains that perception is necessarily valid cognition. As stated in the perception chapter of his *Treasure of the Science of Valid Cognition*, "Because of being nonmistaken, it is just valid cognition," he takes perception to be a valid cognition by reason of being nonmistaken. He also asserts that subsequent cognition must be conceptual recollection, and he does not accept that perception can include indeterminate perception. This kind of explanation also occurs in Khedrub Je's *Ornament of the Seven Treatises Clearing Mental Obscurity* (*Tshad ma sde bdun gyi rgyan yid kyi mun sel*); however, the logicians of Ngok's tradition, as well as the Gelukpa

epistemologist Gyaltsab Je and his followers, assert that perception includes subsequent cognition and indeterminate perception.

389. Dignāga, *Pramāṇasamuccayavṛtti*, chap. 1. Toh 4204, 15b, Pd 97:60. The so-called *Abhidharma Sūtra* either is no longer extant or is no longer known by that name.
390. Dharmakīrti, *Nyāyabindu*, chap. 1. Toh 4212, 231a, Pd 97:812.
391. A reflexive awareness of a thought, for example, is a perception. Yet it is not free of the substance of thought because the perceived and the perceiver here are the same substance.
392. Dignāga, *Pramāṇasamuccayavṛtti*, chap. 1. Toh 4204, 15b, Pd 97:60.
393. Devendrabuddhi, *Pramāṇavārttikapañjikā*, chap. 3. Toh 4217, 254b, Pd 98:618.
394. Dharmakīrti, *Pramāṇavārttika*, 3.281. Toh 4210, 129a, Pd 97:551.
395. Dharmakīrti, *Pramāṇaviniścaya*, chap. 1. Toh 4211, 167a, Pd 97:658. In responding to the objection against the Cittamātra view that all cognitions involving subject-object duality are deluded, Dharmakīrti draws a distinction between those cognitions that are affected by immediate conditions of illusion versus those affected only by deeply ingrained, stable, latent potencies. On this basis, Dharmakīrti states, a distinction between what is an erroneous cognition versus what is a veridical cognition can be maintained without a problem within the context of everyday convention.
396. Dharmakīrti, *Pramāṇavārttika*, 2:133. Toh 4210, 111b, Pd 97:510.
397. This verse, which His Holiness also quotes at the outset of his introduction, is quoted by Tsongkhapa at the beginning of his *Essential Explanation of the Provisional and the Definitive* (*Drang nges legs bshad snying po*). On its source, see note 36 in volume 1 of this series.
398. *Saṃdhinirmocana-sūtra*. Toh 106, 51a, Pd 49:119.
399. Asaṅga, *Abhidharmasamuccaya*, chap. 2. Toh 4049, 103a, Pd 76:258.
400. Asaṅga, *Śrāvakabhūmi*. Toh 4036, 57b, Pd 73:138.
401. Asaṅga, *Śrāvakabhūmi*. Toh 4036, 57b, Pd 73:139.
402. Asaṅga, *Śrāvakabhūmi*. Toh 4036, 58a, Pd 73:139.
403. Asaṅga, *Śrāvakabhūmi*. Toh 4036, 58a, Pd 73:140.
404. Asaṅga, *Śrāvakabhūmi*. Toh 4036, 135a, Pd 73:331.
405. Three types of valid cognition are mentioned here because the passage from *Śrāvaka Levels* cited above adds scripture as an additional means of valid cognition. The Dharmakīrtian account of valid cognition followed elsewhere in this volume accepts only perception and inference as valid cognition and then subsumes scriptural inference under the rubric of inference itself.
406. This treatise is not included in the Dergé or the Coné editions of the Tengyur, but it is included in both the Peking and the Narthang editions. Dignāga, *Nyāyapraveśaprakaraṇa*. P5706, 183b, Pd 97:442.
407. *Vaidalyasūtra*. Toh 3826, 22b, Pd 57:59.
408. In Tibetan, the term *rgol ba* (Skt., *vādin*) refers to a disputant in the kind of

debate that involves the inferential reasoning being discussed here. In general, one disputant holds a position that is being supported through inferential arguments that are meant to lead the other disputant to understand and adopt that position. The one who is being convinced in this fashion is often called the "opponent" (*pha rol po*). For the sake of simplicity and clarity, the word *opponent* has also been used to translate a related term, namely, the *phyi rgol* (Skt., *uttaravādin*); this is the disputant who literally "speaks later" in response to the arguments laid out by the "proponent" (*snga rgol*; Skt., *pūrvavādin*), or the "one who speaks first." Additionally, it is important to note that there is some potential for confusion in regard to the related terms "former position" (*phyogs snga ma*; Skt. *pūrvapakṣa*) and "latter position" (*phyogs phyi ma*; Skt., *uttarapakṣa*). These terms are used to identify the parts of an argument in a text, where the author lays out the "former position" as the object of the critique presented from the perspective of the "latter position," which is the philosophical position that the author seeks to defend. While the same kinds of inferential argument are used in the context of critiquing a "former position" to establish a "latter position" in a text, these terms do not map directly onto the notion of a "proponent" and "opponent" in a debate.

409. The term *pervasion* (*khyab pa*; Skt., *vyāpti*) is a metaphor that expresses the invariable concomitance that pertains between two properties. For example, to say that the quality of being "produced" is pervaded by the quality of being "impermanent" is to say that anything that is produced is necessarily also impermanent.
410. See volume 1 of this series, pages 65–67. Volume 1 uses the term "reason" instead of "evidence."
411. Mokṣākaragupta, *Tarkabhāṣā*, chap. 4. Toh 4264, 358a, Pd 106:986.
412. When a consequence is put forward, as in "some x is has the predicate-property P because it has the evidence-property E," there is a *pervasion* or logical relation between the property [E] attributed to the subject [x] and another property [P] that is attributed to the subject as a consequence of having the property [E]. The logical relation between [P] and [E] is the pervasion, and on this account, the consequence is convincing when there is some valid logical relation between [P] and [E]. According to the system presented here, there are eight ways in which [P] and [E] can stand in a valid relation.
413. Although this is logically identical to the second of these eight, they are different in terms of which term is being incorrectly negated. In the second, the correct evidence (e.g., "produced") for proving the predicate (e.g., " impermanent") has been replaced with its negation ("not produced"), and in the eighth case, the correct predicate ("impermanent") has been replaced with its negation ("permanent").
414. There is a very famous and extensive Collected Topics text known as *Rato Collected Topics* (*Ra stod bsdur grwa*) by Jamyang Choklha Öser ('Jam dbyang Phyogs lha 'od zer). There is also a suitably condensed text that amalgamates

the points of all three paths of reasoning—small, medium, and great—known as *Yongzin Collected Topics* (*Yongs 'dzin bsdus grwa*) by Je Jampa Gyatso (Rje Byams pa rgya mtsho). A succinct account of the history of the development of the Collected Topics literature in Tibet can be found in the introduction to *Selected Texts on Geluk Epistemology* (*Dpal dge ldan pa'i tshad ma rig pa'i gzhung*), volume 21 of the Bod kyi gtsug lag gces btus series being published in English as *The Library of Tibetan Classics.*

415. Mokṣākaragupta, *Tarkabhāṣā.* Toh 4264, 356b, Pd 106:958.
416. Dignāga, *Pramāṇasamuccaya*, 3.1. Toh 4203, 6a, Pd 97:13.
417. Dharmakīrti, *Pramāṇavārttika*, 4.1. Toh 4210, 139a, Pd 97:574.
418. Dignāga, *Nyāyapraveśa.* Toh 4208, 89a, Pd 97:454.
419. Dharmakīrti, *Pramāṇaviniścaya*, chap. 2. Toh 4211, 184a, Pd 97:688.
420. Dharmakīrti, *Pramāṇavārttika*, 1.27–28. Toh 4210, 95b, Pd 97:472.
421. Dharmakīrti, *Pramāṇavārttika*, 4:20. Toh 4210, 95b, Pd 97:576.
422. Dignāga, *Pramāṇasamuccaya*, 3.2ab. Toh 4203, 6a, Pd 97:13. A Sanskrit version and English translation of these lines is provided by Tillemans 1999, 86: *svarūpeṇaiva nirdeśyaḥ svayaṃ iṣṭo 'nirākṛtaḥ //.*
423. Devendrabuddhi, *Pramāṇavārttikapañjikā*, chap. 4. Toh 4217, 275a, Pd 98:668.
424. Devendrabuddhi, *Pramāṇavārttikapañjikā*, chap. 4. Toh 4217, 275a, Pd 98:668.
425. Devendrabuddhi, *Pramāṇavārttikapañjikā*, chap. 4. Toh 4217, 275a, Pd 98:668.
426. Devendrabuddhi, *Pramāṇavārttikapañjikā*, chap. 4. Toh 4217, 275b, Pd 98:668.
427. Devendrabuddhi, *Pramāṇavārttikapañjikā*, chap. 4. Toh 4217, 276b, Pd 98:669.
428. "Unaccompanied nonoccurrence" (Skt., *avinābhāva*) refers to the kind of relation that must hold between the evidence and the predicate to be inferred. Specifically, it points to the fact that the quality or thing that is being put forward as the evidence cannot occur in the absence of the predicate to be inferred. Thus there is necessarily no arising or occurrence of the evidence without the predicate, and as a result, if the evidence is present in some locus, the predicate to be inferred must also necessarily be present in that locus. See discussion below on pages 316–17.
429. Dharmakīrti, *Pramāṇavārttika*, 1.1. Toh 4210, 94a, Pd 97:469.
430. This list of synonyms reflects the complex historical process of translating terms from Sanskrit into Tibetan. From the perspective of the original Sanskrit, the word *liṅga* (literally, a "mark" or "sign") is the most straightforward term for something that is serving as evidence in the type of inferential reasoning discussed here; it has been translated into English simply as "evidence," and in Tibetan it is translated by the single word *rtags.* In Sanskrit, the synonyms for "evidence" include *sādhana* or "proof" (Tib., *sgrub byed*) and *gamaka* or

"cause of understanding" (*shes byed*). Each of these is translated consistently with one Tibetan term for each, as noted. However, two other common synonyms for evidence in Sanskrit, *hetu* and *kāraṇa*, are *both* translated inconsistently by two Tibetan terms *gtan tshigs* ("inferential evidence") and *rgyu mtshan* ("reason"); in other words, either of these terms could be referring to either *hetu* or *kāraṇa* in the original Sanskrit. Finally the latter Tibetan term (*rgyu mtshan*) additionally translates another Sanskrit term, *nimitta*, that can also be used to refer to the evidence in an inference. This means that, unlike much of the other technical vocabulary in this volume, there is not a one-to-one correspondence between the original Sanskrit synonyms for "evidence" and their Tibetan translations, but as is pointed out here, these terms are fully synonymous, and the lack of correspondence is thus not problematic in this context.

431. Dignāga, *Nyāyapraveśaprakaraṇa*, chap. 1. P5706, 183b, Pd 97:442.
432. Dharmakīrti, *Pramāṇavārttikasvopajñavṛtti*, on *Pramāṇavārttika* 1. Toh 4216, 266b, Pd 97:913.
433. Dharmakīrti, *Pramāṇavārttika*, 1.24–25. Toh 4210, 95b, Pd 97:471.
434. Dharmakīrti, *Nyāyabindu*, chap. 2. Toh 4212, 232a, Pd 97:813.
435. *Daśadharmaka-sūtra*, chap. 1. Toh 53, 167b, Pd 40:469.
436. Dharmakīrti, *Pramāṇavārttika*, 1.2ab. Toh 4210, 95a, Pd 97:469.
437. Dharmakīrti, *Pramāṇavārttika*, 2.36. Toh 4210, 108b, Pd 97:503.
438. Dharmakīrti, *Pramāṇavārttika*, 2.180ab. Toh 4210, 114b, Pd 97:516.
439. Dharmakīrti, *Pramāṇavārttika*, 3.390d–391b. Toh 4210, 133a, Pd 97:560.
440. Dharmakīrti, *Pramāṇavārttika*, 1.9. Toh 4210, 95a, Pd 97:470.
441. *Kanakavarṇapūrvayoga*. Toh 350, 53a, Pd 76.151.
442. Dharmakīrti, *Pramāṇavārttika*, 1.2. Toh 4210, 94a, Pd 97:470.
443. Dharmakīrti, *Pramāṇavārttika*, 1.186. Toh 4210, 101b, Pd 97:486.
444. Correct evidence consisting in nonperception is simply the negative form of the other two types of evidence—correct effect-evidence and correct nature-evidence.
445. *Sarvadharmapravṛttinirdeśa-sūtra*. Toh 180, 275a, Pd 60:734.
446. Dharmakīrti, *Pramāṇavārttika*, 3.85. Toh 4210, 121b, Pd 97:533. Here Dharmakīrti is making the point that if something's existence is established by it being perceived via a valid cognition—i.e., a direct perception or inference—then if a thing does not exist, a valid cognition of it will not occur. Thus, the non-occurrence of any valid cognition in relation to some putative object is evidence for the nonexistence of that object.
447. Dharmakīrti, *Pramāṇavārttika*, 1.3ab. Toh 4210, 94a, Pd 97:470. In these two lines Dharmakīrti explains that there are contexts where the nonperception of a specific thing by a person can serve as evidence not for the actual existence or nonexistence of the object under discussion itself but rather for the absence of any cognitive engagement by that person toward that object, since that

object is effectively nonexistent from the standpoint of that person's cognitive capacities.

448. Dharmakīrti, *Pramāṇavārttika*, 1.3c–4b. Toh 4210, 95a, Pd 97:470. Here Dharmakīrti lists four main categories of this second type of nonperception evidence, whereby what is being negated is in principle perceptible. The first two function by demonstrating that something that is directly contradictory to the negandum or an effect of this contradictory thing can be perceived. In contrast, the latter two function by demonstrating the nonperception of a fact that is closely related to what is being negated, such as its cause or something that is an essential property of the negandum. Dharmakīrti in fact presents several different classifications of this type of nonperception evidence. As cited here, he presents four main types of such evidence in *Exposition of Valid Cognition*. But in his *Ascertainment of Valid Cognition*, he presents a classification of ten, while in *Drop of Reasoning*, he enumerates a list of eleven. Later Buddhist commentators of Dharmakīrti introduced further elaborations on the list such that, by the time of Mokṣākaragupta's *Language of Logic* (Kajiyama 1998, 81–86), the list includes sixteen types of this kind of nonperception evidence.
449. The phrase "immediate, unobstructed cause" (*dngos rgyu nus pa thogs med*) refers to a particular causal context in which a causal complex will produce its effect in the very next moment and thus cannot be obstructed from producing that effect.
450. Dharmakīrti, *Pramāṇavārttika*, 3.95. Toh 4210, 122a, Pd 97:534. The second reasoning referred to here is presented against the Sāṃkhya view that effects already exist in their causes. The refutation takes the following form: "Rice sprouts, for instance, do not exist prior to their arising for they come into existence from their causes." Here prior existence is being refuted by the evidence that they are produced by their causes.
451. Tangible cold is the cause of the presence of bristling hair (which is the predicate of the negandum).
452. Dharmakīrti, *Pramāṇavārttika*, 1.6. Toh 4210, 94a, Pd 97:470.
453. Dharmakīrti, *Pramāṇavārttikasvopajñavṛtti*, chap. 1. Toh 4216, 262a, Pd 97:902.
454. Dharmottara, *Pramāṇaviniścayaṭīkā*, chap. 1. Toh 4227, 29b, Pd 104:67.
455. Dharmakīrti, *Pramāṇavārttika*, 4.116–17. Toh 4210, 143b, Pd 97:584.
456. Asvabhāva, *Mahāyānasaṃgrahopanibandhana*, chap. 3. Toh 4051, 230b, Pd 76:600.
457. Dharmottara, *Pramāṇaviniścayaṭīkā*, chap. 1. Toh 4227, 60b, Pd 104:143.
458. Dharmakīrti, *Pramāṇavārttika*, 1.214/215. Toh 4210, 102b, Pd 97:488.
459. Dharmakīrti, *Pramāṇavārttika*, 1.215/216. Toh 4210, 102b, Pd 97:488.
460. Dharmakīrti, *Pramāṇavārttika*, 1.1. Toh 4210, 1a, Pd 97:469.
461. Literally, "subtle living beings" (*srog chags phra mo*).
462. Dignāga, *Hetucakraḍamaru*, v. 5. Toh 4209, 93a, Pd 97:465.
463. Dignāga, *Hetucakraḍamaru*, v. 5. Toh 4209, 93a, Pd 97:466.

464. Dignāga, *Hetucakraḍamaru*, vv. 9–11. Toh 4209, 93a, Pd 97:466.
465. Dignāga, *Hetucakraḍamaru*, vv. 3–4. Toh 4209, 93a, Pd 97:465.
466. Dignāga, *Hetucakraḍamaru*, vv. 4–5. Toh 4209, 93a, Pd 97:465.
467. There is an excellent commentary on Dignāga's *Drum of a Wheel of Reasons* by Alaksha Tendar Lharampa (2011) entitled *Precious Lamp* (*Rin chen sgron me*). This commentary provides three different tables detailing a wheel that is in accordance with reality, a wheel of concordant and discordant examples, and, additionally, the wheel that is laid out in this chart. Readers are encouraged to refer to this text.
468. Dharmakīrti, *Pramāṇavārttika*, 4.196. Toh 4210, 147a, Pd 97:592. The term "unique" here refers to the unique proof statement where the defining characteristic of the subject is stated as the evidence, for example: "Sound is impermanent because it is audible." The term "general" refers to the proof statement where an overly general quality of the subject is stated as the evidence, for example: "Sound is impermanent because it is an object of knowledge."
469. Asaṅga, *Yogācārabhūmi*. Toh 4035, 83a, Pd 72:870.
470. Asaṅga, *Abhidharmasamuccaya*, chap. 2. Toh 4049, 78b, Pd 76:199.
471. Vasubandhu, *Abhidharmakośabhāṣya*, chap. 5. Toh 4090, 246a, Pd 79:603. The passage is quoting *Treasury of Knowledge* 5.34.
472. Nāgārjuna, *Mūlamadhyamakakārikā*, 18.5bc. Toh 3824, 11a, Pd 57:26.
473. Vasubandhu, *Abhidharmakośa*, 3.28cd. Toh 4089, 8b, Pd 79:17.
474. Vasubandhu, *Abhidharmakośabhāṣya*. Toh 4090, 131b, Pd 79:324.
475. *Tathāgatācintyaguhyanirdeśa-sūtra*, chap. 15. Toh 47, 161b, Pd 39:439.
476. Nāgārjuna, *Mūlamadhyamakakārikā*, 18.5. Toh 3824, 11a; Pd 57:26.
477. Āryadeva, *Catuḥśataka*, 6:10. Toh 3846, 7b, Pd 57:796.
478. Dharmakīrti, *Pramāṇavārttika*, 2.218. Toh 4210, 115b, Pd 97:519.
479. Dharmakīrti, *Pramāṇavārttika*, 2.220. Toh 4210, 116a, Pd 97:520.
480. Candrakīrti, *Madhyamakāvatārabhāṣya*. Toh 3862, 292b, Pd 60:774. This passage is commenting on *Entering the Middle Way* 6.120.
481. Vasubandhu, *Abhidharmakośa*, 5.32cd–33. Toh 4089, 17a, Pd 79:39.
482. Vasubandhu, *Abhidharmakośa*, 5.48–51ab. Toh 4089, 17b, Pd 79:40.
483. This distinction between aspiration-type (*'dun pa'i rigs*) and intelligence-type (*shes rab kyi rigs*) mental states roughly parallels the contemporary cognitive science distinction between affective and cognitive mental states.
484. Āryadeva, *Catuḥśataka*, 14.25. Toh 3846, 16a, Pd 57:815.
485. Dharmakīrti, *Pramāṇavārttika*, 1.49. Toh 4210, 96b, Pd 97:474.
486. Dharmakīrti, *Pramāṇavārttika*, 2.223. Toh 4210, 116a, Pd 97:520.
487. Dharmakīrti, *Pramāṇavārttika*, 2.213. Toh 4210, 115b, Pd 97:519.
488. Śāntideva, *Bodhisattvacaryāvatāra*, 6.7–8ab. Toh 3871, 14b, Pd 61:980.
489. Āryaśūra, *Jātakamālā*. Toh 4150, 122a, Pd 94:292.
490. Vasubandhu, *Abhidharmakośabhāṣya*, chap. 6. Toh 4090, 8a, Pd 79:697. This passage is commenting on *Treasury of Knowledge* 6.5cd.
491. Vasubandhu, *Sūtrālaṃkāravyākhyā*, chap. 19. Toh 4026, 227a, Pd 70:1371.

492. *Lalitavistara-sūtra*, chap. 18. Toh 95, 129b, Pd 46:313.
493. Śāntideva, *Bodhisattvacaryāvatāra*, 5.25. Toh 3871, 11a, Pd 61:972.
494. Śāntideva, *Bodhisattvacaryāvatāra*, 5.24. Toh 3871, 11a, Pd 61:972.
495. Nāgārjuna, *Suhṛllekha*, v. 54. Toh 4182, 43b, Pd 61:675.
496. Śāntideva, *Bodhisattvacaryāvatāra*, 5.19. Toh 3871, 11a, Pd 61:971.
497. Śāntideva, *Bodhisattvacaryāvatāra*, 7.69. Toh 3871, 23a, Pd 61:999. This is verse 68 in the Sanskrit edition.
498. Śāntideva, *Bodhisattvacaryāvatāra*, 5.34. Toh 3871, 11b, Pd 61:973.
499. Śāntideva, *Bodhisattvacaryāvatāra*, 5.108. Toh 3871, 14b, Pd 61:979.
500. Śāntideva, *Bodhisattvacaryāvatāra*, 7.71. Toh 3871, 23a, Pd 61:999. This is verse 70 in the Sanskrit edition.
501. Dharmamitra, *Prasphuṭapadā*, chap. 6. Toh 3796, 85b, Pd 52:107.
502. Asaṅga, *Śrāvakabhūmi*. Toh 4036, 75a, Pd 73:183.
503. For an explanation of these four types of objects of meditation see Sopa and Rochard 2017, 547–50.
504. Contemporary psychologists speak of the "big five": (1) openness, (2) conscientiousness, (3) extroversion, (4) agreeableness, (5) neuroticism. Psychologists give detailed explanations about the differences among these and say that there is no need to consider them to be temperaments of different people, in that one person can have many temperaments.
505. Asaṅga, *Śrāvakabhūmi*. Toh 4036, 71a, Pd 73:173.
506. Asaṅga, *Śrāvakabhūmi*. Toh 4036, 71b, Pd 73:174.
507. Asaṅga, *Śrāvakabhūmi*. Toh 4036, 72a, Pd 73:175.
508. Asaṅga, *Śrāvakabhūmi*. Toh 4036, 72b, Pd 73:176.
509. Asaṅga, *Śrāvakabhūmi*. Toh 4036, 72b, Pd 73:177.
510. Asaṅga, *Śrāvakabhūmi*. Toh 4036, 71a, Pd 73:173.
511. In his *Path of Purification* (*Visuddhimagga* 3.75, in Buddhaghosa 1991, 101–2), Buddhaghosa presents an alternative theory of personalities where he lists only six kinds of temperament. He identifies these as (1) greedy, (2) hating, (3) deluded, (4) faithful, (5) intelligent, and (6) speculative. Buddhaghosa draws parallels between faithful-greedy, intelligent-hating, and speculative-deluded temperaments, and correlates these temperaments of persons with the relative prominence of certain elements, or the lack thereof, as well as with the relative strength of the humors such as phlegm.
512. Kamalaśīla, *Bhāvanākrama (2)*. Toh 3916, 45b, Pd 64:128.
513. Often translated as the "seven-point posture of Vairocana," this traditional instruction pertains to the postures of (1) one's legs, (2) hands, (3) back, (4) shoulders, (5) head and neck, (6) mouth, and (7) eyes. In some sources, an eighth—breath—is added.
514. *Vajramālā-tantra*, 6.53–54. Toh 445, 218a, Pd 39:439.
515. Kamalaśīla, *Bhāvanākrama (2)*. Toh 3916, 46b, Pd 64:131.
516. The vajra-fist gesture is a fist with the thumb tucked inside.
517. Āryaśūra, *Pāramitāsamāsa*, 5.12. Toh 3944, 229a, Pd 64:1610.

518. Atiśa, *Bodhipathapradīpa*, 43. Toh 3947, 240a, Pd 64:1645.
519. Kamalaśīla, *Bhāvanākrama (1)*. Toh 3195, 31a, Pd 64:11.
520. *Saṃdhinirmocana-sūtra*, chap. 8. Toh 106, 26b, Pd 49:61.
521. Maitreya, *Madhyāntavibhāga*, 4.3–4. Toh 4021, 43a, Pd 70:908.
522. Maitreya, *Madhyāntavibhāga*, 4.5. Toh 4021, 43a, Pd 70:909.
523. Maitreya, *Madhyāntavibhāga*, 4.5–6. Toh 4021, 43a, Pd 70:909.
524. Śāntideva, *Bodhisattvacaryāvatāra*, 5.3. Toh 3871, 10a, Pd 61:970.
525. Śāntideva, *Bodhisattvacaryāvatāra*, 5.33. Toh 3871, 11b, Pd 61:973.
526. Śīlapālita, *Āgamakṣudrakavyākhyāna*, 88. Toh 4115, 170a, Pd 11:409. (*Āgamakṣudraka* is an alternate name for the *Vinayakṣudrakavastu*.)
527. *Mahāparinirvāṇa-sūtra*, chap. 7. Toh 119, 272b, Pd 42:622.
528. Kamalaśīla, *Bhāvanākrama (2)*. Toh 3916, 47b, Pd 64:134.
529. Vasubandhu, *Abhidharmakośabhāṣya*, chap. 2. Toh 4090, 65, Pd 79:164. This passage is commenting on *Treasury of Knowledge* 2.26ac.
530. *Saṃdhinirmocana-sūtra*, chap. 8. Toh 106, 35b, Pd 49:82.
531. Candragomin, *Deśanāstava*, v. 30. Toh 1159, 205b, Pd 1:602.
532. Asaṅga, *Yogācārabhūmi*. Toh 4035, 134a, Pd 72:990.
533. Asaṅga, *Yogācārabhūmi*. Toh 4035, 134a, Pd 72:990.
534. Maitreya, *Mahāyānasūtrālaṃkāra*, 15.11. Toh 4020, 19a, Pd 70:845. For a presentation of the remaining lines of this portion of the text see Tsongkhapa 2000, 73–75.
535. Asaṅga, *Śrāvakabhūmi*. Toh 4036, 132b, Pd 73:325.
536. *Mahāśūnyatā-mahāsūtra*. Toh 291, 255b, Pd 71:681.
537. *Saṃdhinirmocana-sūtra*, chap. 8. Toh 106, 26b, Pd 49:61.
538. Maitreya, *Mahāyānasūtrālaṃkāra*, 15.15. Toh 4020, 19a, Pd 70:845.
539. Kamalaśīla, *Bhāvanākrama (2)*. Toh 3916, 48a, Pd 64:134.
540. Asaṅga, *Abhidharmasamuccaya*, chap. 1. Toh 4049, 49a, Pd 76:127.
541. Sthiramati, *Triṃśikābhāṣya*. Toh 4064, 156b, Pd 77:414.
542. However, although it does not appear in most lists of mental factors, it is listed as a virtuous mental factor in Anuruddha's *Compendium of Abhidhamma* (see page 186 above).
543. Sthiramati, *Triṃśikābhāṣya*. Toh 4064, 156b, Pd 77:414.
544. Asaṅga, *Śrāvakabhūmi*. Toh 4036, 163a, Pd 73:403.
545. Asaṅga, *Śrāvakabhūmi*. Toh 4036, 163a, Pd 73:403.
546. This entire sequence of the arising of physical and mental pliancy as well as their two corresponding joys is drawn from Tsongkhapa's *Great Treatise on the Stages of the Path to Enlightenment* (*Lamrim Chenmo*). See Tsongkhapa 2002, 80–84.
547. Asaṅga, *Śrāvakabhūmi*. Toh 4036, 163a, Pd 73:404.
548. Āryaśūra, *Pāramitāsamāsa*, 5.11. Toh 3944, 163a, Pd 73:404.
549. Śāntideva, *Bodhisattvacaryāvatāra*, 8.4. Toh 3871, 23b, Pd 61:1000.
550. *Saṃdhinirmocana-sūtra*, chap. 8. Toh 106, 26b, Pd 49:62.
551. Ratnākaraśānti, *Prajñāpāramitopadeśa*. Toh 4079, 128b, Pd 78:339.

552. Kamalaśīla, *Bhāvanākrama (2)*. Toh 3916, 45a, Pd 64:127.
553. Kamalaśīla, *Bhāvanākrama (3)*. Toh 3917, 55b, Pd 64:158.
554. Vimalamitra, *Kramaprāveśikabhāvanārtha*. Toh 3938, 343b, Pd 64:963.
555. *Brahmaviśeṣacintiparipṛcchā-sūtra*. Toh 160, 42a, Pd 59:108.
556. Kamalaśīla, *Bhāvanākrama (3)*. Toh 3917, 64a, Pd 64:178.
557. Kamalaśīla, *Bhāvanākrama (3)*. Toh 3917, 61b, Pd 64:172.
558. Kamalaśīla, *Bhāvanākrama (3)*. Toh 3917, 62b, Pd 64:174.
559. *Mañjuśrīvikrīḍita-sūtra*, chap. 2. Toh 96, 231b, Pd 46:624.
560. Kamalaśīla, *Bhāvanākrama (3)*. Toh 3917, 63a, Pd 64:176.
561. Maitreya, *Mahāyānasūtrālaṃkāra*, 2.10. Toh 4020, 2b, Pd 70:807.
562. Kamalaśīla, *Bhāvanākrama (3)*. Toh 3917, 64a, Pd 64:178 (also cited above).
563. Vimalamitra, *Kramapraveśikabhāvanārtha*. Toh 3938, 340a, Pd 64:978.
564. Kamalaśīla, *Bhāvanākrama (2)*. Toh 3916, 49b, Pd 64:139.
565. Kamalaśīla, *Bhāvanākrama (1)*. Toh 3915, 34b, Pd 64:96.
566. *Satipaṭṭhāna Sutta*, Majjhima Nikāya 10.2–3. As translated by Bodhi 2005, 145, with slight modification.
567. *Satipaṭṭhāna Sutta*, Majjhima Nikāya 10.31. Bodhi 2005, 149, with slight modification.
568. Asaṅga, *Abhidharmasamuccaya*, 2. Toh 4049, 96b, Pd 76:242. *Śrāvaka Levels* says: "The way to engage in the application of mindfulness on the mind is to focus on twenty aspects: the mind with attachment, without attachment, with anger, without anger, and so on." *Śrāvakabhūmi*. Toh 4036, 108a, Pd 73:265.
569. *Ratnacūḍaparipṛcchā-sūtra*, chap. 2. Toh 91, 230a, Pd 44:638.
570. Asaṅga, *Abhidharmasamuccaya*, 2. Toh 4049, 96b, Pd 76:242.
571. Vasubandhu, *Abhidharmakośa*, 6.14. Toh 4089, 19a, Pd 79:44.
572. Vasubandhu, *Abhidharmakośabhāṣya*, chap. 6. Toh 4090, 12a, Pd 79:707.
573. Asaṅga, *Abhidharmasamuccaya*, chap. 2. Toh 4049, 96b, Pd 76:244.
574. Vasubandhu, *Abhidharmakośa*, 6.14. Toh 4089, 19a, Pd 79:44 (also cited above).
575. *Vinayottaragrantha*, chap. 51. Toh 7, 290b, Pd 13:681.
576. These colors seem to correspond to those of a corpse in various states of decay.
577. *Aṣṭādaśasāhasrikāprajñāpāramitā-sūtra*, chap. 16. Toh 9, 157b, Pd 29:377.
578. Vasubandhu, *Abhidharmakośa*, 6:10–11. Toh 4089, 19a, Pd 79:43.
579. Vasubandhu, *Abhidharmakośabhāṣya*, chap. 6. Toh 4090, 11a, Pd 79:705. This verse summarizes the commentary on *Abhidharmakośa* 6.12.
580. Vasubandhu, *Abhidharmakośabhāṣya*, chap. 6. Toh 4090, 9a, Pd 79:700.
581. *Ratnacūḍaparipṛcchā-sūtra*, chap. 2. Toh 91, 226a, Pd 44:627.
582. Vasubandhu, *Abhidharmakośa*, 5.45. Toh 4089, 17b, Pd 79:44. As Asaṅga clarifies in the *Compendium of Knowledge* (*Abhidharmasamuccaya*, 80b5): "The three poisons are 'bindings' (Skt., *bandhana*) because they bind one to the three types of suffering."
583. *Kāśyapaparivarta-sūtra*, chap. 2. Toh 87, 139a, Pd 44:378.
584. *Kāśyapaparivarta-sūtra*, chap. 2. Toh 87, 139b, Pd 44:379.

585. *Ratnacūḍaparipṛcchā-sūtra*, chap. 2. Toh 91, 229a, Pd 44:635.
586. Asaṅga, *Abhidharmasamuccaya*, 2. Toh 4049, 96b, Pd 76:243.
587. *Ratnacūḍaparipṛcchā-sūtra*, chap. 2. Toh 91, 224a, Pd 44:623.
588. Yaśomitra, *Abhidharmasamuccayabhāṣya*, chap. 2. Toh 4053, 62b, Pd 76:1108.
589. Śāntideva, *Bodhisattvacaryāvatāra*, 8.12–13. Toh 3871, 23b, Pd 61:1001.
590. *Ratnacūḍaparipṛcchā-sūtra*, chap. 1. Toh 91, 153a, Pd 44:428.
591. Nāgārjuna, *Suhṛllekha*, v. 29. Toh 4182, 42a, Pd 96:672.
592. Śāntideva, *Bodhisattvacaryāvatāra*, 6.57–58. Toh 3871, 16b, Pd 61:985.
593. Śāntideva, *Bodhisattvacaryāvatāra*, 8.79. Toh 3871, 26b, Pd 61:1007.
594. Śāntideva, *Bodhisattvacaryāvatāra*, 8.123. Toh 3871, 28a, Pd 61:1011.
595. That is, the craving itself limits one's choices and restricts one's freedom, much like an ox tethered by a nose ring.
596. Śāntideva, *Bodhisattvacaryāvatāra*, 8.88c. Toh 3871, 27a, Pd 61:1008.
597. *Vīradattagṛhapatiparipṛcchā-sūtra*. Toh 72, 200b, Pd 43:567.
598. Śāntideva, *Bodhisattvacaryāvatāra*, 8.23. Toh 3871, 24a, Pd 61:1002.
599. Śāntideva, *Bodhisattvacaryāvatāra*, 6.92. Toh 3871, 18a, Pd 61:988.
600. Mātṛceṭa, *Mahārājakaniṣkalekha*, v. 24. Toh 4184, 54a, Pd 96:709.
601. Śāntideva, *Bodhisattvacaryāvatāra*, 6.53. Toh 3871, 16b, Pd 61:984.
602. Mātṛceṭa, *Mahārājakaniṣkalekha*, v. 22. Toh 4184, 54a, Pd 96:709.
603. Śāntideva, *Bodhisattvacaryāvatāra*, 6.91cd. Toh 3871, 18a, Pd 61:988.
604. Śāntideva, *Bodhisattvacaryāvatāra*, 7.64. Toh 3871, 22b, Pd 61:998. This verse is enumerated as number 63 in the Sanskrit edition because the Tibetan version includes a verse 62 that does not appear in the Sanskrit.
605. Śāntideva, *Bodhisattvacaryāvatāra*, 6.10. Toh 3871, 15a, Pd 61:981.
606. Śāntideva, *Bodhisattvacaryāvatāra*, 6.90–91ab. Toh 3871, 18a, Pd 61:988.
607. Śrī Jaganmitrānanda, *Candrarājalekha*, v. 30. Toh 4189, 72b, Pd 96:767.
608. Śāntideva, *Bodhisattvacaryāvatāra*, 8.21. Toh 3871, 24a, Pd 61:1001.
609. *Udānavarga*, 25.23. Toh 326, 232a, Pd 72:656.
610. Dharmakīrti, *Pramāṇavārttika*, 2.74. Toh 4210, 110a, Pd 97:506.
611. Dharmakīrti, *Pramāṇavārttika*, 2.124–27. Toh 4210, 112a, Pd 97:511.
612. Dharmakīrti, *Pramāṇavārttika*, 1.220. Toh 4210, 103a, Pd 97:489.
613. Nāgārjuna, *Dharmadhātustava*, vv. 20–21. Toh 1118, 64b, Pd 1:180.
614. Śāntideva, *Bodhisattvacaryāvatāra*, 6.14. Toh 3871, 15a, Pd 61:981.
615. *Connected Discourses* (*Saṃyutta Nikāya*), I.5 (*Bhikkhunī Saṃyutta*); cf. Bodhi 2000, 230. A parallel quotation is cited in Candrakīrti's *Entering the Middle Way Autocommentary* (*Madhyamakāvatārabhāṣya*), where he says it is "from a Śrāvaka scripture" but does not name it. Toh 3862, 296b, Pd 60:792.
616. *Akṣayamatinirdeśa-sūtra*, chap. 6. Toh 175, 150b, Pd 60:375.
617. Asaṅga, *Paryāyasaṃgraha*. Toh 4041, 31b, Pd 75:81.

Glossary

absorption (*bsam gtan*; *dhyāna*). *See* meditative absorption.

afflictions (*nyon mongs*; *kleśa*). *Also* emotional afflictions, mental afflictions. Nonvirtuous mental states, such as attachment, hatred, and ignorance, which motivate contaminated actions that cause one to continue to be born in cyclic existence.

aggregates (*phung po*; *skandha*). *See* aggregates of appropriation.

aggregates of appropriation (*nyer len gyi phung po*; *upādānaskandha*). The five appropriated aggregates are the psychophysical constituents that function as the basis of imputation of a conventionally existent person or self. They include the body, the primary consciousness, various mental factors, and all the other causal factors within the continuum of a person, traditionally listed as the five of form, feeling, discernment, conditioning factors, and consciousness. They are what is grasped or appropriated in the continual round of rebirth, which occurs as a result of contaminated actions motivated by mental afflictions, especially grasping. Therefore the aggregates themselves, as well as the rebirth taken, are in the nature of suffering.

analysis (*dpyod pa*, *yongs su dpyod pa*; *vicāra*). This is a mind of analytical wisdom that mentally dissects its object of meditation, which may be either conventional or ultimate reality, in a subtle and precise way. It is more refined than inquiry.

analytical meditation (*dpyad sgom*). The practitioner investigates the aspects and qualities of the object of meditation by analyzing it into its parts or by searching for it in relation to its parts, thereby arriving at its conventional or ultimate nature. Any analytical meditation, whether conceptual or nonconceptual, is included within the category of insight meditation (*vipaśyanā*).

appearing object (*snang yul*; *pratibhāsaviṣaya*). This refers to what appears to a consciousness apprehending an object. In general, to a directly perceiving awareness, the object manifestly appears; to a conceptual awareness, a mental image of the object appears mixed with an appearance of the object itself.

argument (*sbyor ba*; *prayoga*). This is a positively stated argument that has the following form: the *subject* is the *predicate* because it is the *evidence*, as in the *homologous example*. For an argument to be correct it must be characterized by the three modes. However, it need not be an autonomous inference, where the three modes must be known by both parties using a commonly accepted notion of validity.

attribute of the subject (*phyogs chos*; *pakṣadharmatā*). This is the first element of the tri-modal criterion of a valid argument, where the evidence is an attribute of the subject. More precisely, it refers to the ascertainment by valid cognition that the evidence is present in the subject of the thesis sought to be proven. *See also* argument; three modes.

aspect (*rnam pa*; *ākāra*). Image, feature, way of apprehending. In the case of a sense consciousness, this refers to the image that is transferred onto that consciousness by its objective condition. And in the case of any conceptual or nonconceptual consciousness, it refers to that consciousness's way of holding its object (i.e., its object as cognized). It may also refer to any feature of an object or subject of awareness.

awareness (*rig pa*; *saṃvitti*). *Also* intelligence, knowledge. Awareness is stated to be the nature of cognition and the function of consciousness. It is treated as synonymous with cognition and consciousness.

bodhisattva (*byang chub sems pa*). A person who has entered the Mahāyāna path. Such a being (*sattva*) has a continuous, spontaneous wish to attain enlightenment (*bodhi*) in order to bring other sentient beings to the state of enlightenment.

calm abiding (*zhi gnas*; *śamatha*). A special type of single-pointed meditative concentration. Calm abiding arises as a result of having progressively developed nine specific levels of concentration. It is accompanied by bodily and mental pliancy and their concurrent forms of bliss. It is united with special insight so as to subdue or remove the mental afflictions.

Cārvāka (*rgyang 'phen pa*). The Cārvākas are non-Buddhist materialists who do not believe in past and future lives or in liberation. The Tibetan translation of this name suggests that in denying causality and liberation, a follower of this school casts the possibility of enlightenment and higher rebirth far away.

causal relation (*de byung 'brel*; *tadutpattisaṃbandha*). *Also* cause-and-effect relation. This is a relation that involves impermanent things only: any cause and its effect. It is a one-way relation, not mutual. An effect is related to its cause—not the other way around: smoke is causally dependent upon fire, but the reverse is not true. *See also* intrinsic relation; relation.

cause of error (*'khrul pa'i rgyu*; *bhrāntihetu*). In terms of location, there are two types: external and internal causes of error. In terms of subtlety, there are two types: temporary and deep causes of error. A deep cause of error is only ever internal. *See also* temporary cause of error.

classificatory convention (*tha snyad*; *vyavahāra*). This refers to a term as it continues to be used after its initial establishment as a linguistic convention. *See also* linguistic convention.

cognition (*blo*; *buddhi*). *Also* mind, intellect. The nature of cognition is awareness. It is treated as synonymous with awareness and consciousness.

common locus (*gzhi mthun*; *sāmānyādhikāranya*). This is where two particular attributes or classes of object are instantiated in one thing, which means that they are not contradictory.

compassion (*snying rje*; *karuṇā*). A mind wishing beings to be free from suffering.

conceived object (*zhen yul*). A conceived object is an object of thought only. Thought conceives something, and the object that it holds is the conceived object. A thought's conceived object is a mental image that represents its focal object held in a certain way—rightly or wrongly, depending on whether that thought is correct or incorrect.

conceptual cognition (*rtog pa'i shes pa*; *savikalpakajñāna*). *Also* conceptual mind, conceptual consciousness. A conceptual cognition does not arise due to the causal capacity of an observed object nearby. Instead, it arises due to previous habituation and to latent potencies of language-based thought processes. *See also* construing awareness.

conceptual identity (*ldog pa*; *vyāvṛtti*). *Also* distinguisher, abstract entity. The term *ldog pa* indicates an exclusion of that which does not have the expected causal capacities. It may be used in combination with another word to refer, for example, to a thing itself (*rang ldog*) or to a meaning itself (*don ldog*), i.e., a definition.

conceptualization of inappropriate attention (*tshul bzhin ma yin pa'i rnam rtog*; *amanasikārakalpanā*). There are four fundamental conceptualizations of inappropriate attention: holding impure things as pure, holding things in the nature of suffering as pleasurable, holding impermanent things as unchanging, and holding selfless things as a self or as belonging to a self. *See also* inappropriate attention.

conceptually distinct (*ldog pa tha dad*). *Also* different isolates. Conceptually distinguishable. For example, a product and an impermanent thing are inseparable in reality and therefore one individual entity, yet they are conceptually distinct because for a conceptual consciousness the one term does not evoke the other. *See also* different conceptual identities.

condition (*rkyen*; *pratyaya*). Conditions are impermanent factors that function as causes or assist causes in producing their effects. There are four types of conditions listed as giving rise to a sense consciousness: objective condition—what is perceived; dominant condition—a sense faculty; immediately preceding condition—a previous moment of consciousness; and causal condition—any other conditions required, such as a body, karma, or proximity.

conditioned thing (*'dus byas*; *saṃskṛta*). *Also* conditioned phenomena. Whatever is produced by causes and conditions. Also, an object perceived by a sense consciousness.

conditioning action (*'du byed*; *saṃskāra*). The second of the twelve links of dependent arising.

conditioning factors (*'du byed*; *saṃskāra*). *Also* formative factors. The fourth of the five aggregates constituting the basis of imputation of a person. There are conditioning factors concomitant with the mind, such as volition, and conditioning factors not concomitant with the mind, referring to those factors not included in either form or mental phenomena.

confusion (*rmongs pa*; *moha*). *See* ignorance.

consciousness (*shes pa*, *rnam par shes pa*; *jñāna*, *vijñāna*). Consciousness is that which knows its object, or that which knows differentiated aspects of its object. The nature of consciousness is said to be clear (or luminous) and aware. It is treated as synonymous with awareness and cognition. It is also treated as synonymous with mind and mentality.

consequence (*thal 'gyur*; *prasaṅga*). *Also* logical consequence, consequential reasoning. A logical consequence is a formally structured argument designed to uproot a wrong view. It usually has the form "If *A* then *B*" and, correspondingly, "If ~*B* then ~*A*." A logical consequence can be valid without the two parties in a debate having a commonly accepted valid knowledge of its components. There are two types of consequences: affirming consequences and refuting consequences.

construing awareness (*zhen rig*). This is a cognition that apprehends word and referent as suitable to be associated. As such, it defines conceptual cognition.

contradictory (*'gal ba*; *viruddha*). *Also* mutual exclusion. Two things are contradictory if nothing instantiates both; in other words, there is no common locus or shared basis. *See also* directly contradictory; contradictory in the sense of being mutually exclusive; contradictory in the sense of being contrary.

contradictory in the sense of being mutually exclusive (*phan tshun spangs te gnas pa'i brgyud 'gal; parasparaparihārasthitivirodha*). This includes cases that are *directly* contradictory as well those that are *indirectly* contradictory in a strict sense. In each case, being both things is impossible, and being neither is impossible. There is no third ground or possibility at all, whether positive or negative. The difference between them is as follows. *Directly* contradictory in the sense of being mutually exclusive requires that the two are contradictory in terms of reality and that the terminology itself shows this, such as *permanent* and *impermanent*. *Indirectly* contradictory in the sense of being mutually exclusive requires that the two are contradictory in terms of reality but not in terms of terminology or understanding, such as *permanent* and *produced*. If one mentally or verbally cuts out *permanent*, this does not mean one will naturally understand *produced*. Or if one cuts out *produced*, then one does not necessarily understand *permanent*. Therefore they are not directly contradictory according to Buddhist logic. *See also* directly contradictory.

contradictory in the sense of being contrary (*lan cig mi gnas 'gal*; *sahabhāvavirodha*). This is a less strict way of being indirectly contradictory, and in Western philosophy it is described in terms of being "contrary." In such a case, there are no instances that are both, yet there are instances that are neither. An example of this is *circular* and *square*. Whatever is circular cannot be square, and whatever is square cannot be circular. There is nothing that is both. But there is a common locus of their negations: something that is neither circular nor square, such as something that is triangular. Therefore, circular and square are *contraries* but not contradictories.

cooperative condition (*lhan cig byed rkyen*; *sahakāripratyaya*). Although not the primary producer of the object, it functions as a complementary condition to help the substantial cause produce its effects.

correct assumption (*yid dpyod*; *manaḥparīkṣā*). This is defined as construing awareness that with conviction conceives the main thing that is its engaged object but does not obtain a realization of its object of scrutiny.

correct effect-evidence (*'bras rtags yang dag*; *saṃyakkāryahetu*). This is evidence based on a causal relationship; it proves the existence of the cause based on the existence of its effect. For example, "On a smoky mountain pass there is fire, because there is smoke."

correct evidence consisting in nonperception (*ma dmigs pa'i rtags yang dag*; *saṃyaganupalabdhihetu*). This is simply the negative form of the other two types of evidence—correct effect-evidence and correct nature-evidence. It proves the nonexistence or negation of one thing based upon the nonexistence or negation of another, which are related either causally or in terms of being the same nature.

correct nature-evidence (*rang bzhin gyi rtags yang dag*; *saṃyaksvabhāvahetu*). This is evidence based on an intrinsic relationship; it proves one thing on the basis of another that is of the same nature as it. For example, "Sound is impermanent because it is produced."

delusion (*gti mug*; *moha*). *See* ignorance.

denial (*skur 'debs*; *apavāda*). Denigration, deprecation, repudiation, undermining. This refers to the implicit or explicit denial of the existence of something that actually exists. It is the opposite of *reification*.

dependent arising (*rten 'brel*; *pratītyasamutpāda*). According to varying degrees of subtlety, the term *dependent* may mean: dependent on causes, dependent on parts, or dependent upon imputation. A prime example of causal dependence would be the twelve links of dependent arising. The other two types will be discussed more extensively in the philosophy volumes of *Science and Philosophy in the Indian Buddhist Classics*.

different conceptual identities (*ldog pa tha dad*, *ldog pa so so ba*; *bhinnavyāvṛtti*). Things that have different conceptual identities are distinguished through the concepts applied to them, even though they may be the same individual entity. For example, one speaks of "an atom's impermanence"; the quality "impermanence" is conceptually different from the atom, but the atom and its impermanence are actually the same individual entity. *See also* conceptually distinct.

direct perception (*mngon sum*; *pratyakṣa*). A mind that perceives its object directly, or free of conceptuality. Free of any temporary causes of error, it is valid. *See also* valid cognition.

directly contradictory (*dngos 'gal*). This is synonymous with *directly contradictory in the sense of being mutually exclusive* (*phan tshun spangs te gnas pa'i dngos *'gal; parasparaparihārasthitivirodha*). Two things are directly contradictory if the terminology itself shows that it is impossible to be both and impossible to be neither. This means that by negating one term, you affirm the other. If two things are contradictory in terms of reality but not in terms of terminology or understanding, then they are not directly contradictory. *See also* contradictory in the sense of being mutually exclusive.

discernment (*'du shes; saṃjñā*). Translated in other sources as *discrimination, perception, recognition*. This is the third of the five appropriated aggregates. It is that which "apprehends distinguishing marks" (*mtshan mar 'dzin pa; nimittagrahan*) of objects.

distinguishing mark (*mtshan ma; nimitta*). This refers to the distinguishing characteristic of an object that serves as a basis for applying a linguistic convention.

distorted cognition (*log shes; viparītajñāna*). *Also* distorted consciousness. A distorted cognition is erroneous in how it apprehends its object, and as such, it is not epistemically reliable.

dominant condition (*bdag rkyen; adhipatipratyaya*). This refers to the sense organ or faculty that empowers its respective sense consciousness when in contact with an appropriate object.

doubt (*the tshom; vicikitsā*). This is a cognition that vacillates between two standpoints with uncertainty. It may adopt any of three positions: doubt tending away from the fact, balanced doubt, and doubt tending toward the fact.

elements (*khams; dhātu*). When classified as six, these are earth, water, fire, wind, space, and consciousness. When classified as eighteen, these are the six internal bases of sense consciousness (the sense faculties), the six external bases of sense consciousness (their objects), and the six sense consciousnesses themselves, which arise in dependence upon the sense faculties and their objects. The same term is translated as "realm" when referring to the desire, form, and formless realms. *See also* sense base.

emptiness (*stong pa nyid; śūnyatā*). According to the Yogācāra and Madhyamaka schools of philosophy, emptiness is the ultimate nature of all phenomena, but emptiness is interpreted differently in these systems. Emptiness and its varieties will be discussed further in the philosophy volumes of *Science and Philosophy in the Indian Buddhist Classics*.

engaged object (*'jug yul; pravṛttiviṣaya*). *Also* object of activity, object of application. This is generally considered to be the main object of awareness to which cognition applies.

evidence (*rtags; liṅga*). *Also* reason, sign, mark. The evidence or reason posited in a valid argument to show that the predicate applies to the subject. All logical reasoning depends on relations. Dharmakīrti posits only two types of relation—*causal* and *intrinsic*—based on which there are just three types of evidence that can be used in a valid argument: correct effect-evidence, correct nature-evidence, and correct evidence consisting in nonperception.

exclusion (*sel ba*; *apoha*). A conceptual cognition engages its object by way of exclusion.

extensive [mental] factors (*sa mang po ba*; *mahābhūmika*). Vasubandhu's term for omnipresent mental factors.

fine investigation of the ultimate (*yang dag par so sor rtog pa*; *saṃyakpratyavekṣaṇa*). This is a very refined mind of wisdom that analyzes its object in detail, searching for its ultimate nature or final mode of existence. *See also* wisdom of fine investigation.

five aggregates (*phung po lnga*; *pañcaskandha*). *See* aggregates of appropriation.

focal object (*dmigs pa*, *dmigs yul*, *dmigs rten*; *ālambanaviṣaya*). *Also* basic object, objective support. This refers to any main object of perception or thought.

foundation consciousness (*kun gzhi rnam shes*; *ālayavijñāna*). Yogācāra philosophers posit eight primary minds: the six sense consciousnesses, the afflictive mental consciousness, and the foundation consciousness. They consider the foundation consciousness to be the receptacle of the karmic seeds and imprints as well as being the basis of identity of the self.

general characteristics (*spyi mtshan*; *sāmānyalakṣaṇa*). Characteristics shared by things of the same kind or belonging to the same category. *See also* specific characteristics.

hidden (*lkog 'gyur*; *parokṣa*). A hidden or remote object can be known only in dependence upon logical reasoning or some other thought process.

homologous case/example (*mthun dpe*; *udāharaṇa*). The homologous case is provided when setting forth a formally structured argument to help the opponent recognize the relation between the evidence and the predicate in his or her own experience.

ignorance (*ma rig pa*; *avidyā*). *Also* delusion, confusion. This usually refers to afflictive ignorance, which is a distorted awareness. It is a mental factor rather than a primary mind. Ignorance is not simply not knowing the truth; it is the opposite of knowing the truth, a radical misunderstanding of reality.

image (*rnam pa*; *ākāra*). *See* aspect.

immediately preceding condition (*de ma thag rkyen*; *samanantarapratyaya*). The immediately preceding condition of any consciousness is a prior moment of consciousness.

inappropriate attention (*tshul min yid byed*; *ayoniśomanasikāra*). This is a type of mental attention that superimposes on its object additional factors that do not accord with the way in which the thing exists, especially seeing impure things as pure. *See also* conceptualization of inappropriate attention.

indeterminate perception (*snang la ma nges pa*). A cognition that has the clear appearance of a unique particular as its engaged object yet is unable to ascertain it.

inference (*rjes dpag*; *anumāna*). The term *inference* may refer to an inferential understanding or to a formally structured argument designed to engender that understanding through the use of valid reasoning.

indicative conceptions (*rang bzhin gyi rtog pa*). These are the conceptions indicative of the three appearances presented in the tantric texts: thirty-three indicative of the white luminous appearance, forty indicative of the red luminous radiance, and seven indicative of the black luminous imminence.

intention (*sems pa*; *cetanā*). This is the mental factor that directs any action of body, speech, or mind. It is what constitutes karma.

intrinsic relation (*bdag gcig 'brel*; *tādātmyasaṃbandha*). Related as sharing the same intrinsic nature. This is a relation whose components do not exist separately and yet can be distinguished conceptually. *See also* causal relation; relation.

introspection. *See* meta-awareness.

invariable relation. *See* unaccompanied nonoccurrence relation.

latent potencies (*bag chags*; *vāsanā*). These are causally effective predispositions or propensities imprinted on the mindstream by any virtuous, nonvirtuous, or neutral activity of body, speech, or mind.

linguistic convention (*brda*; *saṃketa*). *Also* label. This refers to a term as it is initially applied to an object. *See also* classificatory convention.

love (*byams pa*; *maitrī*). A mind wishing beings to be happy.

main mind (*gtso sems*; *citta*). This refers to any of the six sense consciousnesses, which are always accompanied by a retinue of mental factors. *See also* mental consciousness; primary consciousness.

meditative absorption (*bsam gtan*; *dhyāna*). This is a type of meditative concentration that places the mind in a state beyond the desire realm. It includes four levels of the form realm and four levels of the formless realm.

meditative concentration (*ting nge 'dzin*; *samādhi*). There are many levels of concentration, which may accompany different types of mind. When gradually developed in conjunction with a virtuous mind, meditative concentration eventually gives rise to calm abiding (*śamatha*).

memory (*dran pa*; *smṛti*). *See* mindfulness.

mental afflictions (*nyon mongs*; *kleśa*). *See* afflictions.

mental consciousness (*yid kyi rnam shes*; *manovijñāna*). This is one of the six primary consciousnesses. However, unlike the five sense consciousnesses, it does not arise in dependence on a physical sense organ. It arises in dependence on its object and its dominant condition, a previous moment of consciousness, which functions as the mental sense faculty.

mental factor (*sems 'byung*; *caitta*, *cetasika*). These are secondary consciousnesses that accompany a primary consciousness. Where the main mind cognizes an object, its concomitant mental factors cognize the object's attributes.

mentality (*yid*; *manas*). This refers to the aspect of mind that functions to cognize its object.

meta-awareness (*shes bzhin*; *saṃprajanya*). *Also* introspection. Meta-awareness is the mental factor that monitors one's conduct of body, speech, and mind from moment to moment. In meditation practice, it is used in tandem with mindfulness.

mind (*sems*; *citta*). The nature of mind is luminous clarity and awareness. In this regard it is the same as consciousness. However, when differentiated from mental factors, which are also types of consciousness, mind refers to any of the six types of primary consciousness within the continuum of a living being. *See also* main mind; primary consciousness.

mindfulness (*dran pa*; *smṛti*). *Also* recollection, memory. Mindfulness is the mental factor that prevents distraction. In meditation practice, it is used in tandem with meta-awareness.

mistaken cognition (*'khrul shes*; *bhrāntajñāna*). *Also* mistaken consciousness. A mistaken cognition is mistaken with regard to how its object appears, but it can still be epistemically reliable. A mistaken cognition, whether it is a perception or a thought, need not be distorted. It is valid if it holds its principal object correctly.

nature (*rang bzhin*; *svabhāva*). The term *nature* has at least three meanings: conventional nature, which mainly refers to a thing's conventional characteristics; ultimate nature, which refers to a thing's emptiness of true existence; and inherent nature, which some Buddhist schools argue is not an attribute of anything at all because it does not exist.

negative pervasion (*ldog khyab*; *vyatirekavyāpti*). This is the third element of the trimodal criterion of a valid argument, where the absence of the predicate entails the absence of the evidence. More precisely, it refers to the ascertainment by valid cognition that the evidence is *only* absent in the set of heterogeneous cases owing to its relation to the predicate of the thesis to be proven. *See also* argument; pervasion.

nirvāṇa (*mya ngan las 'das pa*). A state beyond suffering and its causes.

object as cognized (*'dzin stangs kyi yul*). *Also* held object. This may be an object of thought or of direct perception. It is that mind's main object held in a certain way—rightly or wrongly, depending on whether that mind is correct or incorrect.

object universal (*don spyi*; *arthasāmānya*). This is a technical term referring to the appearing object of a conceptual consciousness. Thought engages its focal object via a partly transparent mental image of it, which is a universal. The object appears mixed with this image. *See also* universal.

objective condition (*dmigs rkyen*; *ālambanapratyaya*). The main object of focus.

observed object (*gzung yul*; *grāhya*). This is the same as the appearing object; in the case of direct perception, it is a real thing, and in the case of thought, it is a universal.

ordinary being (*so so'i skye bu*; *pṛthagjana*). Any living being who has not realized directly the ultimate nature of persons or phenomena.

particular (*bye brag*; *viśeṣa*. Also *rang mtshan*; *svalakṣaṇa*). An instance of a general category or universal.

pervaded object (*khyab bya*; *vyāpya*). That which is encompassed by another. An instance is pervaded by any of its universals. For example, impermanent things are encompassed by existent things; they are included within the domain of existent things. In this case, impermanent things are the pervaded.

pervader (*khyab byed*; *vyāpaka*). That which encompasses another. A universal encompasses all its instances. For example, existent things encompass or include impermanent things as well as permanent things. In this case, existent things are the pervader.

pervasion (*khyab pa*; *vyāpti*). To encompass or include; to extend over a domain. Pervasion is an expression of relationship. There are two types of relation—causal relation and intrinsic relation—therefore there are two types of pervasion. As an example of the latter type, a universal pervades all its instances. For example, the universal, horse, encompasses all kinds of horse and all individual horses. Pervasion in the context of logic requires that, in a formally structured argument, the predicate must pervade the evidence.

placement meditation (*'jog sgom*). *See* stabilizing meditation.

pliancy (*shin sbyangs*, *shin tu sbyangs pa*; *praśrabdhi*). This is a mental factor that enables the mind and the body to become serviceable in meditation focused upon a virtuous object. It is a quality especially associated with the development of calm abiding. However, after having attained calm abiding, there is another more special kind of pliancy that arises from special insight meditation.

positive pervasion (*rjes khyab*; *anvayavyāpti*). This is the second element of the trimodal criterion of a valid argument, where the presence of the evidence entails the presence of the predicate. More precisely, it refers to the ascertainment by valid cognition that the evidence is present in *only* the set of homologous cases owing to its relation to the predicate of the thesis to be proven. *See also* argument; pervasion.

prasaṅga (*thal 'gyur*). *See* consequence.

primary consciousness (*rnam par shes pa*; *vijñāna*). This refers to any of the six sense consciousnesses, which arise in dependence upon the six sense faculties. *See also* main mind.

primordial mind/consciousness (*gnyug sems*; *nijacitta*). The very subtle innate mind of clear light, which rides upon the very subtle wind.

primordial wind (*gnyug rlung*; *nijavāyu*). The mount of the primordial mind, which causes the mind to move.

probandum (*bsgrub bya*; *sādhya*). *Also* thesis. Combination of the subject and the predicate of a formally structured argument, which is to be proved by the evidence posited.

proof (*sgrub byed*; *sādhana*). The evidence, or reason, that proves the thesis, or probandum, in a formally structured argument.

proof statement (*sgrub ngag*; *sādhanavākya*). A type of argument that clears away doubt that oscillates between two standpoints.

real thing (*dngos po*; *vastu*, *bhāva*). *Also* thing, causal thing, impermanent thing. A thing that arises and ceases in dependence on causes and functions to produce a result.

reflexive awareness (*rang rig*; *svasaṃvedana*). Self-cognizing consciousness. This is a directly perceiving, inwardly focused mind that observes an outwardly focused, other-cognizing mind. These minds are substantially identical. Prāsaṅgikas do not posit any self-cognizing consciousness.

refutation (*sun 'byin*; *dūṣaṇa, niṣedha*). This is a form of consequential reasoning employed to destroy a distorted conviction or a distorted conception that holds on to one standpoint.

reification (*sgro btags*; *samāropa*). Superimposition, exaggeration. This refers to the implicit or explicit acceptance of the existence of something that does not actually exist. It is the opposite of *denial*.

relation (*'brel ba*; *saṃbandha*). According to earlier Buddhist systems, there are just two ways of being related: causal relation and intrinsic relation (or related as sharing the same nature). A relation holds between two things or attributes and is characterized as follows: if this does not exist, then definitely that does not occur. This definition is known as *unaccompanied nonoccurrence*. It applies to both types of relation and in one direction only, not mutually (unless both components in an intrinsic relation are coextensive). A third type of relation is presented in the Madhyamaka system: related as mutually dependent. *See also* causal relation; intrinsic relation; unaccompanied nonoccurrence relation.

Sāṃkhya (*grangs can pa*). The Sāṃkhya is a non-Buddhist school whose followers believe that things arise from their own intrinsic nature, the universal principle, and are thereby self-producing. There are two divisions of the Sāṃkhya: one accepts a creator god, who works with the universal principle in creating things, and the other does not.

sense base (*skye mched*; *āyatana*). There are twelve bases of sense consciousness: internally, the six sense faculties, and externally, the objects of the six sense consciousnesses. *See also* elements.

sense consciousness (*dbang shes*; *indriyajñāna*). The six sense consciousnesses include the mental sense consciousness, which is often treated separately. The five sense consciousnesses refer to those of the eye, ear, nose, tongue, and body. Each arises in dependence on its respective object together with its own uncommon dominant condition, the subtle sense organ located in its fleshly counterpart. The mental consciousness does not depend on a subtle sense organ as its dominant condition but on a previous moment of mind. *See also* mental consciousness.

sense faculty (*dbang po*; *indriya*). Sense organ. The subtle sense organs are located in their fleshly counterparts, such as the eyeball, and function as the internal bases of their respective sense consciousnesses. However, the mental sense faculty is not an organ but a previous moment of mind.

special insight (*lhag mthong*; *vipaśyanā*). A mind of wisdom engaging in ultimate analysis, which in union with calm abiding sees the ultimate nature of its object.

specific characteristics (*rang mtshan, rang gi mtshan nyid*; *svalakṣaṇa*). A real or causally effective thing has its own specific characteristics that belong to itself as well as more general characteristics, such as impermanence, that belong to all conditioned things. The meditator can use either of these types of characteristics as a focal object for placement or analytical meditation. *See also* unique particular.

stabilizing meditation (*'jog sgom*). Placement meditation. The practitioner, instead of analyzing the aspects and qualities of the object, synthesizes the various aspects into one and focuses on that object single-pointedly. When the attention wanders off, it is brought back to the object and placed there again and again until it remains spontaneously.

subsequent cognition (*bcad shes*; *pṛṣṭhalabdhajñāna*). This is a cognition that follows upon a perception, and it often takes the form of a conceptual judgment that interprets the content of the perception. It is not a newly realizing cognition because its object has already been apprehended by the perception.

substantial cause (*nyer len gyi rgyu*; *upādānahetu*). The substantial cause of something must be a continuum of what is of a similar type as itself. It is the primary cause of that thing, whether it is something material, something in the nature of consciousness, or a conditioning factor that is neither of those two.

temporary cause of error (*'phral gyi 'khrul rgyu*). This mainly applies to sense consciousness and includes both external and internal causes of error, excluding the deeper cause of error—ignorance of ultimate reality. Internal temporary causes of error include faulty sense organs, diseases affecting perception, hallucinogenic drugs, and so on. External causes of error include such things as the blades of an airplane propeller spinning so fast that we see it as one vibrating ring. *See also* cause of error.

thesis (*bsgrub bya*; *sādhya*). *See* probandum.

three modes (*tshul gsum*; *trirūpa*). This is the tri-modal criterion of a valid argument, which requires that (1) the evidence is an attribute of the subject, (2) the presence of the evidence entails the presence of the predicate, and (3) the absence of the predicate entails the absence of the evidence. The latter two modes are called the *positive pervasion* and the *negative pervasion*, respectively. *See also* attribute of the subject; negative pervasion; positive pervasion.

three poisons (*dug gsum*; *triviṣa*). The three root afflictions: attachment, hatred, and delusion/ignorance.

unaccompanied nonoccurrence relation (*med na mi 'byung ba'i 'brel pa*; *avinābhāvaniyama*). *Also* invariable relation. This is a criterion that encapsulates the two types of relation, the causal relation and the intrinsic relation. Most notably, it is crucial in the context of a correct inference because it is the kind of relation that must hold between the evidence and the predicate to be inferred. This means that when the *pervaded* is present, the *pervader* too is necessarily present, and when the *pervader* is absent, the *pervaded* too is necessarily absent. *See also* relation.

unique particular (*rang mtshan*, *rang gi mtshan nyid*; *svalakṣaṇa*). A unique particular is a real thing that is causally effective. *See also* specific characteristics.

universal (*spyi*; *sāmānya*). A universal is an abstract object of thought that encompasses all its instances. Although a universal is not causally effective and is therefore unreal, its instances may be real things. There are different types of universal: an *object universal* (*don spyi*) is a universal that encompasses its specific instances; a *word universal* (*sgra spyi*) is the universal type of a word that is expressed through the specific utterances that are its tokens. *See also* object universal.

valid cognition (*tshad ma*; *pramāṇa*). Some Buddhist sources say it is newly acquired and nondeceptive cognition. Others say it is cognition that correctly knows its principal object. All agree that there are two types of valid cognition: direct perception and inferential understanding.

wisdom (*shes rab*; *prajñā*). *Also* intelligence. One of the five mental factors with a determinate object, it accompanies every instance of cognition. There is innate wisdom and wisdom acquired through learning and mental cultivation.

wisdom of fine investigation (*so sor rtog pa'i shes rab*; *pratyavekṣaṇaprajñā*). Investigative wisdom. This is a very refined mind of wisdom that is employed in analytical meditation searching for the nature of the object scrutinized. *See also* fine investigation of the ultimate.

word universal (*sgra spyi*; *śabdasāmānya*). *See* universal.

wrong view (*log lta*; *mithyādṛṣṭi*). This is an afflictive intelligence that, upon considering something that exists, such as karmic cause and effect, views it to be nonexistent. It is a distorted apprehension that causes one to behave perversely regarding what to accept or reject, whereby one abandons virtue and engages in nonvirtue. The root wrong view is a distorted cognition holding its object to be truly existent.

Bibliography

Sūtras and Tantras

Buddha Nature Sūtra. Tathāgatagarbha-sūtra. De bzhin gshegs pa'i snying po'i mdo. Toh 258, Sūtra, za.

Chapter on the Thirty-Three. Trāyastriṃśatparivarta-sūtra. Sum cu rtsa gsum pa'i le'u gyi mdo. Toh 223, Sūtra, dza.

Collection of Aphorisms. Udānavarga. Ched du brjod pa'i tshoms. Toh 326, Sūtra, sa.

Compendium of Teachings Sūtra. Dharmasaṅgīti-sūtra. Chos yang dag par sdud pa'i mdo. Toh 238, Sūtra, zha.

Concentration Combining All Merit Sūtra. Sarvapuṇyasamuccayasamādhi-sūtra. Bsod nams thams cad bsdus pa'i ting nge 'dzin gyi mdo. Toh 134, Sūtra, na.

Connected Discourses. Saṃyutta Nikāya. Lung yang dag ldan pa. Pali canon.

Descent into Laṅkā Sūtra. Laṅkāvatāra-sūtra. Lang kar gshegs pa'i mdo. Toh 107, Sūtra, ca.

Dhammapada. Dhammapada. Chos kyi tshigs su bcad pa. Pali canon.

Enquiry of the Four Goddesses. Caturdeviparipṛcchā. Lha mo bzhis zhus. Toh 446, Tantra, ca.

Finer Points of Discipline. Vinayakṣudrakavastu. 'Dul ba phran tshegs kyi gzhi. Toh 6, Vinaya, tha.

Great Final Nirvāṇa Sūtra. Mahāparinirvāṇa-sūtra. Mdo myang 'das: Yongs su mya ngan las 'das pa chen po'i mdo. Toh 119, Sūtra, nya.

Great Play Sūtra. Lalitavistara-sūtra. Rgya cher rol pa'i mdo. Toh 95, Sūtra, kha.

Great Sūtra on Great Emptiness. Mahāśūnyatā-mahāsūtra. Mdo chen po stong pa nyid chen po. Toh 291, Sūtra, sha.

Guhyasamāja Compendium of Vajra Wisdom. Guhyasamāja-vajrajñānasammucaya. 'Dus pa'i bzhad rgyud ye shes rdo rje kun las btus pa. Toh 447, Tantra, ca.

Guhyasamāja Subsequent Tantra. Guhyasamāja-uttaratantra. Dpal gsang ba 'dus pa'i rgyud phyi ma. Toh 443, Tantra, ca.

Guhyasamāja Tantra. Guhyasamāja-tantra. Gsang ba 'dus pa'i rgyud. Toh 442, Tantra, ca.

Kanakavarṇa's Past Practice. Kanakavarṇapūrvayoga. Gser mdog gi sngon gyi sbyor ba. Toh 350, Sūtra, ah.

Kāśyapa Chapter Sūtra. Kāśyapaparivarta-sūtra. 'Od srung gi le'u gyi mdo. Toh 87, Ratnakūṭa, cha.

King of Concentrations Sūtra. Samādhirāja-sūtra. Ting nge 'dzin gyi rgyal po'i mdo. Toh 127, Sūtra, da. A.k.a. *Moon Lamp Sūtra* (*Candrapradīpa-sūtra*).

Long Discourses. Dīgha Nikāya. Lung ring po. Pali canon.

Middle-Length Discourses. Majjhima Nikāya. Lung bar ma. Pali canon.

Moon Lamp Sūtra. Candrapradīpa-sūtra. Zla ba sgron me'i mdo. See *King of Concentrations Sūtra* (*Samādhirāja-sūtra*).

Numerical Discourses. Aṅguttara Nikāya. Gcig las 'phros pa'i lung. Pali canon.

Ornament for the Light of Wisdom That Engages the Object of All Buddhas Sūtra. Sarvabuddhaviṣayāvatārajñānālokālaṃkāra-sūtra. Sangs rgyas thams cad kyi yul la 'jug pa'i ye shes snang ba'i rgyan gyi mdo. Toh 100, Sūtra, ga.

Perfection of Wisdom Sūtra in Eight Thousand Lines. Aṣṭasāhasrikāprajñāpāramitā-sūtra. Shes rab kyi pha rol tu phyin pa brgyad stong pa'i mdo. Toh 12, Prajñāpāramitā, ka.

Perfection of Wisdom Sūtra in Eighteen Thousand Lines. Aṣṭādaśasāhasrikāprajñāpāramitā-sūtra. Khri brgyad stong: Shes rab kyi pha rol tu phyin pa khri brgyad stong pa'i mdo. Toh 9, Prajñāpāramitā, ka.

Perfection of Wisdom Sūtra in Twenty-Five Thousand Lines. Pañcaviṃśatisāhasrikāprajñāpāramitā-sūtra. Shes rab kyi pha rol tun phyin pa stong phrag nyi shu lnga pa'i mdo. Toh 9, Prajñāpāramitā, ka, kha, ga.

Pile of Precious Things Collection. Ratnakūṭa. Dkon mchog brtsegs pa chen po'i chos kyi rnam grangs le'u stong phrag brgya pa. Toh 45–93, Ratnakūṭa, ka, kha.

Play of Mañjuśrī Sūtra. Mañjuśrīvikrīḍita-sūtra.'Jam dpal rnam par rol pa'i mdo. Toh 96, Sūtra, kha.

Questions of Gaganagañja Sūtra. Gaganagañjaparipṛcchā-sūtra. Nam mkha' mdzod kyis zhus pa'i mdo. Toh 148, Sūtra, pa.

Questions of Brahma Viśeṣacinti Sūtra. Brahmaviśeṣacintiparipṛcchā-sūtra. Tshangs pa khyad par sems kyis zhus pa'i mdo. Toh 160, Sūtra, ba.

Questions of the Householder Vīradatta Sūtra. Vīradattagṛhapatiparipṛcchā-sūtra. Khyim bdag dpal byin gyis zhus pa'i mdo. Toh 72, Ratnakūṭa, ca.

Questions of Ratnacūḍa Sūtra. Ratnacūḍaparipṛcchā-sūtra. Gtsug na rin po ches zhus pa'i mdo. Toh 91, Ratnakūṭa, cha.

Range of the Bodhisattva Sūtra (Sūtra Teaching the Miracles in the Range of a Bodhisattva's Skillful Means). Bodhisattvagocara-sūtra (Bodhisattvagocaropāyaviṣayavikurvāṇanirdeśa-sūtra). Byang chub sems dpa'i spyod yul gyi thabs kyi yul la rnam par 'phrul pa bstan pa'i mdo. Toh 146, Sūtra, pa.

Sacred Vinaya Scripture. Vinayottaragrantha. 'Dul ba gzhung dam pa. Toh 7, Vinaya, na, pa.

Short Discourses. Khuddaka Nikāya. Lung phran tshegs. Pali canon.

Sūtra Invoking the Supreme Intention. Adhyāśayasañcodana-sūtra. Lhag pa'i bsam pa bskul ba'i mdo. Toh 69, Ratnakūṭa, ca.

Sūtra of Individual Liberation. Prātimokṣa-sūtra. So sor thar pa'i mdo. Toh 2, Vinaya, ca.

Sūtra of the Ten Dharmas. Daśadharmaka-sūtra. Chos bcu pa'i mdo. Toh 53, Ratnakūṭa, kha.

Sūtra on the Application of Mindfulness. Saddharmasmṛtyupasthāna-sūtra. Mdo dran pa nyer bzhag: Dam pa'i chos dran pa nye bar gzhag pa'i mdo. Toh 287, Sūtra, ya, ra, la, sha.

Sūtra Teaching the Great Compassion of the Tathāgata. Tathāgatamahākaruṇanirdeśa-sūtra. De bzhin gshegs pa'i snying rje chen po nges par bstan pa'i mdo. Toh 147, Sūtra, pa.

Sūtra Teaching the Nonarising of All Phenomena. Sarvadharmapravṛttinirdeśa-sūtra. Chos thams cad 'byung ba med par bstan pa'i mdo. Toh 180, Sūtra, ma.

Sūtra Teaching the Tathāgata's Inconceivable Secret. Tathāgatācintyaguhyanirdeśa-sūtra. De bzhin gshegs pa'i gsang ba bsam gyis mi khyab pa bstan pa'i mdo. Toh 47, Ratnakūṭa, ka.

Sutta on the Application of Mindfulness. Satipaṭṭhāna Sutta. Dran pa nye bar bzhag pa'i mdo. Pali canon.

Teachings of Akṣayamati Sūtra. Akṣayamatinirdeśa-sūtra. Blo gros mi zad pas bstan pa'i mdo. Toh 175, Sūtra, ma.

Unraveling the Intention Sūtra. Saṃdhinirmocana-sūtra. Dgongs pa nges par 'grel pa'i mdo. Toh 106, Sūtra, ca.

Vajra Garland Tantra. Vajramālā-tantra. Rdo rje phreng ba'i rgyud. Toh 445, Tantra, ca.

Canonical Treatises

Anuruddha. *Compendium of Abhidhamma. Abhidhammatthasaṅgaha. Chos mngon pa bsdus pa.* See Sempa Dorje 1988 and Wijeratne and Gethin 2007.

Āryadeva. *Discernment of the Self-Consecration Stage. Svādhiṣṭhānakramaprabheda. Bdag byin gyis brlab pa'i rim pa rnam par dbye ba.* Toh 1805, Tantra, ngi.

———. *Four Hundred Stanzas. Catuḥśataka. Bstan bcos bzhi brgya pa.* Toh 3846, Madhyamaka, tsha.

———. *Lamp on the Compendium of Practices. Caryāmelāpakapradīpa. Spyod bsdus pa'i sgron ma.* Toh 1803, Tantra, ngi.

Āryaśūra. *Compendium of the Perfections. Pāramitāsamāsa. Phar rol tu phyin pa bsdus pa.* Toh 3944, Madhyamaka, khi.

———. *Garland of Birth Stories. Jātakamālā. Skyes pa'i rabs kyi rgyud.* Toh 4150, Jātaka, hu.

Asaṅga. *Compendium of Ascertainments. Viniścayasaṃgraha. Rnam par gtan la dbab pa bsdu ba.* Toh 4038, Cittamātra, zhi, zi.

———. *Compendium of Bases. Vastusaṃgraha. Gzhi bsdu ba.* Toh 4039, Cittamātra, zi.

———. *Compendium of Enumerations. Paryāyasaṃgraha. Rnam grang bsdu ba.* Toh 4041, Cittamātra, 'i.

———. *Compendium of Knowledge. Abhidharmasamuccaya. Chos mngon pa kun las btus pa.* Toh 4049, Cittamātra, ri.

———. *Compendium of the Mahāyāna. Mahāyānasaṃgraha. Theg pa chen po bsdus pa.* Toh 4048, Cittamātra, ri.

———. *Levels of Yogic Deeds. Yogācārabhūmi. Rnal 'byor spyod pa'i sa.* Toh 4035, Cittamātra, tshi.

———. *Śrāvaka Levels. Śrāvakabhūmi. Nyan thos kyi sa.* Toh 4036, Cittamātra, dzi.

Asvabhāva. *Explanation of the Compendium of the Mahāyāna. Mahāyānasaṃgrahopanibandhana. Theg pa chen po bsdus pa'i bshad sbyar.* Toh 4051, Cittamātra, ri.

Atiśa. *Lamp for the Path to Enlightenment. Bodhipathapradīpa. Byang chub lam gyi sgron ma.* Toh 3947, Madhyamaka, khi.

Bodhibhadra. *Explanation of the Compendium of the Essence of Wisdom. Jñānasārasamuccayanibandhana. Ye shes snying po kun las btus pa bshad sbyar.* Toh 3852, Madhyamaka, tsha.

Buddhajñāna. *Two Stages of Meditating on Reality. Dvikramatattvabhāvanā. Rim pa gnyis pa'i de kho na nyid bsgom pa.* Toh 1853, Tantra, di.

Candragomin. *Praise of Confession. Deśanāstava. Bshags bstod.* Toh 1159, Stotra, ka.

Candrakīrti. *Classification of the Five Aggregates. Pañcaskandhaprakaraṇa. Phung po lnga'i rab byed.* Toh 3866, Madhyamaka, ya.

———. *Clear Words. Prasannapadā. Tshig gsal ba.* Toh 3860, Madhyamaka, 'a.

———. *Commentary on the Four Hundred Stanzas. Catuḥśatakaṭīkā. Bzhi brgya pa'i rgya cher 'grel pa.* Toh 3865, Madhyamaka, ya.

———. *Entering the Middle Way. Madhyamakāvatāra. Dbu ma la 'jug pa.* Toh 3861, Madhyamaka, 'a.

———. *Entering the Middle Way Autocommentary. Madhyamakāvatārabhāṣya. Dbu ma la 'jug pa'i bshad pa.* Toh 3862, Madhyamaka, 'a.

Devendrabuddhi (Lha dbang blo). *Commentary on Difficult Points in the Exposition of Valid Cognition. Pramāṇavārttikapañjikā. Tshad ma rnam 'grel gyi dka' 'grel.* Toh 4217, Pramāṇa, che.

Dharmakīrti. *Ascertainment of Valid Cognition. Pramāṇaviniścaya. Tshad ma rnam nges.* Toh 4211, Pramāṇa, ce.

———. *Drop of Reasoning. Nyāyabindu. Rigs pa'i thigs pa.* Toh 4212, Pramāṇa, ce.

———. *Drop of Reasons. Hetubindu. Gtan tshigs thigs pa.* Toh 4213, Pramāṇa, ce.

———. *Exposition of Valid Cognition. Pramāṇavārttika. Tshad ma rnam 'grel.* Toh 4210, Pramāṇa, ce.

———. *Exposition of Valid Cognition Autocommentary. Pramāṇavārttikasvopajñavṛtti. Tshad ma rnam 'grel gyi 'grel ba.* Toh 4216, Pramāṇa, ce.

Dharmamitra. *Clarifying Words. Prasphuṭapadā. 'Grel bshad tshig gsal.* Toh 3796, Prajñāpāramitā, nya.

Dharmottara. *Explanatory Commentary on the Ascertainment of Valid Cognition. Pramāṇaviniścayaṭīkā. Tshad ma rnam par nges pa'i 'grel bshad.* Toh 4229, Pramāṇa, dze.

———. *Proof of the Afterlife. Paralokasiddhi. 'Jig rten pha rol grub pa.* Toh 4251, Pramāṇa, zhe.

Dignāga. *Compendium of Valid Cognition. Pramāṇasamuccaya. Tshad ma kun las btus pa.* Toh 4203, Pramāṇa, ce.

———. *Compendium of Valid Cognition Autocommentary. Pramāṇasamuccayavṛtti. Tshad ma kun las btus rang 'grel.* Toh 4204, Pramāṇa, ce.

———. *Drum of a Wheel of Reasons. Hetucakraḍamaru. Gtan tshigs 'khor lo.* Toh 4209, Pramāṇa, ce.

———. *Entering into Valid Reasoning. Nyāyapraveśa. Tshad ma rigs par 'jug pa.* Toh 4208, Pramāṇa, ce.

———. *Introduction to Entering into Valid Reasoning. Nyāyapraveśaprakaraṇa. Tshad ma rigs par 'jug pa'i sgo.* (Not in Derge.) Peking Tengyur 5706, Pramāṇa, ce; Narthang Tengyur, Pramāṇa, ce.

———. *Investigation of the Object. Ālambanaparīkṣā. Dmigs brtag.* Toh 4205, Pramāṇa, ce.

———. *Investigation of the Object Autocommentary. Ālambanaparīkṣāvṛtti. Dmigs brtag rang 'grel.* Toh 4206, Pramāṇa, ce.

Great Explanation. Mahāvibhāṣa. Chos mngon pa bye brag tu bshad pa chen po. (Not in Tibetan canon.) Translated from the Chinese by Lozang Chopak [Hwatsun] (1945–49). Book (*bris ma*) of the Gaden Potang Library. Published by Krung go'i bod rig pa'i dpe skrun khang, 2011.

Haribhadra. *Short Commentary on the Ornament for Clear Knowledge. Abhisamayālaṃkāravṛtti. Mngon par rtogs pa'i rgyangyi 'grel pa.* Toh 3793, Prajñāpāramitā, ja.

Jinamitra. *Commentary on the Levels of Yogic Deeds. Yogācārabhūmivyākhyā. Rnal 'byor spyod pa'i sa rnam par bshad pa.* Toh 4043, Cittamātra, 'i.

Jinaputra. *See* Yaśomitra.

Jitāri. *Distinguishing the Scriptures of the Sugata. Sugatamatavibhaṅga. Bde bar gshegs pa'i gzhung rnam par 'byed pa.* Toh 3899, Madhyamaka, a.

———. *Distinguishing the Scriptures of the Sugata Autocommentary. Sugatamatavibhaṅgabhāṣya. Bde bar gshegs pa'i gzhung rnam par 'byed pa'i bshad pa.* Toh 3900, Madhyamaka, a.

Jñānagarbha. *Distinguishing the Two Truths. Satyadvayavibhaṅga. Bden pa gnyis rnam par 'byed pa.* Toh 3881, Madhyamaka, sa.

Jñānavajra. *Heart Ornament of the Tathāgata: Commentary on Descent into Laṅkā Sūtra. Laṅkāvatārasūtravṛtti. Lang kar gshegs pa'i mdo'i 'grel pa.* Toh 4019, Sūtra, pi.

Kamalaśīla. *Commentary on Difficult Points of the Ornament for the Middle Way. Madhyamakālaṃkārapañjikā. Dbus ma'i rgyan gyi dka' 'grel.* Toh 3886, Madhyamaka, sa.

———. *Stages of Meditation (First, Middle, and Final Volume). Bhāvanākrama (1, 2, 3). Sgom pa'i rim pa.* Toh 3915, 3916, 3917, Madhyamaka, ki.

Lakṣmīṅkarā. *Commentary on the Five Stages. Pañcakramaṭīkā. Rim pa lnga'i 'grel pa: Rim pa lnga'i don gsal bar byed pa.* Toh 1842, Tantra, chi.
Maitreya. *Distinguishing the Middle from the Extremes. Madhyāntavibhāga. Dbus dang mtha' rnam par 'byed pa.* Toh 4021, Cittamātra, phi.
———. *Ornament for Clear Knowledge. Abhisamayālaṃkāra. Mngon par rtogs pa'i rgyan.* Toh 3786, Prajñāpāramitā, ka.
———. *Ornament for the Mahāyāna Sūtras. Mahāyānasūtrālaṃkāra. Theg pa chen po'i mdo sde'i rgyan.* Toh 4020, Cittamātra, phi.
———. *Sublime Continuum. Uttaratantra. Rgyud bla ma.* Toh 4025, Cittamātra, phi.
Mātṛceṭa. *Letter to King Kaniṣka. Mahārājakaniṣkalekha. Rgyal po ka nis ka'i springs yig.* Toh 4184, Lekha, nge.
Mokṣākaragupta. *Language of Logic. Tarkabhāṣā. Rtog ge'i skad.* Toh 4264, Pramāṇa, zhe.
Nāgabodhi. *Clarifying the Meaning of the Five Stages. Pañcakramārthabhāskaraṇa. Rim pa lnga'i don gsal bar byed pa.* Toh 1833, Tantra, ci.
———. *Presentation of the Guhyasamāja Sādhana. Samājasādhanavyavasthāli. Rnam gzhag rim pa: 'Dus pa'i sgrub pa'i thabs rnam par gzhag pa'i rim pa.* Toh 1809, Tantra, ngi.
Nāgārjuna (attributed). *Compendium of Teachings. Dharmasaṃgraha. Chos yang dag par bsdus pa.* (Not in the Tibetan canon.)
Nāgārjuna. *Finely Woven. Vaidalyasūtra. Zhib mo rnam 'thag.* Toh 3826, Madhyamaka, tsa.
———. *Five Stages. Pañcakrama. Rim lnga.* Toh 1802, Tantra, ngi.
———. *Friendly Letter. Suhṛllekha. Bshes pa'i spring yig.* Toh 4182, Lekha, nge.
———. *Fundamental Treatise on the Middle Way. Mūlamadhyamakakārikā. Dbu ma rtsa ba'i tshig le'ur byas pa.* Toh 3824, Madhyamaka, tsa.
———. *Praise of Dharmadhātu. Dharmadhātustava. Chos dbyings bstod pa.* Toh 1118, Stotra, ka.
———. *Precious Garland. Ratnāvalī. Rin po che'i phreng ba.* Toh 4158, Jātaka, ge.
Nāropa. *Clear Compilation of the Five Stages. Pañcakramasaṃgrahaprakāśa. Rim pa lnga bsdus pa gsal ba.* Toh 2333, Tantra, zhi.
Prajñākaramati. *Commentary on Difficult Points in the Engaging in the Bodhisattva's Deeds. Bodhicaryāvatārapañjikā. Byang chub sems dpa'i spyod pa la 'jug pa'i dka' 'grel.* Toh 3872, Madhyamaka, la.
Pṛthivībandhu. *Explanation about the Five Aggregates. Pañcaskandhabhāṣya. Phung po lnga'i bshad pa.* Toh 4068, Cittamātra, si.
Puṇḍarīka. *Stainless Light: Commentary on Kālacakra. Vimalaprabhā-kālacakratantraṭīkā. Dri ma med pa'i 'od dus kyi 'khor lo'i 'grel bshad.* Toh 1347, Tantra, tha, da.
Pūrṇavardhana. *Investigating Characteristics: Explanatory Commentary on the Treasury of Knowledge. Abhidharmakośaṭīkā-lakṣaṇānusāriṇī. Mdzod kyi 'grel bshad mtshan nyid rjes 'brang.* Toh 4093, Abhidharma, cu, chu.

Ratnākaraśānti. *Instructions for the Perfection of Wisdom. Prajñāpāramitopadeśa. Shes rab kyi pha rol tu phyin pa'i man ngag.* Toh 4079, Cittamātra, hi.
Śāntarakṣita. *Compendium of Reality. Tattvasaṃgraha. De kho na nyid bsdus pa.* Toh 4266, Pramāṇa, ze.
———. *Ornament for the Middle Way. Madhyamakālaṃkāra. Dbus ma rgyan.* Toh 3884, Madhyamaka, sa.
———. *Ornament for the Middle Way Autocommentary. Madhyamakālaṃkāravṛtti. Dbu ma rgyan gyi 'grel pa.* Toh 3885, Madhyamaka, sa.
Śāntideva. *Compendium of Training. Śikṣāsamuccaya. Sblab pa kun las btus pa.* Toh 3940, Madhyamaka, khi.
———. *Compendium of Training in Verses. Śikṣāsamuccayakārikā. Bslab pa kun las btus pa'i tshig le'ur byas pa.* Toh 3939, Madhyamaka, khi.
———. *Engaging in the Bodhisattva's Deeds. Bodhisattvacaryāvatāra. Byang chub sems dpa'i spyod la 'jug pa.* Toh 3871, Madhyamaka, la.
Śīlapālita. *Commentary on the Finer Points of Discipline. Āgamakṣudraka-vyākhyāna. Lung phran tshegs kyi rnam par bshad pa.* Toh 4115, Vinaya, dzu.
Śrī Jaganmitrānanda. *Letter to King Candra. Candrarājalekha. Rgyal po zla ba'i springs yig.* Toh 4189, Lekha, nge.
Sthiramati. *Commentary on the Thirty Stanzas. Triṃśikābhāṣya. Sum cu pa'i bshad pa.* Toh 4064, Cittamātra, shi.
———. *Explanatory Commentary on Distinguishing the Middle from the Extremes. Madhyāntavibhāgaṭīkā. Dbus dang mtha' rnam par 'byed pa'i 'grel bshad.* Toh 4032, Cittamātra, bi.
Vasubandhu. *Commentary on Distinguishing the Middle from the Extremes. Madhyāntavibhāgabhāṣya. Dbu dang mtha' rnam par 'byed pa'i 'grel ba.* Toh 4027, Cittamātra, bi.
———. *Explanation of the Ornament for the Mahāyāna Sūtras. Sūtrālaṃkāra-vyākhyā. Mdo sde'i rgyan gyi bshad pa.* Toh 4026, Cittamātra, phi.
———. *Treasury of Knowledge. Abhidharmakośa. Chos mngon pa'i mdzod.* Toh 4089, Abhidharma, ku.
———. *Treasury of Knowledge Autocommentary. Abhidharmakośabhāṣya. Chos mngon pa'i mdzod kyi bshad pa.* Toh 4090, Abhidharma, ku, khu.
———. *Treatise on the Five Aggregates. Pañcaskandhaprakaraṇa. Phung po lnga'i rab byed.* Toh 4059, Cittamātra, shi.
Vimalamitra. *Meaning of the Gradual Approach in Meditation. Kramaprāveśik-abhāvanārtha. Rim gyis 'jug pa'i sgom don.* Toh 3938, Madhyamaka, ki.
Yaśomitra (a.k.a. Jinaputra). *Clarifying the Meaning of the Treasury of Knowledge. Abhidharmakośavyākhyā-sphuṭārthā. Chos mngon pa mdzod kyi 'grel bshad don gsal.* Toh 4092, Abhidharma, gu, ngu.
———. *Explanation of the Compendium of Knowledge. Abhidharmasamuccaya-bhāṣya. Chos mngon pa kun las btus pa'i rnam par bshad pa.* Toh 4053, Citta-mātra, li.

Other Sources Cited

Alaksha Tendar Lharampa (A lag sha Bstan dar lha rams pa, 1759–1831). 2011. *Precious Lamp Illuminating Dignāga's "Drum of a Wheel of Reasons." Phyogs glang gis mdzad pa'i phyogs chos 'khor lo zhes pa'i bstan bcos gsal bar byed pa'i rin chen sgron me*. In Collected Works, 1:95–107. Lanzhou: Kan su'u mi rigs dpe skrun khang.

Anālayo. 2003. *Satipaṭṭhāna: The Direct Path to Realization*. Birmingham, UK: Windhorse Publications.

Block, Ned. 1995. "On a Confusion about a Function of Consciousness." *Behavioral and Brain Sciences* 18: 227–87.

Bodhi, Bhikkhu, ed. 1993. *A Comprehensive Manual of Abhidhamma: The Abhidhammattha Sangaha*. Translated by Nārada Mahāthera and Bhikkhu Bodhi. Kandy, Sri Lanka: Buddhist Publication Society.

———, trans. 2000. *The Connected Discourses of the Buddha: A New Translation of the Saṃyutta Nikāya*. Boston: Wisdom Publications.

———. 2005. *In the Buddha's Words: An Anthology of Discourses from the Pāli Canon*. Boston: Wisdom Publications.

Buddhaghosa. 1991. *The Path of Purification* (*Visuddhimagga*). Translated by Bhikkhu Ñāṇamoli. Kandy, Sri Lanka: Buddhist Publication Society.

Chapa Chökyi Sengé (Phywa pa Chos kyi seng ge, 1109–69). 2006. *Valid Cognition Eliminating the Darkness of the Mind* (*Tshad ma sde bdun yid kyi mun sel*). *Collected Works (Gsungs 'bums) of the Kadampas*, vol. 9. Chengdu: Si khron mi rigs dpe skrun khang.

Chim Jampaiyang (Mchims 'Jam pa'i dbyangs, 13th century). 2009. *Ornament of Abhidharma Commentary* (*Mdzod 'grel mngon pa'i rgyan*). Bod kyi gtsug lag gces btus 23. New Delhi: Institute of Tibetan Classics.

Dalai Lama Tenzin Gyatso. 1999. *Ethics for the New Millennium*. New York: Riverhead Books.

———. 2005. *The Universe in a Single Atom: The Convergence of Science and Spirituality*. New York: Harmony Books.

———. 2011. *Beyond Religion: Ethics for a Whole World*. Boston: Houghton Mifflin Harcourt.

Dreyfus, Georges. 2011. "Self and Subjectivity: A Middle Way Approach." In *Self, No Self? Perspectives from Analytical, Phenomenological, and Indian Traditions*, edited by Mark Siderits, Evan Thompson, and Dan Zahavi, 114–56. Oxford: Oxford University Press.

Dunne, John D. 2004. *Foundation of Dharmakīrti's Philosophy*. Boston: Wisdom Publications.

Eltschinger, Vincent. 2009 [2010]. "Ignorance, Epistemology and Soteriology: Part I." *Journal of the International Association of Buddhist Studies* 32.1–2: 39–83.

Jackson, Roger R. 1993. *Is Enlightenment Possible? Dharmakīrti and rGyal-tshab-*

rje on Knowledge, Rebirth, No-Self, and Liberation. Ithaca, NY: Snow Lion Publications.

Jinpa, Thupten. 2002. *Self, Reality and Reason in Tibetan Philosophy*. London: Routledge Curzon.

Kajiyama, Yuichi. 1998. *Introduction to Buddhist Philosophy*. Vienna: Wiener Studien zur Tibetologie und Buddhismuskunde.

Khedrup Jé Gelek Palsang (Mkhas grub rje Dge legs dpal bzang). 1982. *Ornament of the Seven Treatises Clearing Mental Obscurity* (*Tshad ma sde bdun gyi rgyan yid kyi mun sel*). In *Collected Works (Gsuṅ 'bum) of the Lord Mkhas grub rje Dge legs dpal bzaṅ po: Reproduced from the 1897 Lha sa Old Źol (Dga' ldan Phun tshog Gliṅ) Blocks*, vol. 10. New Delhi: Guru Deva.

Perdue, Daniel. 2014. *The Course in Buddhist Reasoning and Debate*. Boston: Snow Lion Publications.

Potter, Karl H., ed., with Robert E. Buswell Jr., Padmanabh S. Jaini, and Noble Ross Reat. 1996. *Encyclopedia of Indian Philosophies*, vol. 7: *Abhidharma Buddhism to 150 AD*. Delhi: Motilal Banarsidass.

Rogers, Catherine. 2009. *Tibetan Logic*. Ithaca, NY: Snow Lion Publications.

Sempa Dorje (Gsems dpa' rdo rjes), trans. 1988. *Anuruddha's Compendium of Abhidhamma*, vol. 1. Varanasi, India: Central Institute of Higher Tibetan Studies.

Sopa, Geshe Lhundub, with Dechen Rochard. 2017. *Steps on the Path to Enlightenment: A Commentary on Tsongkhapa's Lamrim Chenmo, vol. 5: Insight*. Somerville, MA: Wisdom Publications.

Tillemans, Tom. 1999. *Scripture, Logic, and Language: Essays on Dharmakīrti and His Tibetan Successors*. Boston: Wisdom Publications.

Tsongkhapa. 2000. *The Great Treatise on the Stages of the Path to Enlightenment: Lam Rim Chen Mo*, vol. 1. Translated by the Lamrim Chenmo Translation Committee. Edited by Joshua Cutler and Guy Newland. Ithaca, NY: Snow Lion Publications.

———. 2002. *The Great Treatise on the Stages of the Path to Enlightenment: Lam Rim Chen Mo*, vol. 3. Translated by the Lamrim Chenmo Translation Committee. Edited by Joshua Cutler and Guy Newland. Ithaca, NY: Snow Lion Publications.

Wallace, Vesna A., trans. 2010. *The Kālacakra Tantra: The Chapter on Sādhanā, Together with the Vimalaprabhā Commentary*. New York: American Institute of Buddhist Studies and Columbia University Center for Buddhist Studies.

Wijeratne, R. P., and Rupert Gethin, trans. 2007. *Summary of the Topics of Abhidhamma (Abhidhammatthasaṅgaha) by Anuruddha, and Exposition of the Topics of Abhidhamma (Abhidhammatthavibhāvinī) by Sumaṅgala, Being a Commentary to Anuruddha's Summary of the Topics of Abhidhamma*. Lancaster, UK: Pali Text Society.

Yongzin Yeshe Gyaltsen (Yongs 'dzin Ye shes rgyal mtshan, 1713–93). 2010. *Verse Summary of Mind and Mental Factors* (*Sems dang sems byung gyi sdom tshig*).

In *Mngon pa gong ma sems khams rig pa* (*Abhidharma Psychology and Phenomenology*). Bod kyi gtsug lag gces btus 22. New Delhi: Institute of Tibetan Classics.

Index

A

Abhidharma, 15
  and Dharmakīrtian system, differences between, 236–37
  higher, 87, 176–78, 191, 198 (see also *Compendium of Knowledge*)
  lower, 118, 176, 178–79, 191, 198 (see also *Treasury of Knowledge*)
  mental factors in, motivation for analyzing, 85–87, 91
  mindfulness in, 364
  readings in, 95
  sense and mental consciousness in, 50–51
  as source, 27
  vitarka in, 470–71n60
Abhidharma Sūtra, 273, 483n389
absorptions, four, 104, 166, 212
abstinences, three, 186–87
acquiescence, 190, 191, 196, 198
adherence, 151, 168, 171, 188, 218, 219
adventitious consciousness, 213–14
adventitious stains/pollutants, 12, 41–42, 43, 135, 225, 458–59
afflictions (*nyon mongs*, *kleśa*), 91, 92, 167
  and afflicted persons, distinction between, 134
  analytical meditation and, 426–27
  causes and conditions of, 367–69
  cessation of, 371
  concomitant factors, 153–55
  counteracting, methods of, 353–54
  definition of, 139
  eliminating, 457–59
  equanimity in preventing, 122–23
  extensive afflictive factors, 163, 182–83
  ignorance as basis for, 95, 144, 369–71
  mindfulness and meta-awareness of, 382
  narrower afflictive factors, 164, 183
  secondary, 147, 176–77, 190, 191, 193–94, 197, 198, 373–74
  six root, 11, 139, 146, 176
  in Theravāda tradition, 186
afflictive mental consciousness (*kliṣṭamanas*), 52, 97, 154
aggregates of appropriation, 16, 433
  afflictive views and, 145–46, 147
  arising of, 231
  attachment to, 140
  in *Compendium of Knowledge*, 106
  dissolution of, 227–30
  four forms of rationality and, 295–96
  grossness and subtlety of mind among, 208
  person imputed in dependence on, 462, 463–64
  rebirth and, 62–63
amazement, 219–20
analysis (*dpyod pa*, *vicāra*), 65, 136, 166, 167, 184, 189, 190
  as imputed, 177, 178
  and inquiry, distinction between, 116, 158, 165

in laxity and excitation, 407
in mindfulness of phenomena, 441–42
as particular factor, 185
as variable mental factor, 158, 164–65
wisdom and, 116, 117
analytical meditation, 359–60, 363, 386, 425
inferential cognition in, 270
need for, 426–27
pliancy in, 424
serenity in, 409
in three stages of wisdom, 378, 380, 427
See also special insight (*vipaśyanā*)
anger, 95, 128, 159, 192, 376, 445
antidote to, 374
arising of, 458
cause of error and, 80
forbearance and, 132, 133–36
function of, 92
as indeterminate mental factor, 164–65
mental factors associated with, 148, 150, 153, 154, 169, 177, 178, 179, 373–74
and rage, difference between, 147–48
as root affliction, 140–42, 475n196
three objects giving rise to, 120–21, 140–41
See also hatred
animals, 6, 126, 464
animosity, 131, 148, 153–54, 184, 219, 220
annihilation, 146, 171, 442
antidotes, 374
limitations of, 377–78
possibility of, 43
properly applying, 400, 401–2
See also under individual mental factors
Anuruddha. See *Compendium of Abhidhamma*
apathy, 168, 171, 189
appearing object (*snang yul*), 44, 238, 239, 248
and direct object, distinction between, 247
direct perception of, 273
in engagement via exclusion, 256
mistaken cognition of, 78–80, 81
unmistaken cognition of, 81–82
applications of mindfulness, 363, 364, 442
etymology of phrase, 434
on feelings, 440–41
fruits of, 443
on mind, 441
in Pali sources, 431–33
on phenomena, 441–42
in upper and lower Abhidharma sources, 433–35
See also body, mindfulness of
appreciation, 113, 473n130
apprehensions, four distorted, 433–34. *See also* conceptualization of inappropriate attention
appropriating consciousness, 53, 470n47. *See also* foundation consciousness
arrogance, 149–50, 184, 189, 193, 194
arising of, 373, 374
in *Compendium of Bases*, 170
as imputed, 177, 178
as narrower afflictive factor, 164
as source of fault, 188
Āryadeva, 224. See also *Four Hundred Stanzas*; *Lamp on the Compendium of Practice*
Āryaśūra, 378–79, 398, 419
Asaṅga, 19, 89, 96, 468n15. See also *Compendium of Ascertainments*; *Compendium of Bases*; *Compendium of Enumerations*; *Compen-*

dium of Knowledge; *Compendium of the Mahāyāna*; *Levels of Yogic Deeds*; *Śrāvaka Levels*
ascertainment, 298, 315, 375, 380
Ascertainment of Valid Cognition (Dharmakīrti)
on conceptual and nonconceptual, difference between, 67–69
on direct perception, 54, 275
on nonperception of imperceptible, types of, 487n448
on proof statements, 307
on right cognition, two aspects, 76
on sense and mental consciousness, distinctions between, 51
aspiration, 105, 189, 190, 198
as antidote, 401, 415
in applications of mindfulness, 442
arising of, 420
attention and, 87
changing, 457
distorted, 117
as extensive mental factor, 89, 161, 162, 181, 182
faith and, 119–20
in laxity and excitation, 407
as mental factor with determinate object, 90, 111–12, 118
as particular factor, 185
for special insight, 424
types of, 372
aspiration-type mental states, 375, 376, 377, 488n483
Asvabhāva, *Explanation of the Compendium of the Mahāyāna*, 334
Atiśa, 20, 398
atoms, subtle, 252–53
attachment, 128, 129, 159, 167, 184
antidote to, 374
arising of, 370, 372
to body, 436–38
in *Compendium of Bases*, 168, 169, 171, 172
distorted cognition and, 265
excitation and, 406
as indeterminate mental factor, 164, 165
mental factors associated with, 148, 149, 150, 152, 153, 177, 178, 273, 373, 374
mental factors incompatible with, 154, 374, 376
predominating in temperament, 388
as root of nonvirtue, 169, 192
as source of fault, 188–89
Attainment of Knowledge (Kātyāyanīputra), 105, 473n113
attention, 189, 190
aspiration and, 90
in calm abiding, 400
and concentration, distinction between, 115
cultivation of, 93
distraction involving, 152–53
four types, 412, 413, 415
function of, 101
mental factors and, 90
as omnipresent/extensive/universal factor, 88, 109, 161, 162, 181, 182, 185
regulating, 360, 362, 363–64
in sense perception, 236
in special insight, 426
See also inappropriate attention
"attentional capture," 361. *See also* inappropriate attention
avarice, 149, 154, 189, 192, 193
arising of, 373, 374
in *Compendium of Bases*, 169
as imputed, 177, 178
as narrower afflictive factor, 164, 183
as nonvirtuous factor, 186
as source of fault, 187
aversion
arising of, 370, 372
distorted cognition and, 265

inappropriate attention and, 355–56
predominating in temperament, 388–89
suppressing, 353–54
awakening, full. *See* buddhahood
awareness (*rig pa, saṃvitti*), 12, 29–30, 40, 321
in calm abiding, 420
conceptual and nonconceptual, distinguishing, 90
construing (*zhen rig*), 69–70, 72, 268, 270–71, 273
of main mind, 88–89, 100–101, 103
of mental factors, 185
objective and reflexive, distinction between, 10
quality of, 44–45
in special insight, 424, 427
synonyms of, 41
See also consciousness; meta-awareness (*saṃprajanya*); reflexive awareness

B

balance (P. *tatramajjhattatā*), 186. *See also* equanimity
Barrett, Lisa Feldman, 94, 96
basis/ground, path, and result (*gzhi lam 'bras bu*), 8–9
behavior
and cognition, role of, 240, 244–45
exclusion theory and, 241–42
four types, 384–85, 432
hedonic tone in, 88–89
three gateways of, 141–42, 195
three stages of wisdom and, 380–81
transforming, 30, 86–87, 281
See also ethical conduct
Bhāviveka, *Blaze of Reasoning*, 15
bindings, 190, 191, 192–93, 198, 491n582
birth, 216–17, 226, 227, 231
bliss, 191, 418–19, 423
Block, Ned, 89, 90
Bodhibhadra, *Explanation of the Compendium of the Essence of Wisdom*, 253–54
body, 197
false conception about, 364
gross, arising of, 61
and mind, nonduality of, 202, 203, 215
mind as basis for, 30, 202
qualities contingent on, 455–56
subtle, 203 (*see also* channels; drops (*bindu*))
See also mind-body system
body, mindfulness of, 432, 442
attachment, antidote to, 436–38
conceptualization, antidote to, 438–40
specific and general characteristics in, 435–36
Bohm, David, 3
bonds, 190, 191, 194, 198
brain, 46, 214, 418
bravery, 219, 220
breath meditation
as antidote, 374
counting, 396
in eliminating excitation, 408
in mindfulness of body, 432, 438–40
nine-round exercise, 396–97
brutality, 219, 220
buddha bodies, two, 8–9, 467n5
Buddha Nature Sūtra, 42
Buddha Śākyamuni, 1, 20, 25–26, 36, 201, 283, 293
Buddhaghosa, *Path of Purification*, 477n260, 489n511
buddhahood, 8–9, 204, 260, 467n5
Buddhaśrījñāna, *Two Stages of Meditating on Reality*, 232
Buddhism
authority in, 1–2, 281, 293–94

contemporary science and, 2, 3, 12, 31, 362, 488n483
equality of matter and consciousness in, 46, 469n38
as grounded in causality, 18–19
mind, primacy of, 25, 26, 40
mind training in, 5
three domains of textual tradition, 7–9, 19, 21
universal aspects of, 7–8
See also epistemology, Buddhist; Indian Buddhism; Tibetan Buddhism

C

calm abiding (*śamatha*), 5, 117, 150, 359–60, 386, 422, 423
approximate and genuine, 416–19
balance in, 408–9
definition of, 400
as effortless (*anābhoga*), 122, 123, 359, 412, 416, 429
equanimity and, 123
illustration and meaning of, 414–15
mind as object of, 419–21
mindfulness in, 364
objectless, 363
objects of, 398–99
postures for, 395–97, 489n513
prerequisites of, 394–95
progressive subtlety of mind in, 212
purpose of, 393–94
as remedy to hindrances, 195
sign of, 380
See also concentration (*samādhi*); six ways of settling the mind; special insight (*vipaśyanā*); stabilizing meditation
Candragomin, *Praise of Confession*, 407
Candrakīrti, 17, 76–77, 109, 135. See also *Classification of the Five Aggregates*; *Clear Words*; *Entering the Middle Way*; *Entering the Middle Way Autocommentary*; Prāsaṅgika Madhyamaka
carping, 188, 189, 197
Cārvāka school, 7, 46
causal capacity, 51–52, 67, 241, 261, 378
causality, 18–19
cause of error (*'khrul pa'i rgyu*), 78, 79, 80
certainty, 219, 220, 265, 294, 358
channels, 47, 225, 226, 395, 397
Chapa Chökyi Sengé, 235, 263
cognition typology of, 243
on correct assumption, 482n380
and Dharmakīrti, reinterpreting, 238
on implicit objects, 481n355
object typology of, 239
readings on, 245
Chapter of the Thirty-Three, 41
Chinese translations, 19–20
Chomden Rikral, 481n355
Cittamātra (Mind Only) school, 3
on afflictive mental factors, 177
on direct perception, 275, 483n395
on inquiry and analysis, 165
on matter and consciousness, 469n38
on mind, mentality, and consciousness, 97–98
on mistaken cognition, 82
Nonpluralist view in, 256
on objective condition, 58–60, 470n55
on sensory cognition, images in, 251, 254–55
on subtle mind of death, 210
on two truths, 17–18
Clarifying the Meaning of the Five Stages (Nāgabodhi), 217
Clarifying the Meaning of the Treasury of Knowledge (Yaśomitra), 99, 114, 166–67, 209, 252

clarity (*gsal ba*), 405
in calm abiding, 394, 402
and excitation, relationship of, 362–63, 408–9
in meditating on mind, 420
of mind, 29–30, 41–44, 469n21
and stability, balancing, 409, 421
Classification of the Five Aggregates (Candrakīrti), 190–96, 197, 198
classificatory conventions, 70, 100, 108, 112, 116, 262
clear light (*'od gsal*), 9, 205–6, 441, 457–58, 459
during basic state, arising of, 232
dawning of, 224–25, 230
metaphorical, 206
resting in, 422
and subtle life-sustaining wind, 225–26
clear light of death, 214, 215, 216, 226–27
Clear Words (*Prasannapadā*, Candrakīrti), 76–77, 259, 481n368
Cloud of Jewels Sūtra, 428
cognition, 26
arising of, 230–31
categorizations of, 10, 47–48
correct and distorted, differentiating, 265–66
nonvalid, 10, 77, 266
perceptual and conceptual, distinctions between, 239–40, 357
reliable, 298
of resemblance, 262
sevenfold typology of, 10, 47, 243–45, 246, 263–65, 276–77 (*see also individual types*)
speed of, 259–60
synonyms of, 41
types of, 39, 77, 298
unmistaken, 81–82
See also conceptual cognition; mistaken cognitions; valid cognition (*pramāṇa*)
cognitive distortions, 26, 30, 33, 204, 356–57
cognitive science, contemporary, 12, 31, 362, 488n483
Collection of Aphorisms (*Udānavarga*), 452–53
Commentary on Distinguishing the Middle from the Extremes (Vasubandhu), 101
Commentary on the Difficult Points of the Ornament for the Middle Way (Kamalaśīla), 45
Commentary on the Four Hundred Stanzas (Candrakīrti), 135
Commentary on the Thirty Stanzas (Sthiramati), 417
compassion, 4, 93, 184, 187
as antidote, 374
benefits of, 125–26
definition of, 136
generating, 136–37
as indicative conception, 218
and love, difference between, 130
meditative cultivation of, 354, 426
as nonhatred, 179
nonviolence and, 123
readings on, 96
results of, 137–38
two levels of, 126–27
Compendium of Abhidhamma (Anuruddha), 185–87, 197, 198, 477n247, 490n542
Compendium of Ascertainments (Asaṅga)
on aspiration, 112
on concentration, 115
conventionally existent mental factors in, 177–78
on doubt, 267
on ignorance, 144
on intention, 109

on mental factors with determinate objects, 111
on mind and mental factors, shared features, 104
on mindfulness, 114
on objects, subtle dimension of, 209–10
on resolution, 112
on wisdom, 117
Compendium of Bases (*Vastusaṃgraha*, Asaṅga), 106, 167–73, 178, 197, 198, 477n260
Compendium of Elements (*Dhātukāya*, Pūrṇa/Vasumita), 105, 167, 179, 473n113, 477nn258–59
Compendium of Enumerations (*Paryāyasaṃgraha*, Asaṅga), 464
Compendium of Knowledge (*Abhidharmasamuccaya*, Asaṅga), 20, 87–88, 158–60, 409
on afflictions, 368, 369
on anger, 140
on applications of mindfulness, 433, 435, 442–43
on arrogance, 149
on aspiration, 111–12
on attachment, 140
on attention, 109
on avarice, 149
and *Compendium of Bases*, differences between, 167, 178
on concealment, 148
on concentration, 115
on contact, 110
on diligence, 121
on discernment, 108
on distraction, 152
on doubt, 145
on dullness, 150
on embarrassment, 120
on equanimity, 122–23
on excitation, 150
on faith, 119
on faithlessness, 151
on feeling, 107
on five afflictive views, 145–46, 147
on forgetfulness, 152
on guile, 149
on heedfulness, 122
on heedlessness, 151–52
on ignorance, 143–44
on inquiry and analysis, 158
on intention, 108–9
on jealousy, 148
on laziness, 151
on mental afflictions, definition of, 139
mental factors in, categories of, 106, 196–97
on mental factors with determinate objects, 118, 162
on meta-awareness, 152
on mind, mentality, and consciousness, 97
on mind and mental factors, 103
on mindfulness, 113, 434
on nonattachment, 120
on nondelusion, 121
on nonembarrassment, 150
on nonhatred, 120–21
on nonviolence, 123
omnipresent mental factors in, 89, 107, 477n247
on pliancy, 122, 417
Precious Garland and, 187, 189
on pretense, 149
on pride, 142–43
on rage, 147
on rationality, 294
on regret, 158
on resentment, 148
on resolution, 113
on shame, 120
on shamelessness, 150
on sleep, 157

on spite, 148
on substantial and imputed mental factors, 175, 176, 177
on three poisons, 491n582
and *Treasury of Knowledge*, differences between, 89, 118, 165–66, 167, 477n247
on violence, 150
on wisdom, 116, 434
Compendium of Reality (Śāntarakṣita), 15, 268–69, 482n381
Compendium of Sūtras (Nāgārjuna), 20
Compendium of Teachings (Nāgārjuna), 189–90, 479n295
Compendium of Teachings Sūtra, 131
Compendium of the Mahāyāna (*Mahāyānasaṃgraha*, Asaṅga), 98
Compendium of the Perfections (Āryaśura), 398, 419
Compendium of Training (*Śikṣāsamuccaya*, Śāntideva), 20, 387, 430
Compendium of Training in Verses (*Śikāṣasamuccayakārikā*, Śāntideva), 113
Compendium of Valid Cognition (Dignāga), 20, 66
on conceptualization, 65, 69
on correct thesis, 311
on fallacious perceptions, 81
on inference, 270
on inference for others' purpose, 305
on mental and sense consciousnesses, distinction between, 49–50, 51
Compendium of Valid Cognition Autocommentary (Dignāga), 70–71, 273, 274
competitiveness, 4, 6, 128, 130, 445
completion stage, 205
concealment, 148, 184, 189, 193
arising of, 373, 374
in *Compendium of Bases*, 169
as imputed, 177, 178
as narrower afflictive factor, 164
as source of fault, 187
conceit, 5, 170, 184, 220, 445
conceived object (*zhen yul*), 238, 239, 247–48, 469n33
concentration (*samādhi*), 152, 179, 189, 190
clarity and single-pointedness in, 402
ethics and, 86–87, 394–95
as extensive mental factor, 89, 161, 162, 181, 182
flawless/faultless, 405–6, 421, 429
function of, 101
as mental factor with determinate object, 91, 114–15, 118
mindfulness and, 114
pliancy and, 417–18
and sleep, difference between, 394
strength of, 105
wisdom and, 93, 117
Concentration Combining All Merit Sūtra, 130, 136
concept formation, 27, 34, 36, 240–42, 245, 353
conception (biological), 231
conceptual cognition, 472n86
appearing and observed objects of, 247
categorizations of, 70–72, 240
conceived objects and, 239, 247–48
definition of, 69–70
distorted, 266
engaged objects of, 248
engagement via exclusion by, 256–58, 261, 262
sense cognitions and, 259, 481n368
conceptual constructions (*kun rtog*, *parikalpa*), 65, 471n62
conceptuality/conceptualization (*rtog pa*), 65, 470n63

cessation of, 226
as indicative conceptions, 217, 218, 221–22
meanings of, 236
during mental placement, 409
and nonconceptuality, difference between, 67–69, 89, 90
in sevenfold typology of cognition, 244
suspicion of, 34, 36, 37
conceptualization of inappropriate attention, 364, 370, 443, 471n77
concordant cause, 60
conditioning factors, 372
categories of, 106
in *Compendium of Teachings*, 189–90
and consciousness, levels of, 208
dissolution of, 229–30
mental factors related to, 167, 168, 170, 171, 178, 194
conditions, four, 55
causal, 55, 385–86
immediately preceding, 53–54, 55, 56, 57–58, 60, 79, 210, 321, 369
objective, 49, 51, 55–60, 256, 258–59, 273, 368, 470n55
See also dominant condition
conduct (*samācāra*), 129, 358, 381. *See also* "view, meditation, and conduct"
confidence, 11, 119, 125, 130, 459. *See also* resolution
confusion (*moha*). *See* ignorance (*avidyā*)
consciousness, 39, 101
aggregate of, 60–61, 208, 230
arising of, 230–32
basis of, 46–47
as distinct streams, 100, 472n101
eight classes of, 16
in Indian Buddhism, 85
levels of, 11–12
luminous clarity of, 41–44, 469n21
moments of, 29, 92
overview of, 97–98
phenomenal and access, 89–90
in sense perception, 236–37
substantial cause of, 463
subtle levels of, 201, 203, 207
synonyms of, 41
See also six types of consciousness
consequential reasoning, 15, 285, 290, 291–92, 297–98, 299. *See also* correct consequence; fallacious consequence
contact, 189, 190
function of, 100, 101
as omnipresent/extensive/universal factor, 88, 109–10, 161, 181, 182, 185
in sense perception, 236
contaminants, 190, 191, 194, 198, 457
contemplative practices, 35–36
readings on, 365
taxonomies, role in, 86–87, 92–93
contentment, 5, 167, 184, 188, 447–48
conventional reality, 255
conviction (*yid ches*), 1, 244, 267, 268, 269–70, 386, 427. *See also* distorted conviction
cooperative condition, 60, 322
corpse/skeleton meditation, 436–37, 491n576
correct assumptions, 243, 482n382
definition of, 267–68
five types, 269–70
in sevenfold cognition typology, 264, 276, 277, 482n380
correct consequence, 299–302
correct effect-evidence, 287, 313, 320–22, 331, 347–48, 349
correct evidence
categories of, 319–20
causal relations and, 319–20, 322
definition of, 313–14

on empirical fact, 332, 333, 334
and fallacious, distinguishing between, 349
in homologous cases, 335–36
intrinsic relations and, 319–20, 324
for oneself or for others, 336
on popular convention, 332, 333–34
proving expression, 331
proving thing itself, 331
three modes of, 14–15, 305, 310, 313, 314–17, 468n9
on trustworthy testimony, 332, 335
in wheel of reasons, 345, 346, 348
See also correct effect-evidence; correct evidence consisting in nonperception; correct nature-evidence
correct evidence consisting in nonperception, 288, 313, 320, 348, 349
definition of, 324, 486n444
of imperceptible, 325–26, 330
of perceptible, 325, 326–30
synonyms of, 331
types of, 14–15
correct nature-evidence, 287–88, 289, 313, 322–24, 331, 347–48, 349
covetousness, 195, 436
craving, 167, 371
excitation and, 406
faults arising from, 447, 492n595
inappropriate attention and, 355–56
as indicative conception, 217, 218
suppressing and undermining, 353–54
three types, 140
creators, 463
critical reflection, wisdom arisen from, 116–17, 121, 269–70, 358, 378, 380–81, 425

D

Damasio, Antonio, 94
death, 203
common point of, 229
dissolution at, 214, 226–30, 231–32
experience, reproducing as method, 204–6
feelings at, neutrality of, 209, 210–11
meditation on, 426–27
preparation for, 206
subtle mind/consciousness of, 207, 214
wind dissolution at, 216–17
deathless, thinking oneself to be, 106, 167, 169, 172, 189, 195
debate tradition, 289, 308, 483–84n408
delusion. *See* ignorance (*avidyā*)
denial, 18, 298, 326
dependent arising (*pratītyasamutpāda*), 8, 14–15, 16–17, 19, 442
Descent into Laṅka Sūtra, 42, 53, 66, 80, 97–98, 266
desire, 372
in *Compendium of Bases*, 170–71
engaging by force of, 261
indicative conception of, 218, 219
as nonvirtuous factor, 186
sensory, 167, 189, 195, 218, 411, 433
as source of fault, 188
desire realm, 103–4, 166, 193, 195, 212, 393, 422
despair, 167, 169, 172, 173
Devendrabuddhi, *Commentary on the Difficult Points in the Exposition of Valid Cognition*, 274, 311, 312
Dhammapada, 40, 468n18
dharmakāya, 232
Dharmakīrti, 12, 16, 27, 32, 34
contributions of, 262
on direct perception, 244
on empiricism and scripture, relationship of, 282–83
on evidence and predicate, relationship between, 286–87
object typology of, 238–39
tradition of, 235, 236

on truth of cessation, 18–19
See also *Ascertainment of Valid Cognition*; *Drop of Reasoning*; *Drop of Reasons*; *Exposition of Valid Cognition*; *Exposition of Valid Cognition Autocommentary*
Dharmamitra, *Clarifying Words*, 386
Dharmottara, 62, 481n354, 481n355, 481nn354–55. See also *Explanatory Commentary on the Ascertainment of Valid Cognition*
Dignāga, 12, 27, 32, 78, 236, 468n17
contributions of, 262
inferential reasoning of, 281, 283, 284
See also *Compendium of Valid Cognition*; *Compendium of Valid Cognition Autocommentary*; *Drum of a Wheel of Reasons*; *Entering into Valid Reasoning*; *Introduction to Entering into Valid Reasoning*; *Investigation of the Object*; *Investigation of the Object Autocommentary*
diligence, 177, 184, 190, 442, 443
as antidote, 401
aspiration and, 112, 119–20
concentration and, 115
defining characteristic of, 182
heedfulness and, 122
indicative conception of, 219, 220
in laxity and excitation, 407
as particular factor, 185
power of, 411, 412, 413
as substantial existence, 176
as virtuous mental factor, 121–22, 162
direct perception, 58, 79, 244, 288
apprehension of attributes by, 261
as aspect of right cognition, 76
conceptuality and, 68
and correct assumption, distinction between, 268, 269
definition of, 272–73
four types, 273–75
as free from conceptualization, 65
mental, 273, 274, 276
primacy of, 1–2, 35
scripture and, 33
in sevenfold typology of cognition, 264, 265, 276, 277
subsequent cognition and, 77–78
as unmistaken cognition, 81–82
variant views on, 53–55
yogic, 270, 273, 274–75
discernment, 167, 189, 198
aggregate of, 229
and consciousness, levels of, 208, 211
feeling as concomitant to, 99
function of, 100
as omnipresent/extensive/universal factor, 88, 108, 161, 181, 185
and resolution, distinction between, 113
and wisdom, distinction between, 116
Discernment of the Self-Consecration Stage (Āryadeva), 224
discomposure, 188, 189, 197
discouragement, 106, 151, 166, 167, 169, 171, 179, 184, 189, 195, 442
discourses (*sūtras*), 27, 201
discrimination, 4, 166, 188, 189. *See also* discernment
disenchantment, 184, 190, 191, 197, 220–21, 374, 407, 442
disheartenment, 184, 374
disintegration, 16, 18, 323, 324, 330, 395, 432
dissatisfaction. *See* suffering (*duḥkha*)
Distinguishing the Middle from the Extremes (Maitreya), 65, 98–101, 103, 400–401
distorted cognition, 243
and correct assumption, distinction between, 268

predominating in temperament, 390
in sevenfold cognition typology, 263, 264, 265–66, 276
true existence as, 458–59
distorted conviction, 264, 267, 293, 297–98, 301, 302, 304
distraction, 92, 139, 197
concentration and, 162, 165–66
and excitation, difference between, 406, 407
as extensive affliction, 182
as imputed, 177, 178, 179
mindfulness and, 361
pacification of, 411
during patched placement, 410
as secondary affliction, 152–53
six types, 152–53
dominant condition
direct perception and, 54
in mental and sense consciousnesses, distinguishing, 49, 58
of sense consciousnesses, 31, 52–53, 55–56, 57, 102, 109–10
of three luminosities, 222
uncommon, 49–50, 52, 207, 273, 274, 275
doubt, 192, 243, 244, 374
afflictive, 121, 145, 184
arising of, 373
concomitant factors, 153, 154
in consequential reasoning, eliminating, 304, 307
and correct assumption, distinction between, 268
due to fallacious evidence, 340, 341
as hindrance, 167, 195
as indeterminate mental factor, 164–65
indicative conception of, 220, 221
as nonvirtuous factor, 186
sense direct perceptions and, 274
in sevenfold cognition typology, 263, 264, 267, 276, 277
as source of fault, 189
wisdom and, 116, 118, 121, 162
dreams, 11, 52, 79, 80, 201, 207, 210, 232
Dreyfus, Georges, 90, 238
Drop of Reasoning (Dharmakīrti), 69, 75, 79, 273, 319, 487n448
Drop of Reasons (Dharmakīrti), 56–57
drops (*bindu*), 47, 203, 225, 226
Drum of a Wheel of Reasons (Dignāga), 343–47, 346, 488n467
duḥkha. *See* suffering (*duḥkha*)
dullness, 139, 184, 189
arising of, 373, 374
concomitant factors, 153, 154–55
as extensive afflictive factor, 163
as hindrance, 195
as imputed, 177, 178
and laxity, difference between, 404–5
nonpliancy and, 191
as secondary affliction, 150, 193

E

eight worldly concerns
equanimity toward, 363–64, 445, 446
fame and disrepute, 449–50
material gain and loss, 446–49
pleasant and painful feelings, 450–51
praise and disparagement, 451–53
eighty indicative conceptions, 11, 215, 217–21, 230–31
Ekman, Paul, 95
elements, 216–17, 224, 226–30, 231
embarrassment, 189, 190, 191
as substantially existent, 176
as virtuous mental factor, 120, 162, 163, 183, 184, 186
emotions, 4, 52, 85, 93–95, 96, 456–57
empirical stance, 35, 36, 37
emptiness, 371, 428
energy-wind. *See* wind (*rlung, vāyu*)

engaged object (*'jug yul*), 13, 34, 238, 239–40, 247–48, 469n33, 481n354
conceptual cognitions and, 70
indeterminate perception of, 275–76
mistaken cognition of, 78–80
nonconceptual cognitions and, 72
engagement via affirmation, 258–60, 261
engagement via exclusion, 256–58, 259, 261, 262
Engaging in the Bodhisattva's Deeds (Śāntideva), 27, 446
on anger, 135, 141–42, 376–77
on confusion, 383
on contentment, 448
on disparagement by others, 450, 452
on disparaging the wealthy, 448–49
on equanimity, 137
on fame, 449, 451–52
on forbearance, 133–34
on habituation, 459
on happiness and suffering, sources of, 129
on laziness, three types, 151
on material things, 447
on mind, protecting, 383–84
on mind as source of suffering, 40–41
on mindfulness, 113, 402
on mindfulness and meta-awareness, 114, 382–83, 384, 385, 403
on nonvirtue from worldly life, 445
on pleasure, 450–51
on sentient beings, sameness of, 127
on special insight, 423
on suffering, remedies for, 131–32, 451
Enquiry of the Four Goddesses, 225–26
entanglements, 190, 191, 193, 194, 198
Entering into Valid Reasoning (Dignāga), 307
enthusiasm, 151, 219, 220, 442
epistemology, Buddhist, 12, 235
of Candrakīrti, 76–77
inferential reasoning in, 283–88
logic in, 297–98
perception in, special place of, 242–43
readings on, 37, 245
reliability in, 33–35, 37
sources on, 27, 236
equanimity, 167, 189, 190, 191, 380
as antidote, 401, 402, 415
forbearance and, 131–32
as imputed/conventionally existent, 176, 177
three types, 123
as virtuous mental factor, 122–23, 162, 183, 184, 186
equipoise, 123, 195, 396, 397, 412, 422
error, 34–35, 257–58
causes of, types, 78–80, 82, 275
eliminating, 298
indicative conception of, 220, 221
See also mistaken cognitions
essence, three types, 443
ethical conduct, 5, 6–7
altruistic motivation and, 126
calm abiding and, 394–95
hindrances to, 195
importance of, 125, 358
as mental factor, 184
role of, 86–87, 93
shame and embarrassment in, 120
special insight and, 428–29
evidence (*rgyu mtshan*), 284, 285, 317
as attribute of subject, 308–9
in consequential reasoning, 299, 301, 312
definition of, 313
of direct perception, 1, 2
in inferential proof, 297
and predicate, relationships between, 285–88, 308, 315–16, 319–20, 337, 347–48, 485n428
remainder (*śeṣavat*), 316, 339–40

synonyms, 313, 485n430
See also correct evidence; fallacious evidence
exaggeration, 326, 355–56, 369, 370, 415
exchanging self and others, 137
excitation, 139, 189, 362, 363, 406–7, 410, 411
antidotes to, 123, 374, 401, 402, 442
arising of, 373, 374, 407
concomitant factors, 153, 154–55
distraction and, 152, 153, 406, 407
as extensive afflictive factor, 163, 182, 183
fault of, 400–401
freedom from, 412
as hindrance, 195
as imputed, 177, 178
in meditating on mind, 421
in mindfulness of body, 439
as nonvirtuous factor, 186
pacifying, 115, 408
preventing, 398
as secondary affliction, 150, 193
as source of fault, 189
excitement, 219, 220
exclusion of other, 260, 261
exclusion (*apoha*) theory, 241–42, 245, 262, 324
exhortation, 219, 220
Explanation of the Compendium of Knowledge (Yaśomitra), 104, 136, 139, 157, 158, 443
Explanation of the Ornament for the Mahāyāna Sūtras (Vasubandhu), 117, 382
Explanatory Commentary on Distinguishing the Middle from the Extremes (Sthiramati), 101, 104
Explanatory Commentary on the Ascertainment of Valid Cognition (Dharmottara), 272, 333, 334
Exposition of Valid Cognition (Dharmakīrti), 260
on affirmative engagement, 259, 262
on afflictions, eliminating, 457
on blind person seeing, 54–55
on cognition, 66, 69
on conceptuality, types of, 72
on consciousness, 61, 100
on correct effect-evidence, 321, 322
on correct evidence, 313, 331
on correct nature-evidence, 323
on engagement via exclusion, 261
on evidence as attributes of subject, 347, 488n468
on evidence on mere convention, 334
on evidence on trustworthiness, 335
on fallacious evidence, 337
on fallacious perception, 81
on good qualities, 455, 456
on ignorance, 144, 476n205
on indeterminate perception, 276
on inferential cognition, 270, 271
on mental factors, counteractor and counteracted, 375, 376
on nature of mind, 43
on nonperception of imperceptible, 325, 486n447
on nonperception of perceptible, 326, 328, 329–30, 487n448, 487n450
on nonperceptions, 324–25, 486n446
on proof statements, 305, 307–8, 309
on self and other, 128–29, 371, 372
on sense consciousness, three conditions, 57
on subsequent cognition, 271–72
on unaccompanied nonoccurrence, 317
on valid cognition, 75, 76, 266
on yogic cognition, 274–75
Exposition of Valid Cognition

Autocommentary (Dharmakīrti), 257–58, 316
Extensive Compendium of Sūtras (Atiśa), 20
external objects, 256, 394, 406
indicative conceptions and, 218
luminous clarity and, 41, 43
as objective condition, 59–60
sense consciousnesses and, 51, 53–54
variant views on, 13, 251–53, 254–55
eye consciousness
apprehending form, variant views on, 250–55
location of, 46–47
mental consciousness and, 50–51, 52
three conditions of, 56, 57–58

F

faith, 7–8, 105, 189, 190
as antidote, 401, 415
as aspect of religion, 18–19
compassion as foundation of, 137
defining characteristic of, 182
engaging by way of, 387
meditation for developing, 426
as substantially existent, 176
as virtuous mental factor, 119–20, 162, 183, 184, 186
faithlessness, 151, 189
concomitant factors, 153, 154, 162
as extensive afflictive factor, 163, 182
as imputed, 177
as substantially existent, 178
fallacious consequence, 301, 302, 304
fallacious evidence, 313
contradictory, 337–38, 345, 346, 348
and correct evidence, distinguishing between, 349
definition of, 337
inconclusive, 337, 338–40, 345, 346
unestablished, 337, 340–42
fallacious perception, 80–81, 298. *See also* mistaken cognitions
familiarity, 378
cultivating, 116, 264–65, 359, 386, 459
power of complete, 412, 413
prior, 61, 113, 368
fear, 95, 113, 125, 128, 184, 217, 218
feeling, 167, 189, 198
aggregate of, 228–29
as concomitant to other mental factors, 99, 100, 102, 105
and consciousness, levels of, 208
as indicative conception, 218
mindfulness of, 433, 440–41, 442
as omnipresent/extensive/universal factor, 88–89, 107–8, 161, 181, 185
three types, 208–9
fetters, 190, 191, 192, 198
fine investigation, 114, 121, 136, 218, 229, 264, 386, 423, 426–27
Finely Woven (Nāgārjuna), 298
Finer Points of Discipline (*Vinayakṣudrakavastu*), 106, 166–67, 169, 178–79, 193, 197, 198
five faults, 400–401, 402, 415
Five Stages (Nāgārjuna), 213, 215, 217–21, 222
flattery, 168, 170, 188, 197, 451
floods. *See* streams/floods
fluency, mental and physical, 359. *See also* pliancy
focal object, 146, 248, 249, 362, 367–68, 376, 426, 470n55
foolhardiness, 219, 220
forbearance, 93
as antidote, 374, 376
benefits of, 125, 132–33
cultivating, 134–36
three types, 131–32, 475n175
forgetfulness, 162, 165, 166, 197
as extensive affliction, 182
as imputed, 177, 178, 179
indicative conception of, 220, 221

of meditation instructions, 400–401, 443
recognizing, 410
as secondary affliction, 152
form aggregate, 60–61, 227–28
form realm, 104, 393, 422
formless realm, 207, 212, 393, 422
foundation consciousness (*ālayavijñāna*), 16, 97–98, 154, 226–27, 231, 470n47
four empty states, 11–12, 215, 216
Four Hundred Stanzas (Āryadeva), 61, 107–8, 134–35, 371, 375
four noble truths, 8, 18–19, 433, 479n305
four seals, 16
Friendly Letter (Nāgārjuna), 383, 446
Fundamental Treatise on the Middle Way (Nāgārjuna), 369, 371

G

gain, pursuing with gain, 170, 188, 197
Gandhi, Mohandas, 7
Garland of Birth Stories (Āryaśūra), 378–79
generation stage, 205
generosity, 131, 184, 335, 377
grasping, 140
as indicative conception, 217, 218
at true existence, 371, 377–78, 458–59
See also self-grasping
graspings, 190, 191, 194–95, 198
Great Explanation (*Mahāvibhāṣa*), 105, 167, 473n113, 477n258
Great Final Nirvāṇa Sūtra, 115, 403–4
Great Play Sūtra, 53, 382
Great Sūtra on Great Emptiness, 416
greed, 4, 5, 128, 183, 188, 388, 489n511
grief, 137, 167, 169, 172–73, 179
Guhyasamāja Compendium of Vajra Wisdom, 230–31
Guhyasamāja Subsequent Tantra, 227
Guhyasamāja Tantra, 213, 228, 229, 231
guile, 149, 189, 193, 194
arising of, 374
in *Compendium of Bases*, 169–70
as imputed, 178
as narrower afflictive factor, 164, 183
as source of fault, 187
Gyaltsab Je, 482–83n388

H

habituated patterns, 204–5, 368, 369, 386, 455–57, 458
Half-Eggists, 255
happiness, 212
authentic, 93
cutting root of, 383
from kind attitude toward others, 125–26
mind's role in, 40, 128–29, 457
Haribhadra, *Short Commentary on the Ornament for Clear Knowledge*, 127
hatred, 128, 129, 141–42, 184, 186, 372–73, 458
antidotes to, 130, 134–36, 442
arising of, 373
in *Compendium of Bases*, 169
function of, 100
mental factors incompatible with, 376
as root of nonvirtue, 192
in viewing others, 127
violence imputed on, 179
See also anger
haughtiness, 168, 170, 187–88, 373, 390
heart center, 216, 225, 226
heedfulness, 141, 189, 190
cultivating, 381, 382, 442, 443

as imputed/conventionally existent, 176, 177
as virtuous mental factor, 122, 162, 183, 184
heedlessness, 139, 190
arrogance and, 149, 150
in *Compendium of Bases*, 170
concomitant factors, 153, 154
as extensive affliction, 163, 182, 183
as imputed, 177, 178
as secondary affliction, 151–52
as source of fault, 188
Heisenberg, Werner, 3
hesitation, 218, 219
heterogeneous case/example
in correct evidence, 314–15
in correct proof statements, 306–7, 308
evidence as attribute of subject in, 344–45, 346–47
inconclusive evidence in, 339, 340
hidden objects of knowledge, 71–72, 81, 265, 268–69, 298, 312–13
extremely hidden, 282
slightly hidden, 270–71
very hidden, 271, 332–33, 335
highest yoga tantra, 204
on consciousness, two types, 213–14
on dissolution, 227
four elements in, 216
gross and subtle minds in, 207
on heart, 47
on mind and inner wind, indivisibility of, 46
Hīnayāna, 41–42, 155, 213
hindrances, 167, 171, 189, 190, 191, 195, 198, 433, 478n261
hinting, 168, 170, 188, 197, 477n260
homologous case/example
in correct evidence, 314–15, 319, 335–36
in correct proof statements, 306–7, 308
evidence as attribute of subject in, 344–45, 346–47, 349
inconclusive evidence in, 339, 340
honesty, 184
hunger, as indicative conception, 217, 218
hypocrisy, 168, 170, 188, 197, 443, 479n294

I

ignorance (*avidyā*), 30, 186, 190, 439, 472n91
as basis of afflictions, 108, 369–71, 373
clear-light mind and, 205–6
in *Compendium of Bases*, 168, 169
Consequentialist view of, 290
definition of, 143
as extensive affliction, 163, 182, 183
as facet of luminous imminence, 224
function of, 144–45
grasping at true existence, 371, 377–78, 458–59
inappropriate attention from, 356
mental factors associated with, 148, 149, 150, 151, 152, 157, 158, 164, 177, 178, 373, 404
as most destructive affliction, 95
philosophical interpretations of, 26
predominating in temperament, 389
readings on, 365
reversing, 375
as root affliction, 143
as root of nonvirtue, 192–93
as secondary affliction, 168, 169
suffering and, 18, 25–26, 32, 75
types of, 144, 370
uprooting, 86, 91, 204, 379
ill will, 127, 128, 147
antidote to, 374
bodily tie of, 195
as hindrance, 167, 195
nine causes, 140–41

pacifying, 130
as source of fault, 189, 197
images (*ākāra*), 32, 108, 210, 216, 236, 250–55, 255–56, 260
images, conceptual (*pratibimba*), 34
immediate, unobstructed cause, 327, 487n449
immediately preceding condition. *See under* conditions, four
impermanence
of body, 443
consequential reasoning on, 291
evidence of, 287–88
meditation on, 408, 426–27, 432
as mental perception, 53
products and, 12–13
seal of, 16
imputed existence, four senses of, 176
inappropriate attention, 71, 179, 182, 197, 265, 355–56, 368–69. *See also* conceptualization of inappropriate attention
indeterminate perception, 244–45
sense direct perceptions as, 273–74
in sevenfold cognition typology, 264, 265, 275–76, 277
Indian Buddhism, 19–20, 34, 35–36, 37, 85, 235, 290
Indian traditions, 26, 201, 203, 462
indifference, 220, 221
inference for oneself, 283–84
inference for others, 288–90, 304–5
inference/inferential cognition, 27, 81, 249, 250, 482n374
in consequential reasoning, 297–98
and correct assumption, distinction between, 268, 269
of epistemically remote objects, 282–83
exclusion and, 257–58
readings on, 292
in sevenfold cognition typology, 243–44, 264, 270–71, 276, 277
structure of, 283–84
as subsequent cognition, 272
inferential evidence. *See* evidence (*rgyu mtshan*)
inferential reasoning, 27, 258, 283, 284–88, 292, 293
inquiry (*rtog pa*, *vitarka*), 65, 116, 166, 167, 184, 189, 190, 470n60
as imputed, 177, 178
as particular factor, 185
as variable mental factor, 158, 164–65
wisdom and, 117
insight. *See* special insight (*vipaśyanā*)
instructions, not appreciating, 170, 188
intelligence, 2, 189, 391
afflicted, 145, 146, 152
as extensive mental factor, 161, 162
levels of, 402
loving kindness as enhancing, 126
as root of neutral karma, 192
intelligence-type mental states, 375–76, 377, 403, 488n483
intention, 189, 190, 421
function of, 100–101, 105
in laxity and excitation, 407
as omnipresent/extensive/universal factor, 88, 108–9, 161, 181–82, 185
intermediate state (*bardo*), 214, 231
intolerance, 140, 148, 168, 171, 184, 188, 197
Introduction to Entering into Valid Reasoning (Dignāga), 297–98, 314
Entering the Middle Way (*Madhyamakāvatāra*, Candrakīrti), 131, 133, 134, 144–45
Entering the Middle Way Autocommentary (Candrakīrti), 131, 133, 134, 144–45, 372, 492n615
invariable relation. *See* unaccompanied nonoccurrence relation

investigation, 116, 136, 439, 441–42. *See also* fine investigation
Investigation of the Object (Dignāga), 58–59, 470n55
Investigation of the Object Autocommentary (Dignāga), 59–60

J

jealousy, 128, 189, 192, 445
antidote to, 374
arising of, 373, 374
in *Compendium of Bases*, 169
as imputed, 177, 178
as indicative conception, 218, 219
as narrower afflictive factor, 164, 183
as nonvirtuous factor, 186
pacifying, 130
as secondary affliction, 148–49, 193
as source of fault, 187
Jinamitra, *Commentary on the Levels of Yogic Deeds*, 110
Jinaputra. *See* Yaśomitra
Jinpa, Thupten, 26, 35
Jitāri, *Distinguishing the Scriptures of the Sugata* and *Autocommentary*, 252–53
Jñānagarbha, *Distinguishing the Two Truths*, 266
Jñānavajra, *Heart Ornament of the Tathāgata*, 470n48
joy, 141, 142, 184, 190
arrogance and, 149
and bliss, differences between, 191
bodily pliancy and, 417
in calm abiding, 410
complete, 191
in eliminating laxity, 408
freedom from, 212
immeasurable, 187, 197
indicative conception of, 219
as particular factor, 185–86, 197
in pliancy, 419

K

Kālacakra Tantra, 226
Kālacakra tradition, 231, 397
Kamalaśīla, 45, 481n355, 482n381. See also *Stages of Meditation*
Kanakavarṇa's Past Practice, 323
Kangyur, 9, 19
karma
arising of, 369, 371
conditioning of, 355
environment and, 46
ignorance and, 144
mental factors and, 92
ripening of, 121
roots of neutral, 190, 191, 192, 198
Kāśyapa Chapter Sūtra, 441
Khedrub Je, *Ornament of the Seven Treatise Clearing Mental Obscurity*, 482n388
King of Concentrations Sūtra, 138, 424–25
knowledge, 190, 191, 198
empirical, role of, 282–83
ten categories of, 195–96

L

Lakṣmī, *Commentary on the Five Stages*, 222
lamentation, 169, 172, 197, 431
Lamp for the Path to Enlightenment (Atiśa), 398
Lamp on the Compendium of Practice (Āryadeva), 47, 213, 215, 216
language
-based popular convention, 332, 333–34
engagement via exclusion of, 258, 261
readings on, 37
See also linguistic convention/referent
latent potencies, 80, 138
afflictions and, 368, 369

in Cittamātra view, 251
for conceptualization, 51
direct perception and, 275, 483n395
in engagement via exclusion, 256
learning and, 117
as mind, 97
law of excluded middle, 14
laxity, 123, 362, 363, 410–11, 415
antidotes to, 374, 401, 402, 442
categories of, 405
causes of, 407
definition of, 404–5
eliminating, 408
and excitation, balancing, 408–9
fault of, 400–401
freedom from, 412
gross and subtle, 152, 153
in meditating on mind, 421
in mindfulness of body, 439
nonpliancy and, 191
as nonvirtuous factor, 186, 197
preventing, 398
laziness, 188, 189, 400, 415
antidotes to, 401, 442
concomitant factors of, 153, 154
as extensive afflictive factor, 163, 182
as imputed, 177
indicative conception of, 220, 221
as substantially existent, 178
three types, 151
learning, 121, 474n 141
as antidote, 374
power of, 409, 413
wisdom arisen from, 116–17, 269, 357, 378–80, 425
Levels of Yogic Deeds (*Yogācārabhūmi*, Asaṅga), 429
on afflictions, 367–68
on error, 80
on five hindrances, 171, 478n261
on laxity and excitation, 407
on mental factors, 100, 159
on minds during death and transference, 211
liberation, mental factor of, 190, 191–92, 198
life-force, 185, 197, 210–11, 224, 477n247
life-sustaining wind (*prāṇa*), 47, 214, 216, 217, 222, 225
linguistic convention/referent, 68, 69, 71, 72–73
and classificatory convention, difference between, 70
correct assumptions and, 268–69
as direct object, 247
engagement via exclusion of, 260, 262
listening (*śruti*), 357, 368, 381, 474n141. *See also* learning
livelihood
essence of, 443
right, 186–87, 197
wrong, 149, 189, 477n260
locus of debate, 310
logic, 13–15, 286, 297–98. *See also* reasoning
love, 93, 184
affectionate, three degrees of, 218–19
benefits of, 125–26
and compassion, difference between, 136
definition of, 130
increasing, 455–56
meditating on, 442
mental factors incompatible with, 376
as nonhatred, 179
society and, 129–30
two levels of, 126–27
loving kindness, 93, 136
benefits of, 125, 130–31
cultivation of, 354
nonviolence and, 123

potential for, 126
unbiased, 127
violence and, 150
luminous appearance, 215
defining characteristic of, 222–23
dissolution and, 214
indicative conceptions of, 217–19
winds and, 216
luminous imminence, 215
at death, 227
defining characteristic of, 222, 223–24
dissolution and, 214
indicative conceptions of, 220–21
luminous radiance, 215
defining characteristic of, 222, 223
dissolution and, 214
indicative conceptions of, 219–20

M

Madhyamaka (Middle Way) school, 26
contemporary science and, 3, 17, 19
and lower schools, commonalities with, 27–28
Nonpluralist view in, 256
scientific method and, 10, 467n6
on subtle mind of death, 210
on two truths, 17–18
See also Prāsaṅgika Madhyamaka; Yogācāra-Madhyamaka
Mahāyāna, 8–9, 41–43, 207–12, 213, 463
main mind (*gtso sems*), 88, 98–99, 100, 102, 105, 472n98
Maitreya, 127. See also *Distinguishing the Middle from the Extremes*; *Ornament for the Mahāyāna Sūtras*; *Sublime Continuum*
malice, 166, 169, 171, 219, 220
material gain and loss, equanimity toward, 446–49
Mātṛceṭa, *Letter to King Kaniṣka*, 449, 450
matter, 3–4, 28, 30, 39, 40, 44–46, 213
meditation, 353, 420
environment for, five attributes of, 394
length of sessions, 398
terms for, 358–59, 360
types of, 363–64, 386–87
wisdom arisen from, 116–17, 121, 358, 378, 380, 381
See also analytical meditation; breath meditation; stabilizing meditation
meditation objects, 393
breath as, 438–40
deity visualization as, 398, 399
equipoise on, 412
focusing on, 399
four types, 387
mind as, 419–21
mindfulness and meta-awareness of, 402–3
in nine stages of stabilization, 409
number to focus on, 398–99
in special insight, 423
stability and clarity of, 405
memory, 3, 90, 114, 229
mental afflictions. *See* afflictions (*nyon mongs, kleśa*)
mental consciousness (*manovijñāna*)
causal bases of, 46
as conceptual or nonconceptual, 66, 68
conditions, number of, 55
definition of, 52
as direct perception, views on, 53–55, 470n48
falling asleep and, 210
feelings and, 209
gross and subtle types, 47, 213, 215
ignorance concomitant with, 154
intention and, 109

narrower afflictive factors in, 164
rebirth and, 61
and sense consciousness, distinction between, 31–32, 49–52, 68, 72–73, 201, 207
types of, 52
mental factors, 10–11, 85, 106
Abhidharma sources on, 105–6
arising of, 105
definition of, 102
with determinate objects, 90–91, 111, 162, 176
divergent accounts of, 87–88, 93–94
extensive, 161–62, 166, 181–82, 478n271
mutual incompatibility of, 374–78
omnipresent mental factors and, 107–8
substantial and imputed, distinctions between, 175–79
in taxonomies, roles of, 86, 198
total number of, determining, 196–98
variable/indeterminate, 92, 157–58, 164–65, 176
See also afflictions (*nyon mongs, kleśa*); mind and mental factors; omnipresent mental factors
mental going and coming, 217, 218, 219
mental sense faculty, 49, 56, 79, 80, 213, 222, 274, 469n41
mentality (*manas*), 97–98, 101
mere experience, 45–46
meta-awareness (*saṃprajanya*), 40, 91, 114, 184, 381, 407, 442
as antidote, 401, 415, 443
applying, 384–85, 411
of beginners, 383
in calm abiding, 393, 398
definition of, 403
degeneration of, 383
function of, 362–63, 382, 402, 403–4
inferential cognition and, 270
lack of, 152, 165, 166, 177, 179, 197
of laxity and excitation, 409
meanings of term, 361–62
in meditating on mind, 420–21
power of, 410–11, 413
readings on, 365
method, 8, 202
Mīmāṃsā school, 268–69, 482n381
mind (*sems, citta*), 39, 40, 101, 203
analysis of, 28
and body, nonduality of, 202, 203, 215
Buddhist and Western understandings of, 28–29
classifications of, 10
dissolution of, 226–28, 229–30
extremely subtle (*shin tu phra ba'i sems*), 203
good qualities of, 455–57
gross and subtle types, 208, 222, 230–32
guarding, by beginners, 383
and happiness, role in, 5
innate/nature of, 9, 11–12, 29–30, 40, 43, 422, 457–58
mindfulness of, 433, 441, 442
neutral, developing, 396
overview of, 97–98
plural, use of, 29, 31, 85
readings on, 37
self-knowing, 218
sixteen types of, 433
steadiness in realizing objects, 249
suffering produced by, 26, 40–41, 441
See also clear light (*'od gsal*); main mind (*gtso sems*)
Mind, The, sources and methods, 26–28
mind and cognition genre (*blo rig*), 3, 4, 10–12, 13, 235, 481n353
mind and mental factors, 27, 48

accompanying applications of mindfulness, 435
distinctions between, 98–101, 105, 473n111
five shared features, 101, 102–5
literature of, 87–88, 235
in Theravāda tradition, 185–87
mind training (*blo sbyong*), 363, 365
mind-body system, 86, 287–88
mindfulness, 40, 105, 189, 190, 363–64, 381
as antidote, 401, 415, 442–43
applying, 384, 385, 411
attention and, 87
of beginners, 383
in calm abiding, 393, 398
concentration and, 115
definition of, 403
degeneration of, 383
dissolution of, 224, 225, 230
as extensive mental factor, 161, 162, 165–66, 181, 182
forgetfulness and, 179
function of, 101, 382, 402–3
inferential cognition and, 270
meanings of term, 360–61, 362
in meditating on mind, 420–21
as mental factor with determinate object, 90–91, 113–14, 118
nonvirtuous instances of, 114, 474n133
"of mere nondistraction," 363
power of, 410, 413
readings on, 365
in special insight, 426
as virtuous factor, 186
See also applications of mindfulness
mirror-like wisdom, 228
mistaken cognitions, 48
definition and types, 78–80
epistemic reliability of, 33–34
synonyms for, 80–81
variant views on, 82
modularity thesis, 31
Mokṣākaragupta, *Language of Logic*, 299–300, 304, 487n448
motivation, 126, 129, 138, 198, 384

N

Nāgabodhi, 213, 217, 228–30
Nāgārjuna, 15, 16, 19, 298, 458, 468n12, 468n15. See also *Compendium of Sūtras*; *Compendium of Teachings*; *Five Stages*; *Friendly Letter*; *Fundamental Treatise on the Middle Way*; *Precious Garland*
Nālandā masters, 1–2, 9, 19, 294
Nāropa, *Clear Compilation of the Five Stages*, 213, 214–15
nature of reality, 20–21, 131, 132, 475n175
direct perception, authority of, 1–2
as ground of textual tradition, 7–9
ignorance obscuring, 370
insight into, 15–16
wisdom realizing, 144, 204, 356–57
negandum (antithesis), 288, 300, 315–16, 325, 326, 327, 328, 329, 330, 345, 487n448, 487n451
Ngok Loden Sherab, 482n380, 482n388
nihilism. *See* annihilation
nine stages of mental stabilization, 409–12, 415
equanimity and, 122
four types of attention in, 412–13
sequential generation of, 416
nirvāṇa, 16, 19, 296
nonattachment, 177, 189
equanimity and, 122
as root of virtue, 192
as substantially existent, 176
as virtuous mental factor, 119, 120, 162, 163, 183, 186
nonconceptual cognition
affirmative engagement and, 258–59

appearing and observed objects of, 247
distorted, 266
engagement via affirmation by, 260, 261
fallacious, 81
temporary cause of error in, 79
types of, 72
unmistaken, 82
nonconceptuality, 66, 67–69, 89, 90, 221–22, 237, 428
nondelusion, 165, 166, 197
equanimity and, 122
as imputed, 176, 177, 179
as root of virtue, 192
as substantially existent, 178
as virtuous mental factor, 121
nondesire, 217, 218
nondistraction, 114, 363
nonembarrassment, 150, 167, 184, 189, 193
arising of, 374
in *Compendium of Bases*, 170
concomitant factors, 153, 154
as extensive nonvirtuous factor, 163–64
as imputed, 177
as nonvirtuous factor, 186
as source of fault, 187
as substantially existent, 178
nonexistence, 463
conception of, 429
correct knowledge of, 14
establishing by implication, 325, 486nn446–47
view of, 168, 171
nongrasping, 219, 220
nonhatred (*adveṣa*), 136, 189
equanimity and, 122
forbearance and, 131
function of, 92
love as, 130
nonviolence and, 123, 177, 179
as root of virtue, 192
as substantially existent, 176
as virtuous mental factor, 120, 162, 163, 183, 186
nonperception. *See* correct evidence consisting in nonperception
nonpliancy, 190, 191
Nonpluralists, 255–56
nonrevelation, 183, 197
nonspeaking, 220, 221
nonviolence, 190, 191
compassion and, 136
as imputed/conventionally existent, 176, 177, 179
as virtuous mental factor, 123, 162, 163, 183, 184
nonvirtue
at death, views on, 210–11
from distorted cognition, 265
eliminating, 457
indicative conception of, 217, 218
mindfulness and, 91, 114
motivation and, 94, 396
in resolutions, 112
roots of, 190, 191, 192, 198
transforming, 86–87
wrong view and, 146
nonvirtuous mental factors, 185, 186, 195, 197. *See also* afflictions (*nyon mongs, kleśa*)
no-self, variant views on, 16
Numerical Discourses (*Aṅguttara Nikāya*), 42, 431

O

object image (*grāhyākāra*), 13, 236–38, 239, 250, 253, 424
object universal (*don spyi*), 66–67, 69, 70, 72, 82, 471n67, 472n86
objective condition. *See under* conditions, four
objects, 270
attributes of, 99–100, 101, 208

awareness of, number, 248–49
cognitive accuracy and, 33–34
definition of, 247
of desire, 404, 406
distinguishing marks of, 100, 108
engaging, modes of, 12–13, 30, 51–52, 215–16, 217, 248–50, 256–63
explicit and implicit, 249, 481n355
grossness or subtlety of, 209–10
luminosity of, 41, 43–44
types of, 238–39, 469n33
See also appearing object (*snang yul*); conceived object (*zhen yul*); engaged object (*'jug yul*); focal object; meditation objects; observed object (*gzung yul*); sensory objects
objects of knowledge, 39, 116, 176, 343, 347
Cittamātra approach to, 58–60
correct ascertainment of, 298
sound as, example of, 339, 340, 346, 348–49, 488n468
See also hidden objects of knowledge
observed object (*gzung yul*), 51, 53–55, 60, 238–39, 247–48, 469n33, 481n354
obstructivity, 39, 41, 44, 45, 252
offensive instigation, 106, 170, 173, 188, 189, 197
omnipresent mental factors, 88, 92, 107–8, 167, 477n247
implications of, 89–90
as omnipresent, reason for, 110
as substantially existent, 176
See also mental factors, extensive
opponents, 299, 483n408
Ornament for Clear Knowledge (Maitreya), 127
Ornament for the Light of Wisdom That Engages the Object of All Buddhas Sūtra, 42
Ornament for the Mahāyāna Sūtras (Maitreya), 115, 409, 412, 416, 427
Ornament for the Middle Way and *Autocommentary* (Śāntarakṣita), 44–45, 80, 254–55

P

Pali canon, 19, 431, 468n18
path of knowledge (*jñānamārga*), 26
Path of Purification (*Visuddhimagga*, Buddhaghosa), 477n260
peace
from cherishing others, 128
as indicative conception, 217, 218
loving kindness as expression of, 126
physical health and, 130
perceiver (*bhoktṛ*), 32, 33
perception, 27, 242–43
affirmative engagement and, 259
in consequential reasoning, 297–98
directly cognizing objects, 249–50
fallacious, 80–81 (*see also* mistaken cognitions)
indeterminate, 82
inferential cognition and, 270
nonconceptual, 471n85
object typologies of, 239
readings on, 245
as valid cognition, Sakya Paṇḍita on, 482n388
variant views on, 250–55
See also indeterminate perception; sense perception
perceptual cognitions, 239–40, 242–43, 357
Perfection of Wisdom Sūtra in Eight Thousand Lines, 42
Perfection of Wisdom Sūtra in Eighteen Thousand Lines, 436–37
Perfection of Wisdom Sūtra in Twenty-Five Thousand Lines, 137
Perfection of Wisdom sūtras, 16
Perfection Vehicle, 207–12

permanence, 146, 171, 175, 176, 442
persons
  conceived objects and, 247–48
  definition of, 463
  as dependent imputation, 462
  engaged objects and, 248
  engagement via exclusion by, views on, 261
  ignorance and, 26
  investigating, 461–62
  mental afflictions and, 95
  terms for, 464
  See also self, autonomous (*ātman*)
pervasion (*vyāpti*), 285–86, 290, 484n409
  in consequential reasoning, 299
  eight modes of, 291, 302–4, 484nn412–13
  entailment required by, 288
  in faulty proof statements, 306
  positive and negative, 286, 289, 308–9, 314–17, 337, 338, 348–49
  in proof statement, 289
  valid cognition in, 300
philosophy
  Buddhist, 7, 8, 15, 16–18, 19, 20–21, 27–28, 76, 105
  Western, 237–38, 241
physics, contemporary, 17
Pile of Precious Things Collection, 53
placement meditation. *See* stabilizing meditation
Play of Mañjuśrī Sūtra, 131, 426
pliancy, 117, 189, 190, 191
  as antidote, 401, 415
  bliss of, 418–19
  in calm abiding, 416, 417
  gross and delicate, 419
  importance of, 422
  posture and, 395
  in special insight, 423–24
  as substantially existent, 176
  as virtuous mental factor, 122, 162, 183, 184, 186
  wisdom from meditation and, 380
Popper, Karl, 2–3
Praise of Dharmadhātu (Nāgārjuna), 458
Prajñākaramati, *Commentary on Difficult Points in Engaging in the Bodhisattva's Deeds*, 131
Prāsaṅgika Madhyamaka, 77–78, 82, 290–91, 476n209
Precious Garland (Nāgārjuna), 106, 140–41, 143, 187–89, 197, 198
Precious Lamp (Alaksha Tendar Lharampa), 488n467
predicate, 284, 317
  in consequential reasoning, 299, 301
  correct evidence types in, 319, 330–31
  and evidence, relationships between, 285–88, 308, 315–16, 319–20, 337, 347–48, 485n428
  homologous cases and, 344–45
  and thesis, relationship of, 310
Presentation of the Guhyasamāja Sādhana (Nāgabodhi), 213, 228–30
pretense, 149, 189, 193, 194
  arising of, 373, 374
  in *Compendium of Bases*, 169
  as imputed, 178
  indicative conception of, 219, 220
  as narrower afflictive factor, 164, 183
  as source of fault, 187
pride, 128, 184, 192, 445
  antidotes to, 142, 374, 442
  arising of, 373
  in *Compendium of Bases*, 170
  concomitant factors, 153–54
  distraction and, 152
  as indeterminate mental factor, 164–65
  indicative conception of, 219, 220

as narrower affliction, 183, 184
as nonvirtuous factor, 186
predominating in temperament, 389–90
root affliction of, 142–43
as source of fault, 188
pride of superiority, 143, 168, 170, 183, 184
primordial body-mind, 225
primordial consciousness, 213, 214, 222, 224–25, 232
probandum (thesis), 284, 285, 317
in consequential reasoning, 299, 301
correct and fallacious, 310
correct evidence types in, 319, 332–35
explicit and implicit, 312
in faulty proof statements, 306
meanings of term, 309–10
negation of, 288, 300
for one's own purpose, 310
for others' purpose, 310–12
proof statements and, 288, 309
See also negandum (antithesis)
proclivities, 190, 191, 193, 198
Proof of the Afterlife (Dharmottara), 62
proof statements, 267, 288–89
in consequential reasoning, 297, 298
correct, four features of, 304–5
faults in, 305–6
function of, 289–90, 298, 309
homologous and heterogeneous, 306–8
positing object in, 307–8
Proponents of an Equal Number of Subjects and Objects, 255
Pṛthivībandhu, *Explanation about the Five Aggregates*, 130, 159, 160, 176–77
psychology
Buddhist, 40, 213
contemporary, 3, 4, 5, 489n504
Puṇḍarīka, *Stainless Light*, 226–27
Pūrṇavardhana, *Investigating Characteristics*, 102–3, 164–65, 208, 209

Q

Questions of Brahma Viśeṣacinti Sūtra, 425
Questions of Gaganagañja Sūtra, 115
Questions of King Menander, 15
Questions of Ratnacūḍa Sūtra, 433–34, 440, 441–42, 443, 445–46
Questions of the Householder Vīradatta Sūtra, 448

R

rage, 147–48, 154, 162, 189, 193
arising of, 373, 374
in *Compendium of Bases*, 169, 171
as imputed, 177, 178
as narrower afflictive factor, 164, 183
as source of fault, 187
Range of the Bodhisattva Sūtra, 117
rapture, 219
rationality
of dependence, 294–95, 296–97
engaging by, 387
of functionality, 294, 295, 297
of inferential proof, 294, 295, 297
of nature, 294, 295–96, 297
Ratnākaraśānti, *Instructions for the Perfection of Wisdom*, 424
reasoning
in anger prevention, 136
authority of, 1–2, 293–94
cognizing by means of, 250
and emotions, cultural divide in, 94, 95
in establishing substantial existence, 175–76
role of, 281–83
subtle mental states and, 39
wisdom and, 117
See also consequential reasoning; inferential reasoning

rebirth, 60–63, 193, 209, 210. *See also* transmigration
recollection, 52, 72, 91, 272, 305
reflection, power of, 410, 413
reflexive awareness, 43–45, 250, 273, 274, 276, 483n391
refuge, 7, 137
refutation, 267, 297–98, 487n450
regret, 184
  concealment and, 148
  as hindrance, 195
  as imputed, 177, 178
  mental factors concomitant with, 166
  as nonvirtuous factor, 186
  as secondary affliction, 193, 373, 374
  as source of fault, 189
  as variable mental factor, 158, 164–65
reification, 18, 298. *See also* exaggeration
rejoicing, 184, 374, 378
resentment, 148, 189, 193, 194
  arising of, 373, 374
  in *Compendium of Bases*, 169
  as imputed, 177, 178
  as narrower afflictive factor, 164, 183
  pacifying, 130
  as source of fault, 187
resolution, 189, 190, 198
  attention and, 90
  as extensive mental factor, 161, 162, 181, 182
  as mental factor with determinate object, 90, 112–13, 118
  as particular factor, 185
  wrong, 179, 182–83, 197

S

Sacred Vinaya Scripture, 435
sadness, 95, 184
Sakya Paṇḍita Kunga Gyaltsen
  correct assumption, refutation of, 482n380, 482n382
  critiques by, 235
  on objects of cognition, 481n355
  on perception, 238
  See also *Treasure of the Science of Valid Cognition*
sameness, 127–28, 240–42, 357
Sāṃkhya school, 482n381, 487n450
saṃsāra, 191, 197
Sanskrit, 19, 26–27, 35–36
Śāntarakṣita, 481n355. See also *Compendium of Reality*; *Ornament for the Middle Way* and *Autocommentary*
Śāntideva, 20, 93, 363. See also *Compendium of Training*; *Compendium of Training in Verses*; *Engaging in the Bodhisattva's Deeds*
Sarvāstivāda Abhidharma, seven scriptures of, 105, 179, 473n113
satisfaction, 167, 184, 219, 220
Sautrāntika school, 27, 28
  on consciousness, categories of, 48
  on direct perception, 81–82, 275
  on indeterminate perception, 275–76
  on inquiry and analysis, 165
  on sense consciousness, three conditions for, 58
  on sensory cognition, images in, 250, 251, 252, 253–54
Sautrāntika-Madhyamaka, 81–82
scattering, 362, 404, 408, 415
  concentration in pacifying, 115
  as distraction, 152
  and excitation, distinction between, 406
  in meditating on mind, 421
  preventing, 411, 413
science
  Buddhist, 9, 19, 20–21, 35–36
  Dalai Lama's goals concerning, 3–5

Dalai Lama's interest in, 1–3
purpose of, 4–5
understandings of, 9–10, 35–36
Science and Philosophy in the Indian Buddhist Classics, 15, 461
scientific method, 2, 10, 35
scriptures
authority of, 1–2
and empiricism, relationship between, 282–83
nondelusion and, 121
proof statements and, 305
threefold analysis of, 332–33, 335
wisdom from learning and, 380
seeing (*darśana*), 108, 251–52, 356–57
self, autonomous (*ātman*), 288
Abhidharma critique of, 85–86
cognition not requiring, 32, 33
ignorance grasping at, 370, 371–72
rebirth and, 62–63
as target of meditation, 363–64
See also persons
self and other, 371–72
self-grasping, 128–29, 143, 377–78
selflessness, 442, 462
direct experience, difficulty of, 281
personal, sevenfold typology of cognition of, 243–45
two types, 16–17
wisdom realizing, 457
sense consciousness, 39
as adventitious, 213
causal bases of, 46–47, 49
cessation of, 227
in Cittamātra, 59–60
during death process, 214
definition of, 52
distorted cognitions in, 266
as dumb, 259, 481n368
first moment of, 31–32, 90, 274
five types, 52–53
grossness of, 215
images appearing in, variant views on, 250–55
and mental consciousness, distinction between, 31–32, 49–52, 68, 72–73, 201, 207
natural distraction and, 152
as nonconceptual, 66, 68–69, 72
omnipresent mental factors and, 107–8
sleep and, 157, 210
three conditions of, 55–58
as weak modularity, 31–32
See also eye consciousness
sense faculties/organs
conditions that distort, 79–80
direct perception and, 273, 275
feelings and, 108, 209
intention and, 109
and sense consciousness, relationship between, 49, 109–10, 213
subtle, 469n41
Vaibhāṣika view of, 250
sense perception, 11, 88, 109–10, 236–38, 258–59, 273–74, 275, 276
sensory objects, 32, 415, 422
serviceability, 115, 117, 197, 417. *See also* pliancy
shame, 150, 189, 190, 191
and embarrassment, difference between, 120
as indicative conception, 218
mindfulness and, 113
as substantially existent, 176
from three abstinences, 187
as virtuous factor, 120, 162, 163, 183–84, 186
shamelessness, 150, 167, 184, 189, 193
arising of, 373, 374
in *Compendium of Bases*, 170
concomitant factors, 153, 154
as extensive nonvirtuous factor, 163–64
as imputed, 177

indicative conception of, 219, 220
as nonvirtuous factor, 186
as source of fault, 187
as substantially existent, 178
Siddhārtha, spiritual crisis of, 25–26
signlessness, yoga of, 428
Śīlapālita, *Commentary on the Finer Points of Discipline*, 403
single-pointedness, 11, 115, 182, 185, 402, 418, 429, 477n247
six bases, 231, 433
six powers, 409–12, 413, 415
six types of consciousness, 16, 32, 49, 53, 97–98, 100, 102
six ways of settling the mind, 422
slander, 169, 171
sleep, 184
afflictive, 193, 373, 374
anger and, 141, 142
in *Compendium of Bases*, 171
consciousness at, 210
deep/dreamless, 201, 207, 210, 232, 393–94
as hindrance, 195
as imputed, 177, 178
laziness and, 151
as nonvirtuous factor, 186
as source of fault, 189
as variable mental factor, 157, 164–65
social harmony, 125–26, 128, 129–30
soliciting, 170, 188, 197
sorrow, 167, 169, 172, 179, 184, 217, 218, 379
special insight (*vipaśyanā*), 5, 117, 359–60, 380, 386
approximate and genuine, 423–24
and calm abiding, relationship between, 274–75, 380, 424–25, 426–27
convention and ultimate focus in, 424
definition of, 423
equanimity and, 123
spiritual masters, 141–42, 170, 188, 380
spite, 148, 177, 178, 189, 193, 194
arising of, 374
as narrower afflictive factor, 164, 183
as source of fault, 187
Śrāvaka Levels (Asaṅga), 409, 430
on application of mindfulness, 491n568
on equanimity, 123
on meditation objects, 387
on nine stages of stabilization, 412–13
on pliancy, joy in, 419
on pliancy, order of arising, 418
on rationality, four forms of, 294–96
on temperaments, types of, 387, 388–91
on valid cognition, 483n405
Śrāvaka path, 176–77
Śrī Jaganmitrānanda, *Letter to King Candra*, 452
stability
in calm abiding, 394
and clarity, balancing, 409, 421
indicative conception of, 219, 220
and laxity, relationship of, 362–63, 405
obstructions to, 404
during patched placement, 410
stabilizing meditation, 359, 360, 378, 380, 386, 425, 426–29. *See also* calm abiding (*śamatha*); nine stages of mental stabilization
Stages of Meditation (Kamalaśīla), 387, 430
on calm abiding, prerequisites of, 394
on calm abiding and special insight, relationship between, 424–25, 427–29
on fine investigation, 427
on laxity, 404

on meditation objects, 399
on pliancy, 416–17
on posture of Vairocana, 397
on wisdom, 425–26
stealing, 129, 219, 220
Sthiramati, 101, 104, 417
Stoicism, 94
store consciousness. *See* foundation consciousness (*ālayavijñāna*)
streams/floods, 190, 191, 194, 198
subject (*dharmin*), 284, 317
absence of (*āśrayāsiddha*), 285, 291
in consequential reasoning, 299
evidence as attribute of, 314, 316, 337, 343–49
subject image (*grāhakākāra*), 236–37, 250
Sublime Continuum (Maitreya), 42, 116
subsequent cognition (*bcad shes*), 244, 482n388
and correct assumption, distinction between, 268
direct perception and, 54–55, 77–78, 471n85
sense direct perceptions as, 273–74
in sense perception, 237
in sevenfold cognition typology, 264, 265, 271–72, 276, 277
as unmistaken cognition, 82
valid cognition and, 76
substance dualism, 202
substantial existence, four senses of, 175–76
suffering (*duḥkha*), 184
from anger/hatred, 140–42
approaches to, 25, 353
arising of, 25–26, 43, 369, 371
from attachment, 140
contemplating, 408
of distorted cognitions, 265
forbearance and, 131
future, avoiding, 134
ignorance as cause of, 18, 32–33, 144
karma and, 92
meanings of, 25
mind's role in, 26, 40–41, 441
physical and mental, 4–5
from pride, 142–43
of saṃsāra, 191
seal of, 16
three levels of, 192, 479n305, 491n582
transformation, possibility of, 30
valid cognition and, 75
Sūtra Invoking the Supreme Intention, 149–50
Sūtra of the Ten Dharmas, 320
Sūtra on the Application of Mindfulness, 94, 95, 197, 198
on anger, 141
on compassion, 137
on extensive mental afflictions, 182–83, 478n275
on forbearance, 132
on health benefits of benefiting beings, 131
on laziness, 151
on mental factors arising simultaneously with mind, 181–82, 478n271
mental factors scattered in, 184
on narrower mental afflictions, 183
omnipresent mental factors in, 477n247
on virtuous mental factors, 183–84
Sūtra Teaching the Great Compassion of the Tathāgata, 132
Sūtra Teaching the Nonarising of All Phenomena, 324
Sūtra Teaching the Tathāgata's Inconceivable Secret, 371
Sutta on the Application of Mindfulness (*Satipaṭṭhāna Sutta*), 431–32
syllogisms, 14, 15. *See also* predicate; probandum (thesis); subject (*dharmin*)

T

tantra, Buddhist, 201, 206, 469n39. *See also* Vajrayāna
Teaching on the Undifferentiated Nature of the Sphere of Reality, 41
Teachings of Akṣayamati Sūtra, 130, 464
temperaments
  changing, 456–57
  in contemporary psychology, 489n504
  seven types (*Śrāvaka Levels*), 387, 388–91
  six types (*Path of Purification*), 489n511
Tengyur, 9, 19
Tenzin Gyatso, Fourteenth Dalai Lama, 243, 245
terror, 184
Theravāda tradition, 19
  Abhidhamma in, 473n113
  applications of mindfulness in, 431–33
  on luminous clarity, 42
  mindfulness in, 474n133
  See also *Compendium of Abhidhamma*
thesis. *See* probandum
thirst, 217, 218
three luminosities, 231
  as adventitious consciousness, 214
  at death, 216
  at dissolution, 227, 230
  three levels of wind movement in, 221
three poisons, 122, 150, 151, 152, 177, 406, 422, 491n582
three realms, 207, 212. *See also individual realm*
three stages of wisdom, 116–17, 244, 357–58, 378–81, 386
  mindfulness and meta-awareness in, 381–83
  nondelusion from, 121
  transformation due to, 455–57
  types of meditation and, 427
Tibetan Buddhism
  canon of, 19, 431
  higher Abhidharma in, 87–88
  and Indian, interchange between, 201–2
  nondual traditions of, 364
Tibetan language, 26–27
ties, 190, 191, 195, 197
Topic Divisions (Vasumitra), 105, 473n113
torment, 52, 133, 168, 169, 172, 218
tranquility. *See* calm abiding (*śamatha*)
transformation, 386
  and contemplative practices, role of, 92–93
  direct perception/experience and, 244, 281
  due to good qualities, 455–57
  perception in, 243
  in tantric practice, 206
  taxonomies, role in, 86–87, 94, 198
  through reasoning, 292, 353
  through wisdom, 357, 381
transmigration, 209, 210, 211
Treasure of the Science of Valid Cognition (Sakya Paṇḍita), 482n380, 482n388
Treasury of Knowledge (*Abhidharmakośa*, Vasubandhu), 27, 87, 190, 387, 430
  on afflictions, arising of, 373, 374
  afflictive factors in, 163, 164
  on applications of mindfulness, 434, 435
  and *Compendium of Knowledge*, differences between, 89, 165–66
  extensive mental factors in, 89, 118, 161–62, 477n247
  on ignorance, 369–70
  on images, apprehension of, 251–52

indeterminate mental factors in, 164–65
on inquiry and analysis, 65, 165
on intention, 109
mental factor categories in, 161, 196–97
on mental factors, substantial and imputed existence of, 178–79
on mind and mental factors, relationships between, 101, 103, 104, 105, 166
on mindfulness of body, 437, 438
on mindfulness of feelings, 440–41, 491n582
on neutral feeling and death processes, 210
nonvirtuous factors in, 163–64
Precious Garland and, 187, 189
source for, 167
virtuous factors in, 162–63
Treasury of Knowledge Autocommentary (Vasubandhu), 387, 430
on afflictions, 369
on application of mindfulness, 434–35
on compassion, 136
on death, 210–11
on dullness, 405
on four conditions, 55
on ignorance, 370
on mindfulness of breath, 439, 440
on peak of cyclic existence, 212
on wisdom, three stages, 116, 380
Treatise on the Five Aggregates (Vasubandhu), 159, 160, 162
trickery, 219, 220
Tsongkhapa, 481n355, 482n380
two truths, 8, 17–18

U

unaccompanied nonoccurrence relation, 61–62, 286–87, 288, 313, 316–17, 347–48, 485n428
ungentleness, 169, 171, 197
unique particular, 48, 256, 259, 275–76
universals (*spyi, sāmānya*)
ascertaining, 51
in Buddhist epistemology, 240–41
conceptualization and, 65–66, 68
dependent on linguistic conventions, 260
in engagement via exclusion, 256, 259, 261, 262
Unraveling the Intention Sūtra, 53, 120, 294, 399, 400, 405, 416, 424
unruliness, 184

V

Vaibhāṣika school, 27, 28
on conditions for perception, 58
on inquiry and analysis, 165
on joy, complete, 191
on mindfulness, 434
on sensory cognition, images in, 250, 251–53
on subtle mind of death, 210, 211
Vairocana, seven-point posture of, 395–96, 397, 438, 489n513
Vaiśeṣika school, 262, 347
Vajra Garland Tantra, 216, 217, 397
vajra-fist gesture, 397, 489n516
Vajrayāna, 202, 204, 469n39. *See also* tantra, Buddhist
valid cognition (*pramāṇa*), 33, 275, 482n381, 483n395
correct assumptions and, 268–69
in correct consequences, 300
in correct evidence, 314–16
definitions of, 75–77
in Dharmakīrtian system, 237
doubt and, 267
effects of, two kinds, 77
exclusion and, 257–58
in inferential proof, 297
linguistic convention and, 260

nonperception and, 325, 326, 327, 486n446
perception as, 273–74, 482n388
in proof statements, 305
scripture and, 333
in sevenfold cognition typology, 264
subsequent cognition and, 271–72
as support, 265–66
in thesis for others' purpose, 310, 311, 312
types of, 298, 483n405
unmistaken cognition and, 82
wisdom and, 117
van Fraassen, Bas, 36
Vasubandhu, 101, 144. See also *Explanation of the Ornament for the Mahāyāna Sūtras*; *Treasury of Knowledge Autocommentary*; *Treatise on the Five Aggregates*
vehemence, 219, 220
verbal clarity, indicative conception of, 219, 220
Verse Summary of Mind and Mental Factors (Yongzin Yeshe Gyaltsen), 107, 111, 119, 139, 147, 157
vibrancy, 394, 402, 404, 405, 406. *See also* clarity (*gsal ba*)
view, 184, 186, 192, 358
"view, meditation, and conduct," 357, 358, 385–87
views, five afflictive, 165, 166, 167, 176, 197
arising of, 373–74
concomitant factors, 153–54
holding ethics and vows to be supreme, 146–47, 192, 195
holding views to be supreme, 146, 195
as imputed/conventional existence, 177, 178, 179
view holding to an extreme, 146, 171
view of perishable collection, 142–43, 144, 145–46, 171, 371, 372, 377–78, 476n209
wrong view, 146, 184
Vimalamitra, *Meaning of the Gradual Approach in Meditation*, 425, 427
violence, 189, 190, 191, 193, 194
arising of, 373, 374
compassion as counteraction to, 136
as imputed, 177, 178, 179
as narrower afflictive factor, 164
as secondary affliction, 150
virtue
delighting in, 122
five types, 138
increasing, 455–57, 459
indicative conception of, 219, 220
laziness in reducing, 151
roots of, 121, 190, 191, 192, 198, 441
unsuitable basis of, 141
virtuous mental factors, 92, 93, 119, 166
imputed on wisdom, 177
and nonvirtuous, distinguishing, 87, 94–95
substantial and imputed existence, distinguished, 176
in *Sūtra on the Application of Mindfulness*, 183–84
in Theravāda tradition, 186–87
in *Treasury of Knowledge*, 162–63
as virtue by essential nature, 138
visual consciousness, 98–99, 100, 102, 103, 104–5
von Weizsäcker, Carl Friedrich, 3

W

waking state, 201, 207, 232
wind (*rlung*, *vāyu*), 12
dissolution of, 226, 227, 230
extremely subtle (*shin tu phra ba'i rlung*), 203
five-branched, 47
fluctuation of, 204

manipulating, 205
as mount of consciousness, 213, 214, 215, 217, 231
and nature of mind, indivisibility of, 46
pliancy and, 418–19
posture and, 397
subtle and very subtle, distinctions in, 222
tantric understanding of, 202–3, 207, 480n319
three levels of movement of, 221
very subtle (primordial), 225–26
wisdom (*prajñā*), 105, 166, 190, 198
concentration as basis, 115
cultivating, 356–57
doubt and, 116, 118, 121, 162
as extensive mental factor, 161–62, 181, 182
faulty/flawed, 178, 182
fruits of, 117–18
and ignorance, distinction between, 95, 144
increasing, 455–56
as mental factor with determinate object, 91, 115–17, 118
and nondelusion, difference between, 121
perfect, root of, 426
role of, 25–26, 33, 92–93
subtle level of, 204
in Theravāda tradition, 187
types of, 116–17
wisdom of accomplishing activities, 229
wisdom of equality, 228
wisdom of fine investigation, 229
word referent (*sgra don*), 70, 471n75
word universal (*sgra spyi*), 70, 471n67
wrong view. *See* views, five afflictive

Y

Yaśomitra (a.k.a. Jinaputra). See *Clarifying the Meaning of the Treasury of Knowledge*; *Explanation of the Compendium of Knowledge*
Yogācāra, 27, 28, 254, 476n209
Yogācāra-Madhyamaka, 82, 255, 275
Yongzin Yeshe Gyaltsen. See *Verse Summary of Mind and Mental Factors*

About the Authors

HIS HOLINESS THE DALAI LAMA is the spiritual leader of the Tibetan people, a Nobel Peace Prize recipient, and a beacon of inspiration for Buddhists and non-Buddhists alike. He is admired also for his more than four decades of systematic dialogues with scientists exploring ways to developing new evidence-based approaches to alleviation of suffering and promoting human flourishing. He is the co-founder of the Mind and Life Institute and has helped to revolutionize traditional Tibetan monastic curriculum by incorporating the teaching of modern science. He is a great champion of the great Indian Nalanda tradition of science, philosophy, and wisdom practices.

THUPTEN JINPA is a well-known Buddhist scholar and has been the principal English-language translator for His Holiness the Dalai Lama for more than three decades. A former monk and a Geshe Lharampa, he also holds a BA in philosophy and a PhD in religious studies, both from Cambridge University. He is the author and translator of many books and teaches at McGill University in Montreal.

JOHN D. DUNNE holds the Distinguished Chair in Contemplative Humanities at the University of Wisconsin–Madison. He received his PhD in Sanskrit and Tibetan studies from Harvard University. Before his current position, John was a professor at Emory University, where he helped to establish what is now the Center for Contemplative Science and Compassion-Based Ethics. He is a fellow and former board member of the Mind and Life Institute and the author of *Foundations of Dharmakīrti's Philosophy*.

DECHEN ROCHARD has a BA in philosophy from the University of London and a PhD in Buddhist philosophy from the University of

Cambridge. She also completed the first ten years of the geshe degree program at the Institute of Buddhist Dialectics in Dharamsala, India, including the study of Madhyamaka. She is currently translating texts for The Gaden Phodrang Foundation and is a fellow of the Dalai Lama Centre for Compassion (Oxford) and an honorary fellow of the University of Bristol.

What to Read Next from the Dalai Lama

Buddhism
One Teacher, Many Traditions

The Compassionate Life

Ecology, Ethics, and Interdependence
The Dalai Lama in Conversation with Leading Thinkers on Climate Change

Essence of the Heart Sutra
The Dalai Lama's Heart of Wisdom Teachings

The Essence of Tsongkhapa's Teachings
The Dalai Lama on the Three Principal Aspects of the Path

The Good Heart
A Buddhist Perspective on the Teachings of Jesus

Imagine All the People
A Conversation with the Dalai Lama on Money, Politics, and Life as It Could Be

Kalachakra Tantra
Rite of Initiation

The Library of Wisdom and Compassion series:

1. **Approaching the Buddhist Path**
2. **The Foundation of Buddhist Practice**
3. **Saṃsāra, Nirvāṇa, and Buddha Nature**
4. **Following in the Buddha's Footsteps**
5. **In Praise of Great Compassion**

The Life of My Teacher
A Biography of Kyabjé Ling Rinpoché

Meditation on the Nature of Mind

The Middle Way
Faith Grounded in Reason

Mind in Comfort and Ease
The Vision of Enlightenment in the Great Perfection

MindScience
An East-West Dialogue

Opening the Eye of New Awareness

Practicing Wisdom
The Perfection of Shantideva's Bodhisattva Way

Science and Philosophy in the Indian Buddhist Classics
Volume 1: The Physical World

Sleeping, Dreaming, and Dying
An Exploration of Consciousness

The Wheel of Life
Buddhist Perspectives on Cause and Effect

The World of Tibetan Buddhism
An Overview of Its Philosophy and Practice

About Wisdom Publications

Wisdom Publications is the leading publisher of classic and contemporary Buddhist books and practical works on mindfulness. To learn more about us or to explore our other books, please visit our website at wisdomexperience.org or contact us at the address below.

Wisdom Publications
199 Elm Street
Somerville, MA 02144 USA

We are a 501(c)(3) organization, and donations in support of our mission are tax deductible.

Wisdom Publications is affiliated with the Foundation for the Preservation of the Mahayana Tradition (FPMT).